INTELLIGENT AUTONOMY OF UAVS

Advanced Missions and Future Use

Chapman & Hall/CRC Artificial Intelligence and Robotics Series

Series Editor: Roman V. Yampolskiy

The Virtual Mind
Designing the Logic to Approximate Human Thinking
Niklas Hageback

Artificial Intelligence, Second Edition
With an Introduction to Machine Learning
Richard E. Neapolitan and Xia Jiang

Intelligent Autonomy of UAVs
Advanced Missions and Future Use
Yasmina Bestaoui Sebbane

For more information about this series please visit:
https://www.crcpress.com/Chapman--HallCRC-Artificial-Intelligence-and-Robotics-Series/book-series/ARTILRO

INTELLIGENT AUTONOMY OF UAVS

Advanced Missions and Future Use

Yasmina Bestaoui Sebbane

CRC Press
Taylor & Francis Group
Boca Raton London New York

CRC Press is an imprint of the
Taylor & Francis Group, an **informa** business
A CHAPMAN & HALL BOOK

CRC Press
Taylor & Francis Group
6000 Broken Sound Parkway NW, Suite 300
Boca Raton, FL 33487-2742

Printed on acid-free paper
Version Date: 20180220

International Standard Book Number-13: 978-1-138-56849-5 (Hardback)

Library of Congress Cataloging-in-Publication Data

Names: Bestaoui Sebbane, Yasmina, author.
Title: Intelligent Autonomy of UAVs : advanced missions and future use / Yasmina Bestaoui Sebbane.
Description: First edition. | Boca Raton, FL : Taylor & Francis Group, 2018. | Series: Chapman & Hall/CRC artificial intelligence and robotics ; no. 3 | "A CRC title, part of the Taylor & Francis imprint, a member of the Taylor & Francis Group, the academic division of T&F Informa plc." | Includes bibliographical references and index.
Identifiers: LCCN 2017053553 | ISBN 9781138568495 (hardback : acid-free paper)
Subjects: LCSH: Drone aircraft. | Automatic pilot (Airplanes)
Classification: LCC TL685.35 .B47 2018 | DDC 629.133/39--dc23
LC record available at https://lccn.loc.gov/2017053553

Visit the Taylor & Francis Web site at
http://www.taylorandfrancis.com

and the CRC Press Web site at
http://www.crcpress.com

To my family.

Contents

Preface

Unmanned aviation is a fascinating and rapidly growing industry which is expected to become a fundamental part of aviation. The main topic of the present book is civilian operations (or missions) for **unmanned aerial vehicles** (UAVs). UAVs have potentially innovative applications across industries. As UAVs become an economically feasible option, it becomes important to enhance their autonomy so as to increase productivity. **Unmanned aerial systems** (UAS) are changing the nature of aviation, which is a major driver for economic growth.

UAVs have the ability to carry multiple sensors, transmitters, and imaging equipment. Currently the majority of implemented applications of UAVs are in the field of remote sensing, such as surveying with cameras or extra-spectral imaging sensors. As the use of UAVs continues to increase, they impact industries ranging from entertainment to agriculture, from construction to delivery markets, tracking of moving targets for law enforcement, tracking of wild-life, disaster site exploration, or search and rescue missions, object manipulation tasks, global networking and telecommunications, remote monitoring and data gathering, etc. The ability of UAVs is the key factor of success. The role of UAVs is likely to expand still for many years, thus driving the need to understand mission types that are being established today and also those yet to come.

The aim of this book is to provide an approach to the formulation of the fundamental task typical to any mission and guidelines on how this task can be solved by different generic robotic problems, leading to the mission-method pairs. As such, this book aims to provide a systems engineering approach to UAV projects, discovering the real problems that need to be resolved independently of the application. After an introduction is made in the first chapter to enable the reader to enter the rapidly evolving field of aerial robotics, the second chapter presents the mission framework introducing topics such as autonomy, mission analysis, human-UAV team, then homogeneous and heterogeneous UAV team, and finally UAV-UGV (unmanned ground vehicle) teams. The third chapter considers generic robotic problems such as orienteering and coverage. The fourth chapter introduces deployment, patrolling and foraging, and the fifth chapter tackles an important application: aerial search, tracking and surveillance.

This book is meant for both practitioners and scientists:

1. For practitioners, it presents existing solutions which are categorized

according to the missions: surveillance and reconnaissance, 3D mapping, urban monitoring, precision agriculture, forestry, disaster assessment and monitoring, security, industrial plant inspection, etc.

2. For scientists, it provides an overview on generic robotic problems such as coverage and orienteering; deployment, patrolling and foraging; and search, tracking and surveillance. The design and analysis of algorithms raise a unique combination of questions from many fields, including robotics, operational research, control theory and computer science.

This book is organized as follows:

1. Chapter 1: **Introduction**: This chapter explains the book's purpose and scope. US and European regulations are introduced, based on the federal aviation administration (FAA) and European aviation safety agency (EASA) requirements. This chapter examines then the current unmanned aerial vehicles while also presenting some case studies. Specific applications suitable for the use of a unmanned aerial vehicle ranging from precision agriculture and the environment, to mapping coast line or soil erosion, to species and monitoring of construction sites and mines, to monitoring large networks, diagnoses of the conditions of buildings and civil engineering to disasters are considered in this chapter. UAVs equipped with infrared and optical sensors can be used to detect and monitor humans trapped in hazardous areas under critical rescue missions. UAVs equipped with thermal, visual, multi-spectral imaging and chemical detection sensors can be used for remote sensing operations in the industry, precision agriculture, search and rescue missions, security, etc. To increase precision of sensor data, unlike manned aircraft, UAVs can fly very close to ground facilities even at nighttime, when the environment is calm in absence of usual disturbances, taking into account the regulation aspect and without risking the life of the pilot. By the end of this chapter, the reader will have some clues on rapidly changing regulations with some risk analysis and financial potential as well as taking a look at the large variety of possible UAV missions.

2. Chapter 2: **Mission Framework**: The main topics presented in this chapter are autonomy, human-UAV teams, homogeneous and heterogeneous UAV teams, UAV-UGV teams, and finally mission analysis. An autonomous system has two main goals: make out its mission, while remaining operational and react to the uncertainties of the mission, environment or system. A high level of autonomy for a variety of operating scenarios can be enabled by an automated mission planning, identifying the activities to be performed. Collaborative UAV systems are a very interesting way to speed up the sensing or exploration tasks, to increase accuracy by fusing information, and to complete a task impossible with only one UAV. Homogeneous UAV teams as well as heterogeneous are considered in this matter. Another kind of robotic teams are UAV-UGV

teams with their complementary skills, UAV and UGV collectively can accomplish more complex tasks with higher precision. A framework of autonomy is developed in levels of autonomy. Regardless of the applications, different missions have in common characteristics enabling to use some generic algorithms improving performance and dependability. Mission specificity should be analyzed. Humans being involved in the higher level of planning, decision-making and remote operations, human-UAV collaboration is beneficial due to the uncertainties inherent in supervisory control systems. By the end of this chapter, the reader will be provided a systematic approach on analysis of current UAV missions as well as an introduction to teams of different robotic types, be it aerial or grounded.

3. Chapter 3: **Orienteering and Coverage**: The main topics considered in this chapter are generic robotic problems such as orienteering and coverage. The orienteering problem is a generalization of the vehicle routing problem and its avatars. The team orienteering problem is a multi-level combinatorial optimization problem. In aerial robotics, the problem of coverage is defined as follows: The UAV must pass a sensor over all of the available free space within a bounded region. Some sub-problems are tackled such as barrier coverage, area and perimeter coverage. Area and perimeter surveillance are some of the most well-known UAV applications. Such missions may be repetitive or dangerous and as such can be solved in a more efficient and safe way using a controlled fleet of UAVs. In contrast to traditional coverage, in persistent coverage of an environment with a UAVs team, the coverage level of the environment is always changing. For this reason, the UAVs have to continually move to maintain a desired coverage level. UAVs can generally be placed in an area of interest either deterministically or randomly. By the end of this chapter, the reader will have knowledge of the operational research algorithms and their application in UAV orienteering and coverage.

4. Chapter 4: **Deployment, Patrolling, and Foraging**: The choice of the deployment scheme depends highly on the type of UAV sensors, application and the environment that the UAVs will operate in. Controlled node deployment is viable and often necessary when UAVs are expensive or when their operation is significantly affected by their position. In UAV mission planning, some UAVs attached with various sensors may be required to patrol a set of targets. Each target is assigned a profit, and the planner aims to maximize the total received profit. To solve the area patrolling problem with a team of arbitrarily many robots, intelligent strategies for coordination of the team in order to visit all the surveillance points that need vigilance inside a target area are needed. Foraging involves three basic skills: sensing and locating a target, approaching it so that it is within reach, and finally handling it in some manner. It may involve either communication to signal to others where objects may be

found and/or cooperative transport of objects too large for a single individual to transport. By the end of this chapter, the reader will have considered UAV deployment, patrolling and foraging using some generic robotic algorithms.

5. Chapter 5: **Search, Tracking, and Surveillance**: Search and track is the problem of searching for a mobile target and tracking it once it is found. The problem becomes complicated by several factors: multiple UAVs, moving targets, constrained resources, and heterogeneous UAV capabilities. Moreover, tight coupling of search planning and task allocation increases the complexity of the problem. Complex sequences of behavior must be planned throughout to achieve surveillance goals. They generally consist of gathering as much information as possible given the constraints and communicating findings to human operators. Observers usually operate in unpredictable environments with rapidly changing information. They must decide what action to perform and how to coordinate with other observers almost instantaneously and be highly trained to react quickly. At the same time, observers have limited resources, and need to be strategic in deciding what course to follow, looking ahead at their remaining lifespan and fitting their objectives within this time frame. By the end of this chapter, the reader will have gained knowledge on the different types of search, tracking and algorithms methods encountered in real life.

6. Chapter 6: **General Conclusions**: Some general conclusions are given while also surveying the potential future environment of civilian UAV missions.

CHAPTER 1

Introduction

1.1 INTRODUCTION

Unmanned aerial vehicles (UAVs) are aircraft without the on-board presence of a pilot or crew. Any unmanned aircraft must be deemed to have a pilot. The aircraft may be controlled remotely or it may be running autonomously with a program loaded on the aircraft's on-board computer, but the person who loaded the program and initiated the running of the autopilot is still regarded as the pilot.

UAVs range in size from large military drones with a wingspan of nearly 300m to commercially available small and micro-UAVs. Their ranges of flight vary, with some commercial UAVs with a few meters away around the operator to advanced military drones that can fly for over 25000km without having to land. Likewise, there is a huge variation in their maximum flight altitude, which can be anything from a few meters to a maximum of 20000m. In general, aerial vehicles are distinguished for their ability to fly at various speeds, to stabilize their position, to hover over a target, and to perform maneuvers in close proximity to obstacles, while fixed or loitering over a point of interest, and performing flight indoors or outdoors.

UAVs are expected to be used for a wide spectrum of civilian and public domain applications. A vast market is currently emerging from the big number of potential applications and services offered by small, low-cost, and low-flying UAVs. They can carry payloads such as cameras: visual or infrared, and other sensors of interest for the operation. Thus, they are helpful in applications such as surveillance and reconnaissance, environmental issues, disaster assessment and management, dexterous manipulation and grasping, recognition and tracking, search and rescue, urban monitoring and inspection, coast monitoring, infrastructure inspection, traffic management and surveillance, among others. Any commercial use is in connection with a business, including: selling photos or videos taken from an **unmanned aerial system** (UAS), providing contract services such as industrial equipment or factory inspection, providing professional services such as precision agriculture, security or telecommunications, real estate, cinema photography for a film or television production, providing contract services for mapping or land surveys, cargo delivery, etc. UAV manufacturers, service providers, and platform integrators are seriously considering the business potential across these application types [98]. For the professional operator, the equipment and software needed to deliver the end product to clients is extremely important. In some countries, such as Japan, UAVs have been used commercially for at least 20 years. But in other countries, commercial UAVs are just beginning to take off.

There is an increasingly popular variety of applications that justify the development of UAVs in the civil aviation field. In this new aeronautics field, high performance at the lowest cost is the main objective. By 2020, thousands of UAVs are expected to swarm skies. UAV purchases are rising both for hobbyists and for commercial operators conducting inspections, aiding farmers, and monitoring construction sites. The aviation agencies expect that 42 percent of commercial UAVs will end up in industrial inspection, a further 19 percent in agriculture, 15 percent in insurance, 22 percent in real estate or aerial photography, and just 2 percent in government [47]. Since many disciplines across various fields are involved in this area, one needs also to be aware of the authorized standards, rules and regulations. The commercial UAV market is expected to grow at a **compound annual growth rate** (CAGR) of 19 percent between 2016 and 2020, compared with 5 percent growth on the military side.

The two main system architectures are the fixed-wing aircraft and the rotary-wing type UAV. In civil operations, one of the critical advantages of rotor-craft over fixed-wing aircraft is their hovering capability while the airplane has a much larger radius of action than helicopters. Specifically, the helicopter is capable of decelerating, turning with a very small radius, and accelerating back to its cruising velocity. Similarly to other vehicles, UAVs will need to be maintained, repaired or enhanced with slight upgraded components and insured for liability against potential damages. Approximately two thirds of their value is related to maintenance. Indeed, there is a desire to use UAV assets for multiple years, especially for more expensive ones.

1.2 USE OF UNMANNED AERIAL SYSTEMS

An **unmanned aerial system** (UAS), of which the **UAV** is the airborne component, comprising two fundamental types: **remotely piloted aircraft systems** (RPAS), a class of UAS which has a pilot operating the **remotely piloted aircraft** (RPA) from a **ground-control station** (GCS) or a semi-autonomous UAV. RPAS are a set of configurable elements consisting of a remotely piloted aircraft, its associated remote pilot station(s), the required command and control links and any other system elements as may be required, at any point during flight operations. A UAV operator is a person, organization or enterprise engaged in a UAV operation.

1.2.1 Regulations

All unmanned civilian aircraft, whether remotely piloted, fully autonomous or combinations thereof, are subject to the provisions of Article 8 of the **Convention on International Civil Aviation** (Doc 7300), signed in Chicago on 7 December 1944 and amended by the **International Civil Aviation Organization** (ICAO) Assembly.

UASs are inherently different from manned aircraft. Introducing UAS into the airspace is challenging for the FAA, the EASA, and in general national agencies and the aviation community. The UAV community is committed to working together based on the following principles [2]:

1. UAVs need to be treated as new types of aircraft with proportionate rules based on the risk of each operation.
2. Rules for the safe provision of UAV services need to be developed now.
3. Technologies and standards need to be developed for the full integration of UAV in the national airspace.
4. Public acceptance is key to the growth of UAV services.
5. The operator of a UAV is responsible for its use.

The ICAO has considered annexes regarding the use of UAS when directing the integration of UAS in non-segregated airspace: licensing, rules of the air, aircraft operations, aircraft airworthiness, aeronautical communications, security safeguarding international civil aviation against acts of unlawful interferences and the radio frequency spectrum [55]. With no persons on-board the aircraft, airworthiness objective is primarily targeted at the protection of people and property on the ground. UAS control stations and other remote equipment performing functions that can prejudice take-off, continued flight, landing or environmental protection are considered as part of the aircraft and included in the type of certification basis.

Definition 1 ***Airworthiness*** *is the suitability of an aircraft for safe flight. Each aircraft has an operating envelope, defined in terms of its attitude, inside of which it is flyable, and outside of which it is unstable and probably unrecoverable.*

Safe operation of a UAV is dependent on its attitude being kept within its operating envelope, by conducting maneuvers within that envelope, and by taking corrective action when the attitude is changed by external factors, referred to as upset conditions, such as a wind gust or turbulence.

On 2016, the FAA and EASA have issued approval for flying small commercial UAVs. To fly a UAV for commercial work, a certificate is required in most countries. This certificate applies to the individual who would be employed by a UAS operator certificate holder for commercial work operations. If an individual is piloting a UAV which causes damage to property, or injury or death to a person that individual has to be regarded as being responsible for the accident and should be liable to pay compensation [31]. The legislation of many countries emphasizes the responsibility of the pilot and the need for pilots to be trained and licensed.

Remark 2 *Critical parts of the evolving regulation will be the extent to which UAVs can be operated* ***beyond visual line-of-sight*** *(BVLOS), in populated areas and/or without a dedicated pilot per each UAV (i.e., allowing a single pilot to operate or monitor more than one UAV at the same time).*

1.2.1.1 U.S. regulations

The FAA is taking the following approach to safe UAS integration [3]. Different types of UAS operations are defined:

1. **Governmental operations**: For public aircraft operations, the FAA issues a **certificate of waiver or authorization** (COA) that permits public agencies and organizations to operate a particular aircraft, for a particular purpose, in a particular area. The COA allows an operator to use a defined block of airspace and includes special safety provisions unique to the proposed operation. COAs usually are issued for a specific period.

2. **Civil operations**: Any operation that does not meet the statutory criteria for a public aircraft operation is considered a civil aircraft operation and must be conducted in accordance with all FAA regulations applicable to the operation.

3. **Model aircraft**: Individuals flying for recreation must follow safety guidelines, which include: fly below 150 m and remain clear of surrounding obstacles, keep the aircraft within visual line-of-sight at all times, do not interfere with manned aircraft operations, do not fly near people or stadiums, do not fly an aircraft that weighs more than 25 kg, etc. Operators must fly under daytime **visual flight rules** (VFR), keep the UAS within **visual line-of-sight** (VLOS) of the pilot, and stay certain distances away from airports or heliports:

   (a) Five nautical miles from an airport having an operational control tower; or

   (b) Three nautical miles from an airport with a published instrument flight procedure, but not an operational tower; or

   (c) Two nautical miles from an airport without a published instrument flight procedure or an operational tower; or

   (d) Two nautical miles from a heliport with a published instrument flight procedure.

Federal law requires that all aircraft (which includes UAS and radio/remote controlled aircraft) flown outdoors must be registered with the FAA and marked with a registration number. Unmanned aircraft flown for work or business must be registered individually by the owner, for each UAS they want to fly. Anyone operating a UAS is responsible for flying within FAA guidelines and regulations. Operators should be aware of where it is and is not safe to fly. The rules for operating an unmanned aircraft for civil applications are:

1. **Pilot requirements**: Must have remote pilot, airman certificate, be 16 years old, pass Security Administration (TSA) background security screening, have no physical or mental condition that would interfere with the safe small UAS operations. The goal of the airman certification process is to ensure the applicant possesses knowledge consistent with the privileges of the remote pilot certificate with a **small UAS** (sUAS) rating being exercised, as well as the ability to manage the risks of flight in order to act as a remote **pilot-in-command** (PIC).

2. **Aircraft requirements**: Must be less than 25 kg, if over 25 kg must be registered online, and must undergo pre-flight check to ensure UAS is in condition for safe operation.

3. **Location requirements**: Class G airspace, typically the airspace very near the ground.

4. **Operating rules**: Must keep the aircraft in sight (visual line-of-sight), fly under 400 feet, fly during the day, fly at or below 180 Km/h, yield right-of-way to manned aircraft, not fly over people, and not fly from a moving vehicle.

5. **Legal or regulatory basis**: Title 14 of the Code of Federal Regulations (14 CFR) Part 107.

Operations include radio communication procedures (through radio communications, UAS pilots give and receive information before, during and at the conclusion of a flight), airport operations with or without an operating control tower, emergency procedures (planning and communication, batteries, loss of aircraft control link and a fly-away's loss of GPS signal, frequency spectrums and associated limitations), aeronautical decision making (effective team communication, task management, crew resource management, situational awareness, hazard identification and risk assessment), physiology (stress and fatigue, drug and alcohol use, vision, fitness for flight), as well as maintenance and inspection procedures (basic maintenance, preflight inspection, techniques to mitigate mechanical failures, appropriate record keeping). The sources of weather (aviation routine weather reports, terminal aerodrome forecasts, weather charts, automated surface observing systems and automated weather observing systems) as well as the effects of weather on UAS performance are also important matters to consider. Some important parameters are as follows: density altitude, wind and currents, atmospheric stability pressure and temperature, air masses and fronts, thunderstorms and microbursts, tornadoes, icing, hail, ceiling and visibility, and finally lightning. Notice to airmen (NOTAM) should also be taken care of as well as operating limitations for UAVs such as maximum ground speed, altitude limitations, minimum visibility and cloud clearance requirements.

Remark 3 *In* 2013, *the FAA certified the first UAVs to fly in U.S. airspace for commercial purposes: the AeroVironment* $RQ - 20$ *Puma and the Boeing Insitu Scan Eagle* $X200$. *The Puma is a small (*$2.8m$ *span,* $5.9kg$ *maximum take-off weight (MTOW)), surveillance UAV. It is hand-launched and battery powered, with a standard endurance of* 2 *hours and range of* $15km$. *The ScanEagle is also a small surveillance UAV (*$3.1m$ *span,* $22kg$ *MTOW). It uses a pneumatic launcher and its* $1.12kW$ *single-piston engine allows for endurance of* 20 *hours [77].*

Remark 4 *A new kind of radar is currently installed in the U.S. It will allow air traffic controllers to see a combination of planes and UAVs. The U.S. Air Force and the state of Ohio are footing the* $5 *million radar bill in a radar testing. Right now, UAVs can only fly with visual line-of-sight, or an unobstructed path between the UAV and the controller; with beyond line-of-sight that would increase to 321 km and an altitude of up to 3 km. FAA approval for the state means private companies could test in Ohio. Guidance for UAV operators would come from an air traffic controller who is looking at the special radar on the ground in a trailer. Air traffic controllers in the tower would continue to focus on planes. Knowing where they are and what direction they are going, the operator can tell the pilot in charge of a small UAV to turn in a new direction or airspeed or new altitude in order to stay away or avoid a collision.*

1.2.1.2 European regulations

In terms of regulation, a framework around the operations of UAVs was proposed in 2016 by the European Union as a common basis to harmonize regulation across Europe and enable more applications. The EASA Commission has proposed a risk-based framework for all types of UAV operations [2]. This framework ensures the safe use of UAVs in civil airspace and will create legal certainty for the industry. In this context, concerns related to privacy and data protection, security, liability and insurance or environment must also be taken into account.

Considering the broad range of operations and types of UAVs, three categories of operations and their associated regulatory regime are established:

1. The **open operation** category of UAVs should not require an authorization by an aviation authority for the flight, but stays within defined boundaries for the operation. In the open, the UAV must be flown under

direct **visual line-of-sight** (VLOS): 500 m, at an altitude not exceeding 150 m above the ground or water and outside of specified reserved areas (airport, environmental, security). The UAV's weight should be less than 25 kg.

2. The **specific operation** category will require a risk assessment that will lead to an operation's authorization with specific limitations adapted to the operation. The specific category should cover operations that do not meet the characteristics of the open category where a certain risk needs to be mitigated by additional operational limitations or higher capability of the involved equipment and personnel.

3. The **certified operation** is required for operations with a higher associated risk or might be requested on a voluntary basis by organizations providing services such as remote piloting or equipment such as **detect and avoid**.

Included in this legislation are initial permissions for **beyond visual line-of-sight** (BVLOS) UAV operations that are critical for many of these operations to be economically viable opportunities. Examples of such BVLOS permissions include Spain allowing BVLOS for UAVs under 2kg and France allowing these flights for UAVs under 2 kg with no lateral limitation and additionally for UAVs under 25 kg that operate within 1 km of the remote pilot.

Remark 5 *Other countries offer examples of allowing operations around BV-LOS, including Australia and Japan that require case-by-case permission and Switzerland which requires a total hazard and risk assessment.*

As remotely piloted aircraft systems are very difficult to categorize due to the large variety of shapes, sizes, performance and operations; different traffic classes have been developed to support the management of large numbers of RPAS operations. A class of RPAS traffic is a set of flying rules, operational procedures and system capabilities applicable to the RPAS, the RPAS operator when operating the RPAS in a portion of the airspace, and the services applicable in that airspace. The rules of the air will not be adapted for low-level RPAS operations at this altitude, thereby maintaining the 1 km boundary as implemented around the world already. A **command-control** (C2) service is provided. Detect and avoid and self-separation capability is implemented. RPAS classes of traffic are proposed:

1. Class I: Reserved for RPAS EASA category A (VLOS only);

2. Class II: Free route (VLOS and BVLOS);

3. Class III: Organized commercial medium/long haul traffic (BVLOS only);

4. Class IV: Special operations (VLOS and BVLOS).

Insurance is required for most operators of aircraft, irrespective of the purposes for which they fly, to hold adequate levels of insurance in order to meet their liabilities in the event of an accident. This European regulation specifies amongst other things the minimum levels of third party accidents and risk insurance for aircraft operating into, over or within the European Union (including UAS) depending on their **maximum take-off mass** (MTOM).

The aim is to provide for UAV users a low-level airspace that is safe, secure and environmentally friendly. The **U-space** covers altitudes up to 150 m. Registration of UAVs and their operators, their e-identification and geo-fencing should be in place by 2019. The concept is to develop a system similar to that of Air Traffic Management for manned aviation. The system will provide information for highly automated or autonomous UAVs to fly safely and avoid obstacles or collisions.

1.2.1.3 U.K. regulations

It is **Civil Aviation Authority** (CAA) policy that UAS operating in the U.K. must meet at least the same safety and operational standards as manned aircraft. The Civil Aviation Authority introduces a **Concept of Operations** (ConOps) approach for UAS. The **Chicago Convention** requires that each contracting state undertake to ensure that the flight of such aircraft without a pilot in regions open to civil aircraft shall be so controlled so as to avoid danger to civil aircraft. The term **technical complexity** is used to describe how complex the system; whereas the term **operating environment complexity** describes how complex the environment is. There are scenarios that are not described, which will require differing levels of assessment (for example, a very light UAS operating in a complex environment with an extremely complex flight management system). The concept of operations is the following:

1. Operating within **VLOS** means that the remote pilot is able to maintain direct, unaided visual contact with the UAV which is sufficient to monitor its flight path in relation to other aircraft, persons, vessels, vehicles and structures for the purpose of avoiding collisions. VLOS are normally accepted out to a maximum distance of 500 m horizontally and 150 m vertically from the remote pilot. Operations at a greater distance from the remote pilot may be permitted if an acceptable safety case is submitted. For example, if the aircraft is large, it may be justifiable that its flight path can be monitored visually at a greater distance than 500 m. Conversely, for some small UAV, operations out to a distance of 500 m may mean it is not possible to assure or maintain adequate visual contact.

2. **Extended VLOS** (EVLOS) operations are operations, either within or beyond 500 m / 150 m, where the remote pilot is still able to comply

with his collision avoidance responsibilities, but the requirement for the remote pilot to maintain direct visual contact with the UAV is addressed via other methods or procedures. Collision avoidance is still achieved through **visual observation** (by the remote pilot and/or **remotely piloted aircraft** (RPA) observers). The operator must submit a safety case including a risk assessment for the operation. Factors taken into consideration must include:

(a) the procedures for avoiding collisions;

(b) aircraft size;

(c) aircraft color and markings;

(d) aircraft aids to observation;

(e) meteorological conditions and visibility, including background conditions (cloud/blue sky);

(f) the use of deployed observers; and

(g) Operating range limits: suitable radio equipment must be fitted in order to be able to effect positive control over the unmanned aircraft at all times.

3. Operation of an unmanned aircraft beyond a distance where the remote pilot is able to respond to or avoid other airspace users by visual means is considered to be a **Beyond VLOS** (BVLOS). Unmanned aircraft intended for operation beyond visual range of the pilot will require an approved method of aerial separation and collision avoidance. This requirement to avoid collisions applies to all flights conducted under **instrument flight rules** (IFR) and to flights made with an **Air Traffic Control** clearance, as well as to flights under **visual flight rules** (VFR).

1.2.2 Risk analysis

Every new technology introduces risk with its benefits. It is important to apply UAV technologies effectively and safely.

Definition 6 *Safety is an accumulation of knowledge about risk converted into practice. Safety risk management is the active management and mitigation of forseeable risks the UAS may cause or encounter. It includes assurances from design, operational limitation, pilot action, weather, maintenance or geographic/airspace restriction.*

Safety risk management can be used to ensure regulatory compliance and also to identify, assess, and address unique risks that are not covered by regulation. The safety risks considered must take into account: Mid-air collision with manned aircraft, harm to people, and damage to property in particular critical and sensitive infrastructure. The safety risk assessment has to address airworthiness, operating procedures and environment, competence of involved personnel and organizations as well as airspace issues. The minimum level of safety for airworthiness is based on the results of the safety risk assessment. It may be defined and demonstrated through compliance to acceptable industry standards. It may be acceptable to compensate certain airworthiness risk factors by operational risk mitigating factors (specific limitations on the operations, special qualifications for the personnel, etc.). The airworthiness assessment is closely linked to the operational environment and procedures; e.g., the operation close to crowds could be acceptable when the vehicle has some additional functionality (e.g., automatic loss of link procedures, impact energy limiting devices) and the operation procedures are adequate.

The risk posed by a UAV to people on the ground can be described in terms of the expected number of fatalities associated with a given flight, which can be determined by identifying the possible crash locations and multiplying the probability of a UAV crash by the number of people present in the potential crash location. Typically, this is quantified as a two-dimensional probability distribution representing the likelihood of crashing at a certain distance away from the point of the failure.

Remark 7 *Over the past few years, more and more civil UAVs have come dangerously close to airports across the world, which could damage an aircraft during its take-off or landing. For example, China is currently working on a high-tech solution to keep UAVs at a safe distance, to contribute to aircraft safety. The technology called Electronic Wall, is aimed at breaking communication between a UAV and its operator in the event that the UAV gets closer than ten km from the airport. As a result, the UAV will be unable to receive any signals, and it will only be capable of returning to the place from where it was launched.*

One challenging aspect is the modeling of UAS safety risk for civil applications given the scarcity of actual operational data. With the creation of a probabilistic model, inferences about changes to the states of the accident shaping or causal factors can be drawn quantitatively. These predictive safety inferences derive from qualitative reasoning to plausible conclusions based on data, assumptions, and/or premises and enable an analyst to identify the most prominent causal factors leading to a risk factor prioritization. Such an approach also facilitates the study of possible mitigation effects. In [76], is illustrated the development of an **object-oriented Bayesian network** (OOBN) to UAS with the mission of aerial surveying for a bridge inspection. As a system of systems approach, an object-oriented Bayesian network facilitates decomposition at the sub-system level, yet enables synthesis at a higher-order systems level. The methodology serves as a predictive safety analytic platform to support reasoning to plausible conclusions from assumptions or premises. With the creation of a probabilistic safety risk model, inferences about changes to the states of the causal factors or the presence or absence of mitigation can be made. These inferences may be built on either quantitative or qualitative reasoning, or both, and enable an analyst to identify the most prominent causal factor groupings (i.e., Vehicle or UAS, Operations, Environment or Human) leading to a prioritization of the most influential causal factors. A systematic approach to risk factor sensitivity may lead to vulnerability discovery of emerging hazard causal factors for which mitigation does not yet exist.

For UAS that typically fly low and slow, the possibility of a mid-air collision with a nearby general aviation aircraft needs to be studied by identifying possible hazards and assessing mitigation. The **aviation system risk model** (ASRM) is a first-generation socio-technical model that uses a **Bayesian belief network** (BBN) methodology to integrate possible hazards to assess a non-linear safety risk metric. The ASRM may be used to evaluate underlying causal factors linked to the vehicle and/or to the systems and procedures that led to the unsafe state and the interactions among these factors that contributed to the safety risk. The ASRM can also assess the projected impact of mitigation. The ASRM facilitates robust inductive reasoning on the hypothesized accident scenarios. Recently, it has been updated with the use of the **hazard classification and analysis system** (HCAS) that provides analytic structure for categorizing hazards related to the UAS, Airmen, Operations and the Environment. The safety risk analysis involves systemically following these six steps:

1. Selecting and analyzing an accident/incident scenario,
2. Identifying the case-based causal factors,
3. Constructing an influence diagram depicting causal factor interactions,
4. Building a Bayesian belief network,
5. Inserting mitigation and value functions, and
6. Evaluating the relative risk associated with the insertions.

The dynamic planning and development of a large collection of systems or a system of systems pose significant challenges due to the complex interactions that exist between its constituent systems. The work conducted in [49] develops a tool that adopts an operations research-based perspective to system of systems level planning based on metrics of cost, performance, schedule and risk. This approach allows for identification of near-optimal multi-stage decisions in evolving system of systems architectures. In addition to the logical procedures, an analytic solution framework to objectively quantify the state and outcome of consequent actions to evolve a system of systems architecture is required by practitioners. Actions may involve a sequence of decisions that include adding new systems, retiring old systems, upgrading system, etc. Typical questions could include: how to deal with the interactions between decisions from multiple independent organizations, how to deal with the diverse time scales occurring in a system of systems, how to deal with the complexity resulting from the number of uncertain variables involved, etc.

The goal of risk management is to proactively identify safety-related hazards and mitigate the associated risks. FAA steps for good decision making are as follows:

1. Identifying personal attitudes hazardous to safe flight.
2. Learning behavior modification techniques.
3. Learning how to recognize and cope with stress.
4. Developing risk assessment skills.
5. Using all resources.
6. Evaluating the effectiveness of one's aeronautical decision making skills.

The **perceive, process, perform** (3P) model for aeronautical decision making offers a systematic approach that can be used during all phases of flight. To use it, the UAV pilot will:

1. Perceive the given set of circumstances for a flight,
2. Process by evaluating their impact on flight safety, and
3. Perform by implementing the best course of action.

EASA has proposed a risk-based approach to settle a performance-based framework for regulation related to UAVs. Traffic management solutions associated to required UAV technologies (e.g., detect and avoid, data-link, geofencing) are key enablers for safety. A European Task Force decided to focus on the UAVs available on the mass market that correspond to the proposed EASA Open Category (i.e., less than 25 kg), limiting the assessment to four classes of UAVs that represent the vast majority of the UAVs in this category flying today: large (3.5 kg), medium (1.5 kg), small (0.5 kg) and the smallest (0.25 kg). A simplified model of the UAV threat has been established, considering certain parameters that are assumed to contribute to the potential severity of an impact, which led to the batteries and the motors of UAVs being identified as key critical components. For each product type, the vulnerability of selected aircraft components has been assessed against the four classes of UAV defined. As expected, large UAVs and large rotorcraft are by the nature of their scale and design requirements generally more resilient to collisions with UAVs and the severity level is limited for the smallest UAV categories (small and harmless). For smaller aeroplanes and light rotorcraft, more components are vulnerable and the severity level is higher. The landing gear and the landing lights are expected to be components with the lowest vulnerability.

More specifically, for the case of a collision with a medium UAV, only an impact above 3 km at cruise speed is believed to lead to high severity effects. At lower altitudes, the severity level of a collision with a UAV of this category is expected to be low due to the lower kinetic energy at impact. The use of altitude protection, as defined in the UAV threat specification, which might be implemented in certain UAV designs, is perceived to be a means to mitigate the consequences for large aircraft airframe components of a collision with a medium-sized UAV. Little or no benefit is expected from the use of altitude protection for rotating components (i.e., engines, propellers, and rotors) and the airframe components of rotorcraft or general aviation aircraft. A collision with the smallest UAV category is expected to be harmless, at least for large airplane product types. Further research is needed to determine the consequences for other aircraft product types.

1.2.3 Financial potential

Worldwide UAV markets are poised to achieve significant growth with the use of cameras on stable flying platforms that are used to help implement aerial entertainment and advertising. Entertainment light shows advertising UAVs use LED technology to do innovative skywriting. Aerial visualization lets advertising firms achieve new ways of reaching large numbers of people with a relatively low cost, effective means and lets the UAVs do the work in an automated manner [12]. Tech giants such as Facebook, Inc. and Google, Inc. are planning to use solar-powered UAVs that hover around the atmosphere of the Earth providing Internet access to the remotest of places acting as flying Internet access points or hotspots. Other applications include imaging and data collection activities such as environmental monitoring and mapping, natural hazards research and monitoring, atmospheric monitoring, hyperspectral imaging, plume dispersion and tracking, soil moisture imaging, and aerosol source determinations. AeroVironment Inc., BAE Systems PLC, DJI, Draganfly, Elbit Systems Ltd., General Atomics, Israel Aerospace Industries, Lockheed Martin Corporation, Northrop Grumman, Parrot SA, Textron Inc., and The Boeing Company currently dominate commercial UAV market share [11]. The UAV market is expected to exceed $60 billion by 2020. Analysts said they expect 9 million consumer-grade UAVs and 6 million commercial-quality UAVs flying in the skies by 2020.

The growing UAV marketplace shows significant potential, with European demand suggestive of a valuation in excess of EUR 10 billion annually, in nominal terms, by 2035 and over EUR 15 billion annually by 2050. The impact of civilian missions is expected to generate the majority of this value as related services are anticipated to represent more than EUR 5 billion of annual value by 2035, highlighting their importance within the marketplace. The other main sectors, defense and leisure, will continue contributing to this marketplace and remain the largest sources of value in the near term. Both together represent nearly EUR 2 billion in annual product-related turnover in

Europe over the long term. An estimated total of at least EUR 200 million in additional research and development over the next 5-10 years, based on expectations of the market, is required to address remaining gaps related to **very low level** (VLL) activities that represent the majority of future UAV operations.

Examples of some of the most influential missions, in terms of the potential number of UAVs and economic impact, include the following [13]:

1. **Agriculture sector** where over 100000 UAVs are forecasted to enable precision agriculture to help drive increased levels of productivity that are required and support greener farmer practices. Agriculture is expected to emerge as the dominant application sector over the forecast period on account of several benefits attained such as yield increase due to crop monitoring.

2. **Energy sector** where close to 10000 UAVs limit risk of personnel and infrastructure by performing preventive maintenance inspections, limiting the amount of downtime that is a heavy importer of resources. The environment can be protected by properly maintaining assets.

3. **Delivery and E-commerce** purposes where there is potential for a fleet of nearly 100000 UAVs to provide society with some kind of urgent service capabilities, such as transporting emergency medical supplies, and premium deliveries. Both urban and remote areas will profit of this capability.

4. **Public safety and security** where approximately 50000 UAVs would provide authorities like police and fire forces the means to more efficiently and effectively locate endangered citizens and assess hazards, as they carry out civil protection and humanitarian missions such as search and rescue missions.

1.2.4 Privacy issue

UASs bring with them very substantial increases in the intrusiveness of surveillance generally and of visual surveillance in particular. One of the major concerns related to the use of UAVs is privacy. It is easy to mount a camera or a device to capture information, which may occasionally violate the privacy of people. To overcome such concerns, the U.S. **Center of Democracy and Technology** (CDT) asked the FAA to issue rules on privacy and recommend using data collection statement to know whether the information collected will be retained, used or disclosed. The most suitable means to maintain privacy was considered as **privacy by design** (PbD), which helps in maintaining standards and provides remedies for security breaches. By adopting PbD principles, privacy intrusion becomes limited and privacy can be ensured at an early stage. As part of a privacy education campaign, the agency provides all UAV users with recommended privacy guidelines as part of the UAS registration process.

Privacy is considered a major concern when a UAV goes beyond line-of-sight, but non-line-of-sight operation is also typically required for UAVs, such as for search and rescue or taking a survey of an area. For example, a considerable amount of U.S. operators of UAS should also be aware that, depending upon their actions and intent, they could be charged with criminal video indiscretion.

Countries need to revise or extend the existing regulatory framework or establish a coherent, comprehensive and balanced regulatory framework. There are two principal privacy issues associated with commercial use of UAS: defining what privacy means in the context of aerial systems and selecting a government entity to oversee such issues.

1.3 UNMANNED AERIAL SYSTEM

A **UAS** is a **system of systems** (SoS): a set of complementary technologies brought together to fulfill a specific task. The component systems that make up the UAS include the air vehicle, the ground-control station, payloads, data link, and support equipment [21, 90]. UAS capabilities, such as remote human crew interaction enabled by wireless communication and autonomous vehicle functionality, are central to these systems. The basic design has a micro-controller that acts as a **flight control system** (FCS), usually with actuators, a radio receiver, electronic speed control, and a battery. In addition, inertial measurement unit, gyroscopes and other sensors are added to increase the mid-air stability of the UAV and a GPS device can be used for navigation. Most UAVs also carry at least one camera for aerial imagery, and a gimbal for added image stability. Additionally, other sensors can be attached, though there is a trade-off with increased functionality and weight. Ultrasonic sensors or lidars can be directly integrated in obstacle avoidance operations, while laser range finders provide range measurements for obstacle detection and mapping of 3-D environments. Visual stereo or monocular camera systems are able to provide depth measurements for obstacle avoidance and mapping tasks. High-performance cameras are mostly used, many with useful zooms. Gimbal technology is necessary to capture quality aerial photos, film or 3-D imagery. The gimbal allows for any vibration from the UAV to not reach the camera. Additionally, the gimbal can be tightly coupled with the **inertial measurement unit** (IMU) for visual-inertial ego-motion estimation and the raw image stream is also required for infrastructure inspection [65].

Some UAVs have a radar positioning and return home, as well as a no fly zone UAV technology. **First person view** (FPV) technology is also widespread; it means a video camera is mounted on the UAV broadcasting the live video to the pilot at the UAV from the pilot's actual ground position. FPV control allows for more precise flying around obstacles especially with UAVs which can easily fly indoors and through forests. Another live video uses the 4G/LTE providing an unlimited range and a low latency video, comprising a camera module, a data module and a 4G/LTE model. LED flight indicators are for indicating where the nose of the UAV is. Remote control system and receiver are generally delivered by the manufacturer. Multi-spectral, Lidar, photogrammetry and thermal sensors are now being used on UAVs to provide 3-D models of buildings and landscape; **digital elevation maps** (DEMS) of land, and provide precision data on the health of crops, flowers, fauna, shrubs and trees. Thermal detection allows access to very high-resolution, geo-rectified thermal imagery at local spatial scales enabling a wide array of industry's needs to be met. Applications can include precision agriculture, search and rescue missions, aerial firefighting, and even analysis of heat losses for industrial buildings or electrical utilities. Infrared imagery can improve response time and focus maintenance efforts saving time, money, and lives. On big UAVs, **automatic dependent surveillance broadcast** (ADS-B) can also be implemented. UAV technology is constantly evolving as new innovation and big investment is bringing more advanced UAVs to the market every few months. UAV technology and science covers everything from the aerodynamics, materials in the manufacture of the physical UAV, to the circuit boards, chipset and software.

Key attributes that enable UAV operation are [25]:

1. **Awareness of the UAV location** within the operational space of its attitude and of its direction and speed of movement;

2. **Sensors and/or remote data-feeds** that enable maintenance of the awareness of location, attitude and movement in a sufficiently timely manner;

3. **A set of controls** over the UAV attitude, direction and speed of movement, to enable flight to be sustained under a wide variety of atmospheric conditions;

4. **Maneuverability**: sufficiently rapid response of the UAV to the controls;

5. **Sufficient power** to maintain movement, to implement the controls, and to operate sensors and data-feeds, for the duration of the flight;

6. **Ability to navigate** to destination locations within the operational space;

7. **Situational awareness**: the ability to monitor the operational space;

8. **Collision avoidance**: the ability to navigate with respect to obstacles;

9. **Robustness** to withstand threatening events, such as wind-shear, turbulence, lightning and bird-strike.

Recent developments increase the use of small UAVs in mission applications [50]. It is important to develop tools applicable at all levels of autonomous operation, from inner loop control at a lower level to interaction with human operators and/or mission management systems at a higher level [51]. Challenges of UAS integration with manned aviation in **national airspace system** (NAS) include detect-and-avoid systems, robust and fault-tolerant flight control systems, communications, levels of autonomy, network-controlled teams, as well as challenges related to regulations, safety, certification issues, operational constraints, and frequency management.

1.3.1 Ground control station

Ground control station provides the work environment for the UAV operator. A challenge for remote crews is to create a controlled work environment despite the fact that they may be operating in a wide variety of conditions. The following factors must be considered:

1. Controlled airspace
2. Restricted airspace
3. Nearby airports and heliports
4. Sensitive areas on the ground
5. Obstructions to the aircraft's flight path or to the visual line-of-sight from the ground control system location to the aircraft
6. Hazards to the aircraft in the take-off and landing
7. Unprotected people beneath the flight path of the aircraft.

Current UAVs are controlled in either autonomous or manual control modes depending on the type of UAV and the mission phase. The most common control modes are:

1. Manual Control:
    (a) **Radio control** (RC) aircraft by an **external pilot** (EP) using a third-person remote view of the UAV is very common with small UAVs;
    (b) Flight console, similar to a cockpit, using a forward fixed camera view to allow an external pilot to fly the UAV as in a simulator;
    (c) **Virtual reality** (VR) methods employing various forms of **FPV** flying, including head-tracking techniques.
2. Autonomous Control:

(a) Autopilot control, usually using **global positioning system** (GPS) waypoints to define a flight plan;

(b) Inertial, airspeed and pressure sensors used for inner-loop airframe control;

(c) **Automatic take-off and landing** (ATOL) capabilities offered by some autopilots.

Manual control is still common during landing and take-off. This is in spite of the fact that the majority of UAV accidents occur during take-off and landing, especially by UAVs which rely on a manual pilot to accomplish these tasks [114].

Categories based on the operating ranges between the remote pilot and the UAV can be presented as:

1. **Visual line-of-sight** (VLOS) operation: the operator is always able to see the UAV without visual aids. It should always be in unobstructed view of the pilot.

2. **Extended visual line-of-sight** (EVLOS) operation: the operator commanding the UAV may rely on other remote observers who are in visual line-of-sight of the UAV. The remote observers must be able to relay critical flight information to the operator in real time.

3. **Beyond visual line-of-sight** (BVLOS) operation: the UAV is operated remotely based on instrumentation between the UAV and a remote ground-piloting station. The UAV is allowed to go beyond visual range. An on-board camera-based system is usually employed but not sufficient to allow BVLOS operations. Additional levels of autonomy like detect-and-avoid technologies are deployed on these systems for safety.

4. **First-person view** (FPV) operation: the operator utilizes on-board video cameras to provide a real-time view from the UAV and operates it based on this video. It is also used to collect sensor and imagery data while in flight.

The development of the civil UAV industry is dependent on the ability of UAVs to operate in various areas of the airspace, especially at **very low levels** that today are generally defined as being below 150 m.

1.3.2 UAV operator

The role of the human operator in unmanned aircraft systems depends on the mission being performed. Identifying similarities between mission types helps to leverage research from one domain of UAV operations to another. UAVs are required to complete multiple missions in order to achieve certain objectives. Each mission type consists of three phases: mission planning, mission management, and mission re-planning. The three phases are chosen so as to reflect the changes in operator tasking in time. Each phase consists of a number of steps, called phase goals. The phase goals can be used to derive the functional and information requirements needed for an interface to support the corresponding goals. The specific functions required from the human operator can also be derived from the phase goals of each mission type. Depending on the decision support tools and the level of automation aboard the UAV, the operator functions can be undertaken in several ways. These mission types are [87]:

1. **Monitoring health and status of the UAV**: The health and status of a UAV are essential attributes to the proper functioning of the vehicle.
2. **Notifying relevant stakeholders**: In many UAV missions, the operator is a member of a hierarchy of stakeholders. In these missions, it is essential for the operator to be able to communicate with the relevant stakeholders in order to notify them of the success or failure of the mission objective.
3. **Optimal position supervision**: Many UAV missions require one or more activities to be performed at a predetermined location. Mission types require often optimal position supervision, a function in which the human operator must take part since it is based on a mission goal.
4. **Path planning supervision**: The path taken by a UAV to reach a location is an important part of mission success, particularly in congested or dangerous operating environments. Thus, careful planning and re-planning of a path is an important operator function of many missions.
5. **Resource allocation and scheduling**: Many UAV mission types involve distributing the resources of the UAV on multiple sub-objectives.

1.3.2.1 Training environment

Currently, two main emerging topics in UAS are how to reduce the number of operators to optimize the human capabilities as much as possible, and how to improve the training process to obtain more qualified UAS pilots. The main goal is to make UAV operations more accessible [105]. To ensure that a training process is effective, the most relevant outcomes, such as **knowledge, skills and attitudes** (KSA), must be defined. To extract this information, the analysis of training processes is mostly focused on two different skills: mission monitoring and mission planning.

The data extracted from training processes can be used to generate pilot profiles. User profile extraction is a methodology in **human-machine interaction** (HMI) systems. Given an environment where the user interacts, a user profile can be generated based on its situation awareness. This process is performed by collecting data from the interactions between the user and the environment. It is highly dependent on the mission, the environment and the sensors, architecture and UAV setup.

The data information can be summarized as a set of **performance metrics**, describing the quality of the interactions, and helping to define the user profile. UAV training processes usually require expensive tools to ensure that the operator can deal with as much real issues as possible. However, the skills of an operator can still be evaluated with a simpler environment. The main problem is to collect the right data for defining their profiles. To be able to extract the knowledge to be evaluated, a UAV simulator can be developed to simplify the training process focusing on the most relevant tasks of the missions and ensuring good portability and usability to be more accessible. The operator must introduce an initial mission plan. In many situations, the environment is not fully known or understood, and UAV routes must be planned and executed under time pressure. Effective route planning is a difficult task, and operators report high workload and low accuracy when planning and executing routes in a remotely operated vehicle [108].

Remark 8 *One source of difficulty is that the images on a video feed are perceptually challenging stimuli. Compared to natural vision, the video screen provides less visual information than would be available to a physically present observer. During this plan, unexpected events will occur. That is the moment when the operator has to make decisions in order to complete or abort the mission.*

Decisions give several information about the user skills and attitudes. However, this information needs to be extracted and interpreted; therefore, during the mission, different data related to the user are collected and used later to define the user profile.

In 2035, approximately 250000 pilots in Europe only are expected to support the operation of some 400000 UAV fleet. As technical expertise and responsibility of the pilots varies across mission types, different salary costs are allocated to different types of pilots.

1.3.2.2 Assistant systems

To allow the guidance of a UAV in a manner similar to the guidance of a manned aircraft, the UAV should be furnished with a system able to interpret tasks with respect to the current mission and tactical situation and act upon those tasks in a situation-specific way. These types of systems may use artificial cognition to interpret, act on, and communicate by use of more abstract commands [120].

The introduction of a highly complex automation aboard the UAV poses the challenge of handing the automation to the human operator. Experiences with automation systems for manned aircraft show that complex automation can lead to automation-induced errors and may raise the human workload in situations where the workload is already at a critical level. This problem can be addressed with the introduction of **assistant systems**. To aid the operator in guiding the UAV, the assistant system needs to make decisions on whether, when and how to inform the human operator about any problems concerning the mission progress. Therefore, the assistant system shall be able to anticipate which tasks the human operator should perform in order to achieve the mission goal.

In [37], an approach is described on how to supplement the operators with a cognitive and cooperative assistant system. The issue of detection of the actual operator task and critical operator workload is investigated using models of human behavior. Within different task situations, operator performance and human error during task processing can be observed. To be able to support the operators, the following guidelines for the expected behavior of cognitive assistant systems are followed:

1. Present the full picture of the work situation.
2. Ensure that the operator attention is placed on the most urgent task.
3. If the human operator is overtaxed in performing the most urgent task, then the assistant system should automatically transfer the situation into one that is manageable.
4. If task risk or costs are likely to be too high or the operator is not capable of performing the task, take over or re-allocate the task to an operating supporting mean.

Remark 9 *Cognitive automation provides cognitive capabilities similar to those of a human pilot. These capabilities include planning, decision making and dynamically reacting to changing situation and environment. The operator is able to focus on high-level mission goals and realize the advanced concept of task-based concept.*

The operator will issue a set of abstract tasks to the UAV. These will be processed on the on-board artificial cognitive unit to create a complete and detailed task agenda which will be dynamically adapted in case of unforeseen events [115].

1.3.3 UAV simulation

Realization of complex functionalities of the UAV involves a software platform performing various real-time tasks concurrently. Aircraft environments and techniques for real-time visualization and interaction with a multi-dimensional set of aircraft data are necessary. Development of advanced user interfaces improves completion of the associated tasks. An approach to improve re-usability of data visualization UAV software across different platforms by defining a supporting software architecture based upon common software standards and practices can be proposed [64]. The research produced a set of re-usable object-oriented software, namely, a software tool for UAV mission planning, execution and monitoring. The central module of the system is the visual simulation platform, given the added difficulty of simulating motion through the atmosphere. Some UAV simulators are available such as:

1. X-Plane [10] is a commercially available flight simulator developed by Laminar Research. X-Plane has been used by some aircraft vendors and aerospace agencies as a training simulator because of its high-fidelity simulation of flight model and visualization.

2. Flight Simulator of Microsoft [8] provides expansive environments with real terrain data. It requires a proprietary software development kit for scenery and cockpit creation. However, unmanned aircraft models must be created using third-party three-dimensional (3-D) modeling tools.

3. CAE [1] has developed the synthetic environment as a practical representation of the real world for UAV simulation. A ground control station is being used to operate a simulated UAV in a synthetic environment. The synthetic environment denotes practical representation of the real world comprising the UAV air vehicle simulation, payload simulation, and comprehensive environment.

4. MATLAB [7] presents a numerical simulation environment. It is used for many simulation applications; however, it does not provide the real-time high-fidelity visualization or physical simulation necessary. It can be used successfully in conjunction with environments for high-fidelity visualization.

5. FlightGear [5] is a general-purpose, open-source flight simulator. It provides an extensible scenery base, as well as a set of predefined cockpit environments. Aircraft models for the simulator must be created in an external 3-D modeling application. Additionally, it requires an extensible markup language configuration file describing the various aircraft features that must be created by hand. The entire source code is available for modification and is under continuous development. The simulator can run on different operating systems. In addition to simulation environments that provide a wide range of functionalities, there are specialized solutions covering specific areas of data visualization.

6. OpenGC [9] is an open-source C^{++} environment for developing and implementing high-quality glass cockpit displays for simulation environments. It can be used by FlightGear and Flight Simulator. Atlas is a specialized tool aimed to produce and display high-fidelity maps for users of FlightGear. Except for functional and presentation requirements of UAV data visualization software, there are several issues that must also be considered, such as the ability of the solution to be modified and

re-used to support new scenarios and the level of platform independence from the facets of implementing and running platforms.

7. Robot Operating System (ROS)-GAZEBO [48] is an open-source collection of software packages specialized in robotic applications. One of its main purposes is to provide communication between the user, the computer's operating system and equipment external to the computer. This equipment can include sensors, cameras as well as robots. The benefit of ROS is the hardware abstraction and its ability to control a robot in general, a UAV in particular, without the user having to know all of the details of the robot. UAVs can be simulated using the 3-D Gazebo simulator which includes the physics of the UAV as it moves around in a simulated environment. Gazebo's collision detection engine uses the collision property to identify the boundaries of the object.

1.3.3.1 Architecture

Architecture is the conceptual model that defines the structure and behavior of the system.

Definition 10 *The systems architecture process consists of* ***aggregating****,* ***partitioning****,* ***integrating*** *and finally* ***validating*** *systems architecture. Architecture process is the process by which standards, protocols, rules, system structures, and interfaces are created in order to achieve the requirements of a system, to respond to a given need.*

The UAV architecture assessment model should also consider the complex and dynamic system hardware and software that would enable the UAV to fulfill complex mission requirements. In the systems architecture generation and assessment models made specifically for UAVs, it is essential to have an effective assessment model to determine whether the modeled system architecture is feasible and acceptable [102].

Centralized multiple UAV architectures have been proposed where a single operator interacts with and oversees every UAV in the network. However, a centralized network requires significant operator cognitive resources. Decentralized multiple UAV networks are another, more complex possible architecture where an operator interacts with an automated mission and payload manager, which coordinates a set of tasks for a group of UAVs. While a single operator could maintain effective control of a relatively small network of centralized UAVs, decentralized architectures are more scalable, particularly in terms of operator workload, and more robust to single points of failure. Task-based control of UAV architectures with higher degrees of autonomy as decentralized networks can reduce workload of the operator [32].

1.3.3.2 Human–UAV interface considerations

Current UAVs are frequently equipped with many sensors and controls. Local awareness and control are particularly important. There are many different tasks which would be facilitated by capabilities of a camera-equipped UAV. These tasks range from bridge inspection and news reporting to wilderness search and rescue. An appropriately designed interface should provide a context for interpreting video and support UAV tasking and control, all within a single display screen [28].

Phones and tablets provide currently a means to monitor and control hobby UAVs. The improvement of situational awareness of a UAS pilot to the operational surroundings, state and orientation of the aircraft, and payload data; potentially improve the safety, efficiency, and effectiveness of the application of unmanned systems technology [121]. The fact that the UAS is physically separated from the operator presents unique issues and makes maintaining high degrees of situational awareness difficult. Such systems must compensate by incorporation of varying degrees of automation and autonomy into their design so that operational aides can be implemented to mitigate the effects of sensory isolation and degraded situational awareness.

Human-UAV interfaces must consider several factors that tend to be critical [97]:

1. The dynamics of a small UAV require the interface to support a significant level of autonomy for the UAV to be accessible to many users.

2. Many users, specially the hobbyist, have little to no experience flying UAVs, and can be confused and disoriented by their many degrees of freedom.

3. If the user loses control of the UAV, it may quickly result in significant damage or destruction of the UAV.

4. Since the UAV can fly considerable distances away from its operator, depending on the accessibility of the environment, the UAV may not be recoverable in the event of a crash.

5. Each UAV requires ongoing maintenance. This includes physical maintenance such as blade replacements and routine service inspections, as well as firmware updates. Batteries also require periodic checks.

1.4 CASE STUDIES

The UAV market is expected to grow quickly as applications continue to develop. There exists a wide variety of UAVs in terms of shape, size, features, and configuration. The **National Airspace System** (NAS) is also expected to change significantly over the next years, with the introduction of new technologies and procedures [36]. The aim of this section is to provide an overview of the wide range of potential applications of technology in autonomous aircraft: disaster response, law enforcement, search and rescue, environment protection, geological applications, communication, entertainment, etc. Private companies are now investing and offering photogrammetric products, mainly **digital surface model** (DSM) and ortho-images from UAV-based aerial images as the possibility of using UAVs with variable dimensions, small weight, and high ground resolution. One possible application would be to monitor large areas which are probably difficult to access, either because of terrain or negative environmental influences. Problems and limitations are still existing, but UAVs are a capable source of imaging data for a large variety of applications [88]. Theoretically, these options are also available with satellites, but there arise some important limitations. The bandwidth for satellite transmissions is limited and therefore expensive. Clouds can also obscure the image, thus delaying the monitoring. This does not apply to UAVs; they can fly under the cloud layer, and wireless data transmission can be realized more easily. Another advantage is that due to the lower altitude, much better close-ups can be created [107]. The UAV can be required to visit points of interest and monitoring, surveillance, distribution, coverage tasks, etc. In this context, one of the challenges is to design a policy that assigns and schedules demands for service or targets that are defined over an extended geographical area [24].

Definition 11 *A schedule is a timetable in which each task has a specific time to start, end, or both.*

The following non-exhaustive list of case studies in populated areas or rural areas can be presented in the civil UAV capabilities:

1. Show, communication, publicity, and entertainment;
2. Fisheries and agricultural management;
3. Topography, mapping, geophysics, and geological applications;
4. Inspection and surveillance/security monitoring of infrastructure;
5. Weather and meteorology;
6. Real estate;
7. Environmental protection;
8. Safety and homeland security;
9. Crisis management, search and rescue, and disaster response;
10. logistics, transportation, etc.

Different operational scenarios can be considered such as:

1. Day flight, outside populated area, less than 1 km from the pilot;

2. Day flight in a populated area in sight and within 100 m of the pilot;

3. Analysis of failure modes and effects, and ways to mitigate the associated risks: partial or total loss of the propulsion system; loss of external navigation system such as **global navigation satellite systems** (GNSS); loss of servo actuators; loss of command and control logic; loss of the altimeter; loss of the link command and control.

UAVs with intelligent capabilities ranging from large-scale aircraft to micro-robots can be deployed. Operating costs are typically low, and mission endurance may be high [43]. The operator must have the information on the position of the UAV. The aircraft should be equipped with a recording device of critical flight parameters, allowing an analysis of the last minutes of flight. An automatic device prevents the UAV from exceeding a maximum altitude (virtual ceiling). The main functions are avionics for the autopilot and manual override, GPS navigation, communications link, on-board computing, power and electrical, on-board camera, payload delivery, and ground station instrumentation. Key systems include the autopilot, the on-board computer, the power distribution system, imaging software, mission management and payload delivery system [6].

UAV size, payload, range, and operating mode are typically the main criteria a user defines, by answering the following questions [111]:

1. What is the size of the area to be surveyed?

2. What is the flying altitude the UAV needs to operate?

3. What camera system and what camera mount system is needed?

4. Are there third-party persons working in the area?

5. Are there other physical obstacles present?

6. What take-off/landing space is available?

1.4.1 Industrial applications

Industrial sites often contain areas and facilities that are difficult to access or are hazardous to humans. UAVs can be very useful for monitoring the available infrastructure. A UAV can decrease the operational costs, execute the monitoring process, and can be used in situations where manned inspection is not possible. UAVs are already in use in many industrial domains. Apart from the advantage of lower costs, the UAS does not pose a hazard to aircrews, can operate in more adverse weather conditions, and is less obtrusive to neighboring communities or animals. Hover flight is essential for the inspection tasks with UAV carrying an electro-optic and thermal imaging payload and the data must be available in real time to the operator and also recorded.

UAVs are expected to improve maintenance and be used for inspections, which are segmented into two primary mission types:

1. Local site inspections, performed by multicopters operating today in VLOS and below an altitude of 150 m, and

2. Long-range utility inspections for which the fleet is expected to be composed of BVLOS fixed wing UAVs flying near 150 m of altitude with potentially some certified UAVs operating at higher altitude (likely between 300 m and 3000 m).

1.4.1.1 Infrastructure monitoring

Inspecting and monitoring oil-gas pipelines, roads, and power generation grids is very important in ensuring the reliability and life expectancy of these civil systems. The problem of using a UAV to follow roads, highways and canals can be based on visual feedback. The problem of following a structure using a camera on the vehicle has two main sub-parts [100]: vision-based structure detection and controlling the UAV to follow the structure. Real-time vision sensors allow UAVs to passively interact with their environment in a way that other sensors cannot. With this ability, UAVs become more practical solutions for many suggested applications explained in the sequel.

Power-line monitoring UAS can be suitably employed for power transmission line monitoring by mounted required sensors (hyper-spectral) and cameras (thermal and infrared) to detect many faults, malfunction, and degradation on equipment of power transmission. Monitoring power-line corridors is crucial for the reliability of electricity transmission. Trees and shrubs often create obstructions in corridors and pose risks to power-lines, and therefore utility companies need to scrutinize where and how trees grow in or close to power-line corridors. The need to manage risk motivates the collection of data over power line corridors and the identification of objects of interest in order to assess risk levels and guide field workers for vegetation clearance in the corridors. Remote sensing represents a particularly attractive solution for power-line corridor monitoring. Actually, aerial vehicles have been intensively used in power-line inspection for a long time [74]. Traditionally, in hilly or mountainous terrains, aerial inspections of power lines are carried out with manned helicopters, a costly and dangerous exercise. Rotor-craft UAVs are well suited to power-line inspections. The UAV is required to fly at close quarters to the structure, and is therefore at risk of a collision. The problem is particularly hard for inspections carried out beyond line-of-sight of the UAV operator, as is the case when inspecting long power lines. The power-line inspection task also requires the UAV to fly to a goal or a number of sub-goals, for example to visit a set of transmission towers which have been roughly surveyed. Since the precise locations of the obstacles in the environment are sometimes unknown a priori, the UAV would need to detect these as it flies to the goal, and potentially modify the planned path [61].

Power-plant monitoring The boiler unit of a coal-fired thermal power plant constitutes one example of a risky area. Tremendous precautions have to be taken in order to minimize the risks to human workers. Furthermore, scaffolding often needs to be installed in order to grant access to sections that are otherwise hard to reach. These measures result in significant downtime of the plant and are thus associated with vast economic costs. UAVs can access virtually any area and are not restricted to ferromagnetic surfaces. However, the employment of flying systems poses a number of challenges such as limited payload capabilities, and hence limited computational resources, or reliable state or vehicle ego-motion estimation in the absence of GPS signals, high-quality **inertial measurement units** (IMUs), etc. By arranging the pre-programmed trajectories around an area of interest, an operator can automatically acquire sets of images for visual inspection. In this respect, the autonomy of the system facilitates efficient inspection of large structures and enables remote inspection in areas, where occlusions or impaired radio reception had rendered it infeasible before [89].

Oil and gas applications Pipelines play a critical role in the oil and gas production process because they are used to provide a less expensive and faster means of transporting either the crude oil or the refined products. Aerial and satellite remote sensing integrated with **geographic information system** (GIS) assists pipeline risk assessment to assure the safety of pipeline facilities [106]. The deployment of **supervisory control and data acquisition** (SCADA) systems has also made it possible to monitor in real time the flow rate, pressure, and temperature of crude oil as it moves from one facility to another. Whenever a drop in pressure is detected, the system can be configured to notify the operators and automatically send a **request for inspection** (RFI) to the operator in the flow station covering the pipeline section where the drop is detected. The operator then launches the UAV and guides it to that location using the flight controller, and the UAV provides a video feedback of the pipeline **right of way** (ROW) as it moves to the suspected point of activity. The pipeline networks are to be broken into wide area cells controlled by the processing facility, such as a flow station, with the size of the cells determined by the transmition range of the UAVs located at these flow stations. This system provides accurate information on the type of activity causing the pressure drop [62]. The selection of the UAV and the cell size is made such that the cell sizes are small enough to shorten the required flight time for the UAV to arrive at the location of interest. The pictures taken by the cameras or the video recording is relayed to the ground-control center in real time to enable the operators to determine the state of the pipeline and deploy the appropriate response in the event of any incidents on the pipeline. The desired video- and picture-recording cameras should have pan, tilt, and zoom functionalities that are controllable from the ground-control center. The cameras can also have thermal-imaging capabilities and night-vision capabilities for data capture at night. The UAVs can also be programmed to undertake routine monitoring, with the flight plan and route programmed into the on-board controller. The route of the UAV can be programmed such that the UAV goes from its base through a specific route capturing the video signals along that route [93].

On the basis of a data fusion and evaluation process, the remote monitoring system must be capable of identifying objects and situations representing threats to the pipeline. In terms of its price/performance ratio, it must at least correspond to the methods presently in use, or be better [60]. For teleoperation, UAVs act as a remote system in autopilot mode, with programmed navigation system equipped with GPS and inertial sensors, while transmitting and receiving information signals from ground-level operator [110].

Decision making for improved risk management of oil pipeline infrastructures introduces a set of decision methods for risk management of oil and gas pipelines. A risk-based decision support system reduces the amount of time spent on inspection using **multi-criteria decision analysis** (MCDA) framework. The problem can be stated as a problem of design of monitoring trajectories for a group of UAVs equipped with sensors for damage detection. The UAVs fly autonomously from and to a base and their movement is not restricted by the pipeline. Formally, the problem of oil pipeline monitoring using multiple UAVs can be seen as an optimization problem with multiple mobile agents optimizing a joint criterion function [92].

1.4.1.2 Photovoltaic modules monitoring

After a fast **photovoltaic** (PV) expansion in the past decade, one of the most significant open issues in this sector is to find appropriate inspection methods to evaluate real PV plant performance and failures. Large-scale PV fields are a complicated target for monitoring and performance evaluation [119]. PV modules are the key components affecting the overall system performance; therefore, there is a main concern about the occurrence of any kind of failure. Weather conditions surely affect the system performance over time, and various defects can occur in the PV modules even in the short term. Solar irradiation and temperature are the main stress factors that can degrade PV modules. Other weather and environmental conditions such as wind, hail, snow, dust, and many corrosive gases can indirectly have an effect in increasing and decreasing these two main factors. A concept for monitoring PV plants uses UAS during their operation and maintenance. For this purpose, some thermal imaging cameras and a visual camera can be chosen as monitoring tools to suitably scan PV modules. PV array field monitoring by UAVs has many advantages, such as low cost and large area coverage, precise imagery, the ability to operate in harsh environments, and sensible time reduction. The procedure of utilizing UAV is effective in the detection of different failures of PV modules [96].

Cooperative inspection of PV plants can be carried out by small UAVs and utilize the GPS to find out the optimum route mapping during the solar

PV modules monitoring. Thermal and infrared cameras can be used to detect high-temperature regions in the PV module surface and mounted on-board to perform a fast and reliable monitoring task. Therefore, a thermography technique can be employed to detect different module defects. There are different devices for inspection of energy generation equipment. Typically, PV systems monitoring can be carried out by high precision electro-optic sensors and high-resolution infrared cameras as well. In fact, these latter devices are useful to detect temperature regions in PV modules which may be affected by hot spot. Inspection devices help to find out defects and faults before degradation of modules and it leads to mitigate PV module's defects by early solution recommendation with expert engineers in this area. Thermography analysis is the understanding of the temperature rate of PV modules surface. This inspection method is normally carried out during PV systems working period [58]. The analysis and the visualization of the differences of similar objects is important for scan alignment, nominal/actual value comparison, and surface reconstruction [67].

1.4.2 Civil engineering

Structural health monitoring (SHM) is a key component in maintaining a sound infrastructure. The expansion and development of urban areas, as well as the deterioration of existing infrastructure components, such as bridges and dams, have increased the demand for routine structural integrity assessments. For several decades, various sensors and sensing systems have been developed to monitor and assess the safety condition of structures. SHM is an essential component in civil engineering for safety and integrity of civil structures such as buildings, bridges, power plants, and tunnels. Meanwhile, advanced sensing and robotics technologies greatly facilitated construction automation of infrastructure systems [83]. By capturing very large collections of images and videos, along with methods that process the visual data into 3-D models, UAVs can survey construction sites, monitor work-in-progress, create documents for safety, and inspect existing structures, particularly for hard-to-reach areas [20]. Some methods transform and visualize actionable performance information from the collected data, using **building information models** (BIMs) as a priori information. The goal of a UAV-driven visual performance monitoring procedure is to:

1. Collect images or videos from the most informative views on a project site,
2. Analyze them with or without a priori building information models to reason about performance deviations during construction,
3. Monitor ongoing operations for productivity and safety,
4. Characterize the current existing conditions of civil infrastructure systems,

5. Visualize and communicate the most updated state of work-in-progress with on-site and off-site project participants.

Providing accurate performance information about the state of construction or existing conditions of civil infrastructures requires UAVs to collect visual data in the form of images and videos from the most relevant locations and views on a project site [59]. Image sequences taken from a UAV device have their own specific characteristics: moving camera, high distance from objects, non-constant angle between camera axis and the ground. Consequently, specific techniques have to be developed to analyze their content. Photometric and geometric features can be extracted so that a classification is performed to differentiate the building and non-building. Effective use of multiple features of remotely sensed data including spectral, spatial, and multi-temporal information is especially significant for improving classification accuracy [122].

1.4.2.1 3-D building imaging

Accurate 3-D reconstruction has become essential for non-traditional mapping applications. Low-cost digital cameras, laser scanners, and navigation systems can provide accurate mapping if they are properly integrated at the hardware and software levels. UAVs are emerging as a mobile mapping platform that can provide additional economical and practical advantages [81]. Rotor-craft UAVs, capable of performing the photogrammetric data acquisition with digital cameras, can fly in manual, semi-automated and autonomous modes. With a typical photogrammetric pipeline, 3-D results like **digital surface model** (DSM) or **digital terrain model** (DTM), contour lines, textured 3-D models, vector data, etc. can be produced, in a reasonable automated way [101]. The process of creating a 3-D model concludes the following steps:

1. Creation of a preliminary map,
2. Determination of the unique features/points,
3. Determination of the matching points,
4. Combination of the subsequent images,
5. Specification of the image coordinates, and
6. Creation of a 3-D model.

Climate and condition of the field have a direct impact on the required time for scanning and generation of 3-D models [41]. The quality of a 3-D building model, is directly related to the optimized usage, flight stability and the efficiency of the electronics. When placing diagnostic equipment onboard the UAV it is important to consider the factor of shifted center of gravity of the machine from its ideal positioning in order to generate noise-free control signals [23].

1.4.2.2 Roof insulation inspection

UAVs equipped with high-resolution thermal cameras can graphically depict energy inefficiencies and identify wet insulation in the roof or elsewhere by showing temperature variations within the building surveyed with great efficiency [125]. Roof inspection with thermal infrared cameras is typically performed to detect and locate areas of wet insulation. This non-contact and non-destructive method provides a fast, accurate and inexpensive way to locate areas of wet insulation and potential leaks. Two important imagery specifications to consider for UAV thermal infrared roof inspection to produce usable imagery are thermal sensitivity and image resolution. The thermal sensitivity or **noise-equivalent temperature difference** (NETD) is the measurement of the smallest temperature difference that a thermal imagery can detect in the presence of electronic noise. The lower the thermal sensitivity, the more detailed and less noise present on the thermogram. Highly sensitive (low-NETD) thermal imagery shows more temperature differences, and thus more patterns. Image resolution is important for capturing clear images from a distance. High resolutions are needed when observing the roof from greater distance, such as in a flyover.

The **Property Drone Consortium** (PDC), a collaboration that consists of insurance carriers, roofing industry leaders and supporting enterprises, have a **Cooperative Research and Development Agreement** (CRADA) with the U.S. Department of Homeland Security Science and Technology Directorate. The agreement leverages the knowledge base, capabilities, and resources of the parties to the agreement to advance the understanding of the use of UAS for public safety missions. UAS data can be used by first responders, as well as by insurers and owners of other properties and structures looking to assess damage and take remedial action relative to providing assistance to victims, quick assessment of claims, and urgent repairs.

1.4.2.3 Bridge inspection

While bridges represent only a portion of the national infrastructure, they are a critical element. Visual inspection is required of most structures periodically. In addition to regulating inspection procedures and inspector qualifications, the bridge inspection guidelines standardize a rating system to quantify various structural health levels of three major bridge components: the deck, the superstructure, and the substructure [80]. The structural inspection is a long process that requires the participation of a large staff and a wide variety of specialized equipment. Therefore, the need for better solutions leads to a UAV-based structural inspection. The imagery and data obtained from the UAV is post-processed using a specialized software to reconstruct a virtual model of the structure, taking into account the Venturi effect. Despite the developments in UAV-based structural inspection, there are challenges to be tackled. For instance, the guidance and navigation of the UAV along the structures is an important challenge which consists in the trajectory planning to fly around the entire structure. The inherent nature of a structure inspection mission using UAVs implies the interaction of different domains including, UAV stabilization and control, navigation and obstacle avoidance, wireless communications, computer vision, etc. In these kinds of missions, resources optimization as well as risk and operational cost minimization are fundamental issues which cannot be avoided [20]. A large variety of techniques can be applied such as the potential field method, cell decomposition, the roadmap method, rapidly exploring random trees, mixed integer linear programs and diverse network-based approaches.

1.4.3 Safety and security

UAVs provide a cost-effective solution to other costly equipment; they can provide great return on investment to local public agencies. The aim of this section is to introduce how UAVs are used in hazardous public safety situations, the value of UAVs to public safety organizations and the challenges and rewards of starting a public safety UAVs based business. These needs create three general mission types:

1. Stationary surveying by multicopter UAVs that are operated by on-site forces that carry a UAV in their vehicle;
2. Long-range surveying by future versions of the technology that operate more beyond visual line-of-sight and are operated more centrally at altitudes near or below 150 m, and
3. Higher altitude (i.e., above 150 m) surveying UAVs to screen large areas as part of border security, maritime surveillance, and environmental protection.

1.4.3.1 Traffic monitoring

One promising application of UAVs for safety and security missions is to enhance the traffic monitoring systems which have been widely deployed and serving as a backbone of **intelligent transportation system** (ITS) infrastructure. Despite their importance, the traffic monitoring systems are rare to find in many rural areas or, if they exist, they are to observe simple traffic counts at specific locations rather than be used for comprehensive traffic operations, mainly because of cost effectiveness issues. For this aspect, UAVs provide a cost effective means to meet the need for rural traffic surveillance system. Highway traffic flow is very dynamic and is an uncertain environment that requires instantaneous and accurate information on both accessible and remote areas [104]. Surveillance, travel time estimation, trajectories, lane occupancies, and incidence response are the most required information.

Congestion monitoring Congestion continues to grow on roadway networks. To collect precise and timely information about the traffic state for improved control and response will lead to reduced traveler delay, as well as improved health status of injury victims through faster medical attention. At any given instant, the biggest value comes from monitoring only a small portion of the network. Unfortunately, the specific portion that would provide this largest value is constantly changing and often is not known a priori. For example, the points where queuing will form as a result of an incident depend on where the unpredictable incidents occur. Conventional traffic surveillance uses a set of detectors (including cameras) deployed at fixed locations, requiring a high density of detectors to ensure the ability to respond rapidly under changing conditions throughout the network. When information is needed from beyond the range of these fixed detectors, a person may be deployed to assess conditions [26].

The U.S. **National Consortium for Remote Sensing in Transportation Flows** (NCRST-F) have recognized the potential of UAVs to provide a low-cost means to achieve a **bird's eye view** and a rapid response for transportation operations. The information that would be useful to collect in UAV traffic surveillance is lane changes of the vehicles, average inter-vehicle distances, heavy vehicle counts, accidents, vehicle trajectories and type of vehicles. Even though loop detectors can be used, they provide only local information and cannot provide details like lane changes of vehicles. On the other hand, a UAV equipped with a camera can provide a global view of freeways with the relevant information thus enhancing the real time monitoring aspect. Also, the UAVs have an advantage over manned vehicles as they can fly at lower heights and can be used when weather conditions are not suitable for flying with pilots. Collecting useful information with UAVs in this application has two problems: the first problem is to keep the camera in view of the road and the second is to process the image to collect relevant data [99].

Usually, the traffic data captured by UAV contains much more complex information than those by traditional monitoring system. UAV videos include not only the traditional data such as the traffic flow average speed, density and flow, but also each vehicle's level data, such as vehicle's trajectory data, lane change data and car following data on the road. In addition, the data from a frame of a UAV video contains multiple vehicles and the frame frequency of the UAV video is very high, thus the data size from the UAV video is very large. Considering such features, the data collection, reduction and analysis can be considered as an important component in the big data analysis in transportation. Among the issues that have to be addressed are physical layer issues, communications issues, and network layer issues [112].

The project, presented in [95], serves as a case study for the use of UAVs in remote sensing and multi-modal transportation. The main goals are:

1. Development of reliable software and hardware architectures with both deliberative and reactive components for autonomous control of UAVs;

2. Development of sensory platforms and sensory interpretation techniques with an emphasis on active vision systems to deal with real-time constraints in processing sensory data;

3. Development of efficient inference and algorithmic techniques to access geographic, spatial and temporal information of both a dynamic and static character associated with the operational environment;

4. Development of planning, prediction and chronicle recognition techniques to guide the UAV and predict and act upon behaviors of vehicles on ground; and

5. Development of simulation, specification and verification techniques and modeling tools specific to the complex environments and functionalities associated with the project.

Driving-behavior monitoring In the research of driving behaviors, a detailed and accurate vehicle trajectory data is also necessary. Driving behavior models capture drivers' tactical maneuvering decisions in different traffic conditions, which are essential components in microscopic traffic simulation systems. Compared with traditional traffic surveillance systems, detecting and tracking vehicles through the images captured by a UAV has specific challenges. First of all, the camera of a UAV surveillance platform changes frequently because the camera in a UAV may rotate, shift and roll during video recording. In addition, sudden shakes might also happen due to wind fluctuations, which can cause negative effects in the vehicle tracking. On the other hand, in driver behavior research models, such as car following and lane change models, each car's accurate trajectory data is needed. Missing car data and tracking error could affect the accuracy of the model parameters settings. Therefore, a high resolution of images is crucial for accurately calculating vehicle speed and lateral position of vehicles in the process of vehicle detecting and tracking. A vehicle recognition method can be categorized into optical flow and feature extraction-matching methods [123]. Abnormal driving behaviors have been used as cues to identify **driving while intoxicated** (DWI) drivers and prevent DWI-related accidents. Currently law enforcement officials rely on visual observation for detecting such behaviors and identify potential DWI drivers. This approach, however, is subject to human error and limited to vehicles in a very small region. To overcome these limitations, a UAV can be used for driving-behavior monitoring to prevent accidents and promote highway safety. Dangerous driving activity on highways can be effectively and timely detected and analyzed. These misbehaviors have been commonly used by law enforcement officials to identify potentially drunk drivers as summarized by the U.S. **National Highway Traffic Safety Administration** (NHTSA) in 2010. To observe the eight potential misbehaviors, six key metrics must be identified and quantified. These key metrics consist of vehicle ID, speed, forward distance, lane change, lane change time, and acceleration. The calculation of these key metrics is therefore the first task of the computer vision algorithm to determine and quantify the misbehavior. The calculation of the six key metrics requires both the position of a vehicle relative to a lane line along with the identification and tracking of a vehicle across video frames [126].

1.4.3.2 Nuclear, biological, and chemical accident

Nuclear, Biological, and Chemical (NBC) accident is another example of the teaming of manned and unmanned systems. In disaster recovery projects, it is often necessary to map and quantify radioactive contamination in areas which are difficult or impossible for people to get to. This task is an ideal application for robotic systems. In particular, UAVs appear to offer the promise of a generic solution. UAV projects include emergency response radiological surveys of the site and were also deployed for performing inspections of containment structures. However, UAVs are difficult platforms to use at nuclear sites; they typically rely on GPS for stability and control, which is unreliable or absent near or inside metal clad buildings. It is also difficult to provide quantitative measurements of contamination from data captured by a UAV [109].

In another case of emergency reconnaissance, a UAV should fly into a toxic cloud and bring back a sample of contaminant for analysis. Since human operators do not want to stay close to the accident site, the long range of operation is required as well as the ability to fly against strong opposing winds [54]. Estimation of the plume concentration resulting from a gas source in the atmosphere and the detection of the source location have diverse applications related to environment as well as in search and rescue missions. The real-time estimation of the plume concentration enables the deployment of countermeasures that neutralize the possible adverse effects of the plume and the localization of the source [39]. Plume estimation can be addressed using stationary, surface or aerial sensors in the plume area, with the help of gas transport models. The procedure constructs a map with the possible source localization using measured concentrations at specified locations and an initial assumption of the source location. Sensor mobility control refers to a strategy for gathering sensor measurements to support a sensing objective, such as environmental measurement. When the sensors are installed on UAV platforms, an important part of the problem is planning the sensor path to achieve low working time or low energy consumption, obstacle avoidance in unstructured dynamic environments, or efficiently gather target information.

1.4.3.3 Landmine detection

The use of UAV is clearly suited to covering a minefield without the risk of triggering landmines during the mission. The aerial system enables:

1. The terrain mapping based on a visual stitching method to generate a mosaic image of the covered terrain, and

2. The visual detection of landmine-like objects in real time.

For demining, navigation control is crucial in enabling safer terrain mapping and accurate reconnaissance. In [27], the use of an affordable aerial system is proposed with the following sensors on-board: a CMOS camera, GPS and an IMU, to use an artificial vision-based approach for detecting partially buried landmine-like objects and creating a geo-referenced map of the terrain with the GPS location of the landmines, the system allows:

1. Communication driver to enable the wireless transmission of navigation, sensor and control data between the UAV and the base station.

2. Navigation to handle the flight control. It provides both camera and IMU data to the visual mosaic module.

3. Visual mosaic to build a panoramic image by combining multiple photographic images captured by the camera. It generates a map of the covered terrain.

4. Detection algorithm of partially buried landmine-like objects in real-time using image recognition methods.

5. GPS data: enables the geo-location of the detected landmines based on GPS information.

Vision algorithms are used to analyze the images and to determine whether a landmine has been found. The searching procedure can be developed by analyzing the content of the images obtained from the bottom camera of the UAV.

1.4.4 Environmental applications

The use of cooperative UAVs equipped with simple environmental sensors provides a promising option for safe and cost-effective data collection. Regardless of the number of UAVs employed in an application, the key consideration in the use of UAVs is how to position them in order to gain useful information from their locations and/or path. **Dynamic data driven application systems** (DDDAS) provide a means to position the UAVs in an efficient manner using the real time data obtained from the sensors. The framework of DDDAS are driven by the goal of dynamically incorporating data into a running application (e.g., an environmental simulation) while simultaneously using the application to steer measurement processes. The DDDAS framework is widely used in wildfire simulation, identification of airborne contaminants, and weather forecasting. The DDDAS are a systems-level approach and can be applied to many phenomena where good simulation models are available [94].

1.4.4.1 Geo-referencing

Today a large number of geo-sensors are operational. The available spectrum of geo-sensors ranges from simple static sensors which capture arbitrary physical phenomena such as temperature or humidity to complex sensors mounted on UAS. With real-time close-range imagery, scenarios range from decision support after natural disasters, such as earthquakes, to the virtual piloting of UAV systems. A video stream captured with a small UAV system can be integrated in real-time into a virtual globe technology. Two different integration approaches can be presented, one referred to as augmented monitoring, the other as virtual monitoring. The augmented monitoring approach enables the real-time mapping of arbitrary geo-objects and sharing this information among numerous virtual globe clients with the use of a collaboration framework [44].

Another solution combines state-of-the-art collaborative virtual globe technologies with advanced geo-spatial imaging techniques and wireless data link technologies supporting the combined and highly reliable transmission of digital video, high-resolution still imagery and mission control data over extended operational ranges [85]. The captured video data is geo-referenced in real time, either by direct geo-referencing using the flight control data of the UAV platform or alternatively by an integrated geo-referencing approach. The geo-referenced video streams can then be exploited directly within the virtual globe in either an augmented monitoring or a virtual monitoring scenario.

1.4.4.2 Earth observation and mapping

The primary goal is to provide the user with a top view of the territory without resorting to more expensive classical aerial photogrammetry. The system is designed to collect data for mapping and land monitoring purposes working on areas which represent a difficult task for already existing ground-based mobile mapping systems. According to project specifications presented in [30], the UAV can be used to survey areas of limited extent such as open mines, little rivers, cultivated fields, not only to monitor the land evolution and local changes in the terrain morphology but also to discover illegal uses of land resources. Several kinds of UAVs have been developed for photogrammetric data acquisition and terrain or object modeling. For example, a developed system integrates laser scanner and digital cameras with GPS and with inertial navigation sensors data for constructing digital surface models. UAVs can carry different camera systems like miniature, medium and panorama format cameras and video cameras [29].

Since 2000, the utilization of UAVs in photogrammetry has become common. This scenario can be explained by the spread of the low-cost combined **global positioning system/inertial navigation system** (GPS/INS). As a result, it is possible to obtain geo-referenced products with a spatial accuracy within centimeters. Compared to traditional methods based on **global navigation satellite system** (GNSS) measurements or imagery obtained from aerial or satellite platforms, UAV systems lead to improved cost savings in this task without losing accuracy [78]. **Photogrammetric data** processing is needed to generate a geo-referenced 3-D point cloud from the unordered, overlapping, and airborne image collection of the surface.

Definition 12 ***Structure from motion*** *(SfM) is estimating the locations of 3-D points from multiple images given only a sparse set of correspondences between image features. While this process often involves simultaneously estimating both 3-D geometry (structure) and camera pose (motion), it is commonly known as structure from motion.*

Given three distinct orthographic views of four non-coplanar points in a rigid configuration, the structure and motion compatible with the three views are uniquely determined. Existing SFM algorithms automatically extract features in the images, for example, contour lines, edges, and feature points.

Post-earthquake response Earthquake is one of the natural disasters that cause huge damage, destruction, and loss of life. For emergency logistic system design in crowded urban areas, UAV helicopter is an appropriate vehicle for relief distribution in a multi-modal network. While immediately following an earthquake, aid employees are attempting to open the roads, UAV helicopters are able to transfer water and emergency supplies as well as medical packages to the affected areas. Nowadays, an automatic system which detects the destruction and casualties and sends the special needs to the place is necessary. Immediately after the initial estimation, the UAV helicopters can be sent to impassable places while transporting first-aid commodities and requested supplies. The rotor-crafts return to the center, reload and travel to the same or new ordered locations. The information must be categorized as to the level of injuries and aggregation of injured people in different locations. The flow of information comes from different sources like civilians mobile phone, rescue employees, police reports, social media, buildings motion monitoring system, satellite monitoring and UAV helicopter monitoring system. The **Data Analysis Center** (DAC) must analyze the data in a way to be able to distinguish between the location and quantity of injuries, separate the level of priorities for injuries, the most useful commodities for each location, the distance to emergency centers, and the road condition like traffic congestion or road destruction. The final report must specify a queue, consisting of location of incidents, types of commodities required, and the amount of each commodity needed in place. Such a report also must be sent to the general earthquake response center which will send the aid employees and commodities to the affected areas and controls the whole response activities [86].

Mountain risks Snow avalanches are an inherent consequence of the dynamic and variable snow cover in steep mountainous terrain. Even though avalanches are rare natural hazards, they cause the majority of winter fatalities as well as significant infrastructure loss worldwide. Thus, avalanche research is risk research, dealing with risk reduction by trying to understand avalanche formation in space and time, relative to meteorological and snow pack triggering factors. Traditionally, field-testing of snow properties, field reconnaissance of avalanche activity and dynamics, and modeling of both are used to study avalanche formation. Remote sensing enables objective, safe, and spatial continuous observations of snow avalanches at different spatial scales. Today's abundance of sensor platforms and their sensitivity to a broad range of wavelengths allows for detection of avalanches and associated snow-pack processes. UAV is one of them [38]. Search for victims is resource intensive and time consuming to do if the avalanche spans a large geographical area in rough terrain. Survival decreases rapidly with time to extrication. More than half of avalanche victims are partially buried and visible on the surface. The objective in [14] was to evaluate feasibility of using a rotor-craft UAV to support situation assessment in search and rescue operations in the mountains.

In Switzerland, some 3000 people go missing every year. Some have dementia, others lose their way home and some suffer health problems when out hiking. Currently, teams of rescue workers and dogs operate entirely on the ground, often over a vast area. But now the Swiss Federation of Civil Drones (SVZD) will operate UAVs equipped with thermal imaging and night vision cameras to provide aerial images of search areas. Using UAVs in this way should make searches in difficult, steep or widespread terrain much easier and quicker.

1.4.4.3 Atmospheric monitoring

A UAV that can perform in-situ meteorology in severe meteorological conditions is described in [40]. It is designed to measure electric field and X-rays present within thunderstorms. The operation ability of realization of measures of the ecological monitoring is provided by the presence of radio contact in real time between UAV and surface station of management. High stability and good handling provide exploitation of monitoring complex systems in a wide range of environmental parameters on limited sites [34].

An ability to sample industrial emissions released to the atmosphere is very useful, especially in the view of global warming. In this case, neither long range nor long endurance is required. Atmospheric phenomena can be simulated in near real time with increasing levels of fidelity. Combining autonomous airborne sensors with environmental models enables the collection of data essential for examining the fundamental behavior of the atmosphere. Major challenges for small UAS operating at low to medium altitudes are range and endurance, and operations in adverse weather. Integrated sensing and modeling coupled with autonomous low-level control enable planning to detecting and avoiding mesoscale weather features that degrade performance or pose unacceptable risk to the aircraft. The main challenge for future airborne sampling and surveillance missions is operation with tight integration of physical and computational resources over wireless communication networks, in complex atmospheric conditions. The physical resources include sensor platforms, particularly mobile Doppler radar and unmanned aircraft, the complex conditions in which they operate, and targets or region of interest. Autonomous operation requires distributed computational effort connected by layered wireless communication. On-board decision making and coordination algorithms can be enhanced by atmospheric models that assimilate input from physics-based models and wind fields derived from multiple sources. Finally, the wind field environment drives strong interaction between the computational and physical systems to improve system range and endurance. An **energy-aware airborne dynamic, data-driven application system** (EA-DDDAS) that can perform persistent sampling and surveillance in complex atmospheric conditions is described in [53]. The main challenges that are addressed by the EA-DDDAS include tighter integration of sensor-based processing into online prediction tools; use of these tools inside planning loops that exploit available wind energy; and improved estimation of on-board energy states for higher degrees of autonomous learning.

1.4.4.4 Wildlife evaluation

One of the goals is to develop UAVs that can be used to automate key components of mosquito integrated vector control, in particular, **indoor residual spraying** (IRS), **outdoor residual spraying** (ORS), and **larval source management via larviciding** (LSM-LC). The primary aim is that the integrated socio-technical systems are affordable and sustainable in malaria endemic countries in sub-Saharan Africa, South and South-East Asia, Central and South America, as part of the ongoing global efforts of malaria control, suppression, elimination, and eventual eradication. The current most practical and economic method for both indoor and outdoor spraying in the endemic areas is using manual backpack sprayers. The multi-rotor UAV based solutions are being investigated as replacements for expensive (helicopter and fixed wing aircraft) aerial spraying, ground vehicle spraying, and backpack spraying for mosquito vector control [18].

1.4.5 Precision agriculture

Precision agriculture is an innovative trend in farm management. The integration of UAS for precision agriculture applications offers significant benefits. Precision agriculture involves the application of geo-spatial techniques and sensors (e.g., geographic information systems, remote sensing, GPS) to identify variations in the field and to deal with them using alternative strategies. The use of UAVs for precision agriculture creates two primary mission types:

1. Long range surveying (performed mostly by fixed wing UAVs) to execute remote sensing at an altitude of about 150 m, and

2. Long range light payload UAVs to do precise spraying of chemicals at altitudes below 50 m, as already performed in Japan.

As an example strategy, sensors that can detect field variability, such as **variable rate technology** (VRT) and grain yield monitors have been used together with high position accuracy GPS. Further potential application areas of very high-resolution UAV-based remote sensing might be the detection and mapping of plant diseases [84]:

1. Aerial and ground robotic platforms for soil/crop monitoring, prediction, and decision making;

2. Approaches to low-cost sensing for day/night continuous operation;

3. Long-term autonomy and navigation in unstructured environments;

4. Perception for appearance and geometric change (possibly caused by seasonal and weather changes, or deployment in a new crop) for long-term autonomy;

In agronomic research, new substances and products such as herbicides, pesticides, fungicides or fertilizers are tested on field test sites. Today, these field tests include labor-intensive, typically weekly visual inspections of leaf properties by experienced staff. In this qualitative method the assessment of plant health is often based on number, size and condition of plant leafs. Optical satellite-based remote sensing is successfully used in supporting large scale field tests. However, the prevailing small test plots with sizes around one square meter and the need for short and reliable revisit periods require new solutions. Very high-resolution airborne remote sensing with spatial resolutions in the range of centimeter to decimeter is needed by agronomic research [33]. The use of manned airborne platforms is limited by high operational complexity and costs. UAS are more precise, cost effective and flexible. UAVs embark locating means and perception that allow it to carry a share of the maneuvers in automatic or remote operation and also collect the necessary actions identified in the framework of the mission, measures which may change during the mission [52].

Remark 13 *Users of commercial and recreational UAS should be aware that in remote, rural and agricultural areas, manned aircraft may be operating very close to ground level. Pilots conducting agricultural, firefighting, law enforcement, emergency medical, wildlife survey operations and a variety of other services work in low-level airspace. Operators controlling UAS in these areas should maintain situational awareness, give way to, and remain at a safe distance from these low-level, manned airplanes and helicopters.*

Details of the certification process for agricultural aircraft operations are provided in FAA Advisory Circular $137 - 1A$ [45]. Using the **hazard classification and analysis system** (HCAS) taxonomy, hazards are identified from the causal narrative following the six-step **aviation system risk model process**. The U.S. **National Agricultural Aviation Association** (NAAA) has expressed concerns over the use of UAS for agricultural applications and has also identified a number of potential UAS hazards to the FAA [46]. The aviation system risk model precision agriculture notional scenario provides a systems-level framework for the integration of socio-technical hazards related to the UAS, the crop-duster, operations and the environment. Geo-fencing offers one mitigation strategy in the avoidance of mid-air collisions. A geo-fence, one component in the move towards UAS autonomy, uses the GPS, to check that a UAS is within its designated area of operation.

1.4.5.1 Biomass inspection

Biomass estimation is crucial for yield prediction of crops. Crop parameters, like biomass, are frequently used to assess crop health status, nutrient supply and effects of agricultural management practices. For management optimization, the **nitrogen nutrition index** (NNI) plays a key role. Biomass is needed for calculating the NNI. A method for biomass estimation is the calculation of **vegetation indices** (VIs) in the **near infrared region** (NIR), here defined as the range between 700 and 1300 nm. Field spectro-radiometers are commonly used for the collection of hyper-spectral reflectance data that are used for such calculations. An alternative possibility is to model biomass using plant height information. Plant height information is most useful when it is available at high spatial and temporal resolution. The method of multi-temporal **crop surface models** (CSMs) derived from 3-D point clouds delivers the desired centimeter resolution. Through the availability of high-resolution consumer digital cameras, **red green blue** (RGB) aerial imaging with centimeter resolution can be obtained using UAVs. At the same time, the emergence of **structure from motion** (SfM)-based software has enabled efficient creation of 3-D point clouds and detailed ortho-photos. **Visible band** RGB (VIRGB) may be calculated from the orthophotos. **Near infrared VIs** (VINIR) are more widely used because of the characteristic difference between red and NIR reflection in green vegetation. In addition, smaller, but significant spectral differences in the visible bands exist, which are caused by biochemical plant constituents such as chlorophyll. Crop monitoring by UAV-based RGB imagery enables obtaining the VIRGB and the plant height information from the same data set suggesting to combine both parameters to improve biomass estimation [19].

Crop production Survey of crops is feasible using infra-red and color cameras to detect the onset of disease through changes in crop color. One consociation is a combination of legumes and grasses where nutritive properties, and ecological amplitudes, mutually complement each other. Consociation is an ecological community with a single dominant species. However, in order to be successful, this practice requires supervision and interventions in the sense of fertilizer availability modifications, pest control or composition corrections, preferably immediately and in real time. Use of satellite and aerial images miss adequate temporal resolution which would provide prompt and accurate action in the field, while UAVs are able to provide fast and cheap imagery of the area of interest. The objective of the study presented in [72] is to show possibility of use of aerial imagery acquired by UAV equipped with **visible light** (VIS) and **near infra-red** (NIR) sensors and estimation of botanical composition and status of three component legume-grass mixture. Sampling and determination of the relative share of each component of legume-grass mixture was performed and analysis of floristic composition on sampling points showed share of each component. Use of UAV showed potential in estimation of botanical composition and dynamics status of the three component legume grass mixture, but it is very important to include management data and additional land cover measurements for explanation of the remote sensed data.

Ortho-mosaics for wheat One of the most successful approaches for the acquisition of crop and weed spatial information is through remotely sensed images that can be processed, classified and divided into a series of sub-plots for further adapted applications according to the specific weed emergence. Two of the most important variables in using remote imagery for mapping weeds are the image spatial resolution and the phenological stage of the crop and weeds. Regarding the phenological stage, late-season weed detection maps can be used to design **site-specific weed management** (SSWM), either in-season post-emergence herbicide treatments, or pre-emergence treatments in subsequent years, taking into account that weed infestations can be relatively stable in location from year to year. Thus, the capacity to discriminate weeds at the advanced phenological stage could reduce herbicide use and control costs. In most weed control situations and including **early site-specific weed management** (ESSWM), it is generally necessary to control weeds at an early growth stage of the crop. High spatial resolution images taken by UAVs have been shown to have the potential for monitoring agronomic and environmental variables. It is necessary to capture a large number of overlapped images that must be mosaicked together to produce a single and accurate ortho-image representing the entire area of work. Thus, **ground control points** (GCPs) must be acquired to ensure the accuracy of the mosaicking process. UAV ortho-mosaics are becoming an important tool for early site-specific weed management, as the discrimination of small plants at early growth stages is subject to serious limitations using satellite or piloted aircraft. Small changes in flight altitude are crucial for low-altitude image acquisition because these variations can cause important differences in the spatial resolution of the ortho-images. Furthermore, a decrease of flying altitude reduces the area covered by each single overlapped image, which implies an increase of both the sequence of images and the complexity of the image mosaicking procedure to obtain an ortho-image covering the whole study area. Therefore, the most important parameter to consider when choosing the flying altitude is the ortho-image spatial resolution required rather than the geo-referencing accuracy [56].

1.4.5.2 Soil monitoring

Soil monitoring is important for maximizing crop yields. Accurate and up-to-date maps of water status, nutrient deficiencies and pest infestation allow farmers to take rapid, targeted action that minimizes costs and environmental impact. Manual monitoring involves taking ground samples and estimating the soil water status at random locations. For large areas, this is time consuming and cost intensive; furthermore, it gives only a sparse sampling of the area. Advances in imaging and computation have also made spectral imaging techniques more affordable. By combining these technologies, remote sensing techniques can now be economically applied to make frequent and high resolution surveys of agricultural land. UAS-based **spectral imaging** (SI) acquires spatially resolved images of a measurement sample at different wavelengths and combines them into a 3-D image cube. The two classical approaches for the acquisition of hyper-spectral image data are wavelength scanning and spatial scanning. Both methods have in common that it takes more than one integration time for the acquisition of a hyper-spectral image cube. The image sequence was stitched together and the **normalized difference vegetation index** (NDVI) was calculated and used to classify each pixel in the image [35].

1.4.5.3 Forestry

In natural resources, the urban forest is well suited for small UAV applications [103]. UAV generated products for urban forestry can be used in many ways. Urban forest management objectives are dictated by human use of the areas around trees. To understand how people use the urban forest and to determine tree diversity, it is important to create a spatial tree inventory. Tree diversity across the landscape can be identified with accurate inventories that detail forest characteristics. An indirect benefit of inventory analysis with the UAV is the collection and archive of aerial imagery for future temporal comparison. The affordable repeatability of acquiring UAV imagery offers the opportunity to complete spatio-temporal analysis to detect change over time. Missions are planned using strategic landing/take-off zones to make efficient use of topography and photographic parameters. Geodetic ground control points are established using GPS to aid spatial referencing of images. Results are analyzed to evaluate altitude preferences, radio connectivity, image resolution/detail, and flight parameters.

A **forestry decision support system** (FDSS) is developed in [116] for the management of **short rotation forests** (SRF), FDSS integrates remote sensing data from a UAV with a dynamic crop growth model. UAV missions provide 2 cm resolution of the 3-D structure of the plantations, retrieved by digital stereoscopic restitution of aerial photographs. Similarly, high resolution NIR and NDVI images of the canopy covers are also produced. By means of multi-temporal analysis of these data and their sequential assimilation in a crop growth model, the forestry decision support system can provide accurate information concerning the eco-physiological activity of the plants, the possible occurrence of pest and diseases, as well as biomass yield production.

1.4.6 Disaster relief

Natural disasters happen daily worldwide and represent an important factor that affects human life and development. In order to respond to different types of natural disasters and develop feasible disaster management techniques and methods, it is important to understand the nature of a disaster, its phases and constituents. When a disaster occurs, the most important issue that needs to be solved is to preserve human lives. In this context, the first 72 hours after the disaster hit are the most critical, which means that **search and rescue** (SAR) operations must be conducted quickly and efficiently. UAVs equipped with remote sensing instrumentation offer numerous opportunities in disaster-related situations. UAV airborne surveillance can assist as an essential part at early detection of hazard incidents. During a crisis, airborne surveillance is a needed instrument for real time information support and coordination of rescue forces [124]. A three-stage operational life-cycle can be proposed where UAVs participate in natural disaster management:

1. Pre-disaster preparedness concerning surveying-related events that precede the disaster, static wireless sensor network based threshold sensing and setting up **early warning systems** (EWS),

2. Disaster assessment providing situational awareness during the disaster in real time and completing damage studies for logistical planning, and

3. Disaster response and recovery including SAR missions.

Each stage imposes a set of task demands on the UAVs, lasts different lengths of time, and has varying priority levels [42]. When UAVs acquire photogrammetry data with appropriate imagery data, the capabilities of UAVs for disaster research and management can be further realized. High-resolution images can be analyzed and used to produce hazard maps, dense surface models, detailed building renderings, comprehensive elevation models, and other disaster area characteristics. These data can then be analyzed using remote sensing methods or visual interpretation to coordinate rescue efforts, record building responses to the disaster, detect building failures, investigate access issues, and verify experimental disaster modeling. Collecting photogrammetric-quality imagery with the appropriate meta-data facilitates the rapid creation of UAV imagery-derived maps. These maps can then be used to aid in disaster response efforts as they contain up-to-date spatial information [15].

Because of their exceedingly variable nature, the mission planning has to take a multitude of scenarios into account, considering arbitrary, unknown environments and weather conditions. It is also not feasible to preconceive the large number of unforeseeable events possibly occurring during the mission. In wide area surveillance, low-altitude UAV can provide coverage and investigate events of interest as they manifest themselves. In particular, cases in which close-range information is required on targets detected by other means must be considered and the UAV must proceed to their locations to gather on-site information. Advances in UAV instrumentation have enabled semi-automated and even fully automated map creation, a valuable tool for rapid response.

1.4.6.1 Search and rescue

Search and rescue (SAR) takes many forms, each with its own unique risks and dangers to victim and responder: urban search and rescue, wilderness search and rescue, maritime search, etc.

Urban search and rescue

Definition 14 ***Urban search and rescue*** *(USAR) is defined as the strategy, tactics, and operations for locating, providing medical treatment to, and extrication of entrapped victims.*

USAR is a domain where UAVs have the potential to make a difference. They can determine an approach to deal with SAR, even before humans have gone in. At least seven factors are identified in the problem of prompt rescue in USAR operations in collapsed buildings [113]:

1. Best practices and lessons learned cover the informed evaluation and analysis of past experiences available on demand as the USAR mission develops.

2. Rescue technology involves the development and optimization of enhanced casualty assessment, monitoring and extrication tools.

3. Community involvement establishes collaborative relationships between professional rescue teams and community-based first responders to achieve significant leverage of expertise and resource.

4. Information systems involves identifying, collecting and managing multiple data streams and transforming them into information in order to deliver better high-level planning and time-line situation awareness.

5. Technologies integration deals with the validation and integration of state-of-the-art technologies.

6. Crisis management deals with adaptive and scalable management systems based on the understanding that significant collapses are chaotic.

7. Available budget impacts on purchase but also on the fast deployment and use of systems and technologies. In the long term, it can have a significant impact on the development of new systems and technology.

A USAR team can perform the following tasks: conducting physical SAR in collapsed buildings, providing emergency medical assessments and care to trapped victims, assessment and control of hazards, such as gas or electric service, and evaluation and stabilization of damaged structures [79]. The mission of the human-robot team is to explore a disaster area, and to provide enough information to make a situation assessment. A human-robot team consists of at least one UAV, several humans located at a remote control post, and possibly one or more human operators in-field. The team is thus geographically dispersed. The UAV team largely deploys in the field such as in [68]. This UAV team consists of:

1. The UAV operator, piloting the UAV in-field;
2. A UAV mission specialist, watching the UAV video streams and guiding the UAV operator to mission targets; and
3. A safety command safeguarding the UAV team.

During the deployment, the UAV mission specialist mostly operates with the UAV operator, to provide the UAV operator with an extra pair of eyes on the UAV. The UAV team assesses video material afterward. The information gained from video material is provided directly to the National Fire Corps, and is also used in the briefings for follow-up **unmanned ground vehicles** (UGV) missions. For situation awareness, this requires the approach to be able to integrate different perspectives on the environment, and to facilitate different perspectives and needs [69]. To make this possible, UAV needs more autonomy in perceiving the environment and navigating it. However, disaster areas are harsh places [70]. The persistent performance of robotic systems operating in an urban environment is very challenging. A stable broadband radio link cannot be guaranteed in such environments. The limited availability of computing resources and low-weight sensors operating in harsh environments for mobile systems poses a great challenge in achieving autonomy [117]. Such missions require the UAVs to be modular and flexible in terms of sensor and planning capabilities. The UAVs have to operate in unstructured indoor and outdoor environments, such as collapsed buildings. Navigation systems therefore have to work without external aids, such as GPS, since their availability cannot be guaranteed. UAVs additionally have to provide robust flight capabilities because of the changing local wind conditions in such environments. A key feature to achieving full autonomy in urban disaster areas is on-board processing and decision making. Search assignments also require mission-specific recognition capabilities on the UAVs. Identifying and locating persons, animals, or objects (e.g., landmarks, signs, or a landing zone) is a central issue in USAR missions.

Wilderness search and rescue Wilderness search and rescue (WiSAR) entails searching over large regions in often rugged, remote areas. Because of the large regions and potentially limited mobility of ground searchers, WiSAR is an ideal application for using small UAVs to provide aerial imagery of the search region [57]. UAV-assisted WiSAR requires portability, durability, and operational simplicity if widespread deployment is to be successful. These requirements impose a number of constraints including those that arise from the particular UAV selected, those that result from human factors (particularly from the minimal training requirement), those imposed by the control device used, and those that arise from the specific task at hand, including the necessity of fitting into a pre-existing team structure.

Motivations for path planning: In the priority search phase of WiSAR, the UAV on-board video camera should cover as much of the important areas as possible within a set time:

1. How can such a flight path be generated automatically and quickly?
2. How can the path planning an intuitive, smooth, and effective task for the UAV operator be made to reduce workload and improve search efficiency?

These motivations define the UAV's intelligent path-planning problem to be faced. This problem has three components [75]:

1. The first challenge is to find ways to help the SAR workers generate the probability distribution map, identifying areas where the missing person is likely to be found. In the priority search phase of WiSAR, a probability distribution map for the likely place to find the missing person is created based upon terrain features, profile of the missing person, weather conditions, and subjective judgment of expert searchers. This map is a critical component of the SAR operation because the incident commander uses this map to allocate resources, to direct search, and to coordinate rescue workers. Areas with high probabilities are searched first in the hope of finding the missing person quickly. Such a probability distribution map can also be used by manned or unmanned aerial vehicles for path planning purposes, thus facilitating effective aerial search. First, an automatically generated probability distribution map helps reduce the SAR workers' workload. They can use the generated map as a base, and they can augment the map to incorporate additional information. The systematically generated map may also reduce the chance that the SAR workers might overlook a certain area that should have been allocated higher probabilities. Second, this created map is normally in a very coarse scale reflecting search priorities for various areas but does not precisely represent the desired probability distribution.

2. Then once a probability distribution map is determined, the second challenge is to automatically create flight paths given allowed flight duration. The map can work as a base to be augmented by SAR workers to incorporate additional information. Once a probability distribution map is in place, areas with higher probabilities are searched first in order to find the missing person in the shortest expected time. When using a UAV to support search, the on-board video camera should cover as much of the important areas as possible within a set time. The capability of planning a path with a set destination also enables the UAV operator to plan a path strategically while letting the UAV plan the path locally.

3. The last challenge is to develop an interface that enables the human operator to plan more strategically, while the algorithms plan tactically using allowed flight time as the controlling parameter.

Uncertainty: Search is the process of removing uncertainty regarding the probable location of a missing person. Uncertainty is removed by identifying such things as the **point last seen** (PLS) and the direction of travel, by finding signs that a missing person has recently been in an area, or by covering an area without finding a sign. The map is a representation of the entire search region, but observations are localized to what is visible by the UAV camera, which is controlled by the pose of the UAV and its camera. It is not unusual in mountainous areas that flight is occasionally restricted due to sudden changes in weather and meteorological conditions. With regard to application in mountainous areas, therefore, it is important to obtain imaging information of a broad area with as few flights as possible. Typically, a UAV engaged in a search task requires either two operators or a single operator to fill two roles: a pilot, who flies the UAV, and a sensor operator, who interprets the imagery and other sensors. In WiSAR, the **incident commander** (IC) creates a probability distribution map of the likely location of the missing person. This map is important because it guides the incident commander in allocating search resources and coordinating efforts, but it often depends almost exclusively on prior experience and subjective judgment [75].

Visibility: When conducting remote operations with camera equipped UAVs, remote operators often have difficulty knowing where the camera is or where it is looking due to the limited sensory information coming back from the camera alone. Humans have only a limited view of this system. This keyhole effect has been shown to cause gaps in the space explored by such UAVs. The user's situational awareness can be improved by augmenting the display of the camera's information with other available external information in the form of interfaces. SAR operations rely on thorough coverage of a target observation area. However, coverage is not simply a matter of seeing the area but of seeing it well enough to allow detection of targets of interest. Use of UAV-acquired video geo-registered to terrain and aerial reference imagery creates geo-spatial video coverage quality maps and indices that indicate relative video quality based on detection factors such as image resolution, number of observations, and variety of viewing angles. When used for offline post-analysis of the video, or for online review, these maps also enable geo-spatial quality-filtered or prioritized non-sequential access to the video. Assessing the usability and coverage of aerial video is a matter not only of whether the plane's camera could see a point but how well it saw it. Once the video is geo-registered to the underlying terrain, determining whether the camera saw specific points is a simple matter of viewing geometry, that is **visibility**. But visibility-based coverage alone is not enough to determine how useful the video is; one must consider the viewing resolution as well as the number of times seen, the variation of viewing angle which can often play a role in detection [82].

Maritime search Maritime search can be more efficient using suitable UAV systems. The maritime SAR, used as a **system of systems** (SoS), is domain scenario to both implement and demonstrate the architecture methodology. It utilizes a variety of systems, including UAVs, coordination command control, communication systems and other larger manned vessels. A variety of sensors and data sources are currently in use, such as coastal radar, patrol or surveillance aircraft, ship radar, sporadic report from civilian aircraft or ship, etc. Each of these sensors has their special characteristics. For instance, a coastal radar station has continuous coverage inside its radar horizon but is totally blind outside this horizon. For over-the-horizon and continuous naval surveillance, the most effective, economical and flexible method of observing areas of interest with a revisit time of some tens of minutes would be with a surveillance UAV [63]. The proposed methodology delivers the ability to exploit the interdependence among all systems. Architectures that are robust, efficient, net-centric, and affordable must be generated [16]. Generating architectures for a SoS is a multi-objective optimization problem with a large number of variables and constraints. The information required for architecture generation is the following:

1. Overarching purpose of SoS: A Coast Guard search and rescue capability is selected as the problem.

2. Stakeholders: The Coast Guard has numerous systems with differing capabilities that are available at several stations within the area. Additionally, fishing vessels, civilian craft, and commercial vessels join in an ad hoc SoS to provide assistance when a disaster strikes. The coordination and command control center guides the combination of manned vessels and UAVs in the operation. The communication systems enable coordination of the sensing and rescuing capabilities of each vehicle.

3. **Rough order of magnitude** (ROM) budget.

4. **Key performance attributes** (KPA) of the SoS: performance, affordability, robustness, modularity, and net centricity.

SAR with multiple UAVs on the sea faces several problems. First, commercial UAVs have limited fuel capacity and thus cannot operate indefinitely. Second, the probability that a survivor is in a given area changes with time due to current and wind. This probability is often referred to as the **probability of containment** (POC). Third, a system of many UAVs and fuel service stations should be controlled automatically [73].

Without a proper estimate of the basic drift properties and their associated uncertainties, forecasting the drift and expansion of a search area remains difficult. The direct method measures the object's motion relative to the ambient water using a current meter. The **Search And Rescue Optimal Planning System** (SAROPS) employs an environmental data server that obtains wind and current predictions from a number of sources. It recommends search paths for multiple search units that maximize the increase in probability of detection from an increment of search. The diffusivity of the ocean is an important factor when reconstructing the dispersion of particles either based on observed or modeled vector fields. In many cases one simple stochastic model is sufficient for estimating the dispersion of SAR objects over relatively short time periods. Regional (and possibly seasonal) estimates of diffusivity and the integral time scale should be carefully considered as their impact on the dispersion of SAR objects may be substantial [22].

1.4.6.2 Fires monitoring

In wildland fire fighting, UAVs allow persistent operation and elimination of humans from performing dull and dangerous works. The data generated are used to [4]:

1. Characterize fire behavior in wildland fuels,
2. Document communities damaged by **wildland urban interface** (WUI) fires, and
3. Support future incidents.

UAVs may be used to patrol woods and vulnerable crops to look for hotspots when weather conditions are conducive to the outbreak of fire. They may also be used to assist in directing the application of fire-fighting materials onto any fire, whether rural or urban. The payload would be an electro-optical and thermal video sensors. A team can be defined, using **team definition language**, as a set of aircraft, some capable of carrying large amounts of water or fire retardant, and others equipped with sensors that allow them to detect a fire (such as video cameras, infrared, carbon dioxide concentration or temperature sensors). The **scenario description language** file would contain all static environment elements, such as the description of the base of operations (including airport) that can be used by the team; possible air traffic controllers (with jurisdiction over the base of operations or the mission area); and the existing types of aircraft. The **domain description language** file would specify the characteristics of the fire: how many are present in the environment, their location and creation time, motion pattern, initial size and growth pattern, detailing the growth or dispersion model for each component of the fire, such as temperature, carbon monoxide or carbon dioxide concentrations, smoke density, or others. The **mission description language** file would specify a mission comprised of two stages: in the first stage, the team should detect any possible existing fire (using the vehicles that possess the necessary sensors to do so) in a given area; in the second stage, which is triggered when a fire is detected, the aircraft carrying water or fire retardant should fly over the fire and drop its load in order to put out the fire.

Forest fires: Multiple UAVs can potentially help in fighting wildfires [71]. Two tasks can be considered:

1. The UAVs, equipped with visual and infrared sensors, are used to cooperatively track the fire-front perimeter to provide situational awareness. This implies persistently covering the fire front under the UAVs' footprint. The footprint is the area under the UAV which is within sensing/actuation zone.

2. The UAVs, equipped with fire suppressant, cooperatively extinguish fire.

The task of wildfires fighting with the help of UAVs is challenging because of the dynamic nature of fire growth and propagation which depend on several environmental factors and weather conditions. Models predicting the growth of forest fires aid in the coordination of UAV motion. A forest fire operational scenario is based on the four phases of a typical forest fire: pre-ignition, flaming, smoldering, and glowing phases. Taking into consideration the phases of the forest fire, the scenario can be prepared including five operational stages [66]:

1. Early warning surveillance and fire detection (pre-ignition phase of the fire): during the pre-ignition phase, heat from an ignition source or the flaming front evaporates water and the low molecular weight volatiles from the fuel, and the process of pyrolysis begins.

2. Alert and initial action of forces (flaming phase of the fire at its growth): during the flaming phase, combustion of the pyrolysis products (gases and vapors) with air takes place. The heat from the flaming reaction speeds the rate of pyrolysis and produces greater quantities of combustible gases, which also oxidize, causing increased flaming.

3. Large-scale action planning of forces (flaming phase of the fire at its fully developed stage): the smoldering phase is a very smoky process occurring after the active flaming front has passed. Combustible gases

are still produced by the process of pyrolysis, but the rate of release and the temperatures are not high enough to maintain flaming combustion. Smoldering generally occurs in fuel beds with fine packed fuels and limited oxygen flow.

4. Search and rescue operation in the field and area evacuation (smoldering phase of the fire).

5. Monitoring of the area, ending of forces field deployment (glowing phase of the fire). In the glowing phase, most of the volatile gases have been burned, and oxygen comes into direct contact with the surface of the charred fuel. As the fuel oxidizes, it burns with a characteristic glow, until the temperature is reduced so much that combustion cannot be continued, or until all combustible material is consumed. Cooperation of manned firefighting ground vehicles and unmanned vehicles can take place in order to cope with the risks from a devastating forest fire that burst out in a hot zone and expand towards a territory.

Real-time monitoring of the evolution of the fire involves dynamic information about the fire front and other parameters such as the flame height and the fire front width; as well as houses, roads and utilities. The teleoperation system facilitates the pilot work in low-visibility conditions and makes possible the cooperation with other autonomous and remotely piloted vehicles. Furthermore, the teleoperation systems can be also used as back-up of autonomous navigation systems in case of difficult navigation conditions or difficult tasks that require human intervention [91].

Camping park fires The occurrence of fires in camping parks are relatively frequent and in several cases they lead to fatalities. Fires affecting camping parks may have origin inside the park or may come from the outside since parks are typically located in forested areas. The main ignition sources of fires starting inside the park are electric equipment, use of fire for cooking or other practices, or neglect of some common potentially igniter camping accessories like gas lighting, candles and others. Fires coming from outside the camping area may request the evacuation of the camping park, endangering people and goods. The ignitable camping accessories used such as tents, caravans, sleeping beds, mattresses, etc. can sustain the fire and, when supported by other forest fuels, increase the fire spread through the camping park by conductivity and radiation mechanisms or by spotting. Besides all these risks, the knowledge of fire ignition, spreading and suppression in camping parks is still poor and requires a methodical analysis. Considering that external fires frequently threaten camping parks, the survey of the neighborhood of camping parks is also of great interest. The aerial photographic survey of the camping park has to be carried out in order to produce a fuel map of this area. In order to develop a tool for simulating the fire spread in a camping park, two sets of inputs related to the combustible materials are necessary:

1. The first set of inputs consists of the properties of typical camping accessories that can influence the fire spread.

2. The second set of results is related to the characterization of the fuel cover of the area where the camping park is located.

The determination of the risk of fire in an area requires the characterization of factors such as vegetation type, fuel load, distribution of occupied spaces, neighbor water sources and available access routes. In order to facilitate the characterization procedure, either by doing it remotely or even automatically, a high-resolution digital mosaic may be obtained from aerial images acquired by UAVs flying over the camping park area [17].

1.4.7 Aided communication system

UAVs can be used to establish a communication system for providing connectivity to emergency management field workers [118]. The success of UAV deployment in wireless relay communication systems depends on several factors. Optimal UAV positions can be found by maximizing average data rate, while at the same time keeping the error rate below a certain threshold. The performance of the ground to relay links through control of the UAV heading angle can be optimized in communication systems in which UAVs are used as relays between ground terminals and a base station. The use of multiple networked UAVs as relay nodes can enhance spectral and link connectivity and maintain reliable wireless communication links between users in an obstructed **line-of-sight** (LOS) environment. High uninterrupted bandwidth requires LOS between sender and transmitter to minimize quality degradation. Relays are used to deal with the limited communication range of small UAVs. Quality degradation and limited communication range problems can be solved by creating relay chains for surveillance of a single target, and relay trees for simultaneous surveillance of multiple targets. Wireless data transfer under high mobility in UAV applications is a challenging problem due to varying channel quality and extended link outages. Throughput and reception range can be increased by using multiple transmitters and receivers and by exploiting the resulting antenna beam diversity and parallel transmission effects. The proposed system consists of three main components:

1. End-to-end communication system: the network layer combines a connectivity module, geographic routing protocol, and UAV deployment strategy which achieves these goals. In geographic routing, nodes are capable of localizing themselves.

    (a) Connectivity module is used to detect connectivity problems such as network partitions, and poor communication performance.

    (b) The geographic routing protocol is used to relay messages between nodes.

   (c) The UAV deployment strategy is an algorithm running at the base station producing deployment orders to be transmitted to the UAVs. It accepts as inputs operator directives, the connectivity module information, and current UAV positions.

2. Navigation system plays a key role in the success of the system. It is a layered hybrid reactive software architecture and conceptually can be divided into two sub-systems:

   (a) The low-level sub-system is used for accessing the hardware, implementing continuous control laws, control mode switching, and coordinating the real-time communication between the on-board computer systems.

   (b) The high-level sub-system includes deliberative services of the system such as path planner service, **geographical information system** service and other services.

3. Formation control strategy can be defined as one sub-task of cooperative control of multiple UAVs and can be faced with centralized or decentralized strategies. In decentralized formation control strategy, the control task is divided into two levels. First, the dynamics of each UAV has to be stabilized internally by a local controller which uses the internal states of the UAV. Second, the relative position of the dynamically decoupled UAVs to each other is controlled by the formation controller which needs information acquired from other UAVs and from external references.

1.5 CONCLUSION

The use of UAVs for local surveying that is mostly within visual line-of-sight has the potential to increase rapidly as result of energy infrastructure inspections, public safety and security, mining and construction, insurance (property inspections) and among other medias. These UAVs have a relatively low regulation hurdle to overcome, as many of these operations can be performed within visual line-of-sight.

Beyond visual line-of-sight capabilities yield even greater potential. This includes agriculture remote sensing of crops and livestock, inspection of power line, pipeline, and railway networks. In the future, BVLOS capabilities could be used by public authorities to operate UAVs directly from each station and could complement or replace VLOS units carried in vehicles. Media UAVs used to cover traffic conditions or sporting events are also opportunities along with use at larger construction and mining sites and for conducting new forms of research. Beyond inspections, tethered UAVs converting high-altitude winds into electricity could emerge in European skies.

Bibliography

[1] http://www.cae.com (date of access 2018/01/10)

[2] https://www.easa.europa.eu/easa-and-you/civil-drones-rpas (date of access 2018/01/10)

[3] https://www.faa.gov/uas (date of access 2018/01/10)

[4] https://www.firelab.org (date of access 2018/01/10)

[5] http://www.flightgear.org (date of access 2018/01/10)

[6] http://www.flirtey.com (date of access 2018/01/10)

[7] http://www.mathworks.com (date of access 2018/01/10)

[8] http://www.microsoft.com/games/fsinsider (date of access 2018/01/10)

[9] http://www.opengc.org (date of access 2018/01/10)

[10] http://www.x-plane.com

[11] http://industryarc.com/Report/17144/commercial-drones-market.html

[12] https://goo.gl/n0UeUx (01/06/2017)

[13] SESAR European Drones Outlook Study, http://dronerules.eu/en/, 2016 (date of access 2018/01/10)

[14] Abrahamsen, H. B.: *Use of an unmanned aerial vehicle to support situation assessment and decision-making in search and rescue operations in the mountains*, Scandinavian Journal of Trauma, Resuscitation and Emergency Medicine, pp. 16, 2014.

[15] Adams, S. M.; Friedland, C. J.: *A survey of unmanned aerial vehicle (UAV) usage for imagery collection in disaster research and management*, publisher not identified, 2011.

[16] Agarwal, S.; Pape, L. E.; Dagli, C. H.: *A hybrid genetic algorithm and particle swarm optimization with type-2 fuzzy sets for generating systems of systems architectures*, Procedia Computer Science, vol. **36**, pp. 57-64, 2014.

[17] Almeida, M.; Azinheira, J. R.; Barata, J.; Bousson, K.; Ervilha, R.; Martins, M.; Moutinho, A.; Pereira J. C.; Pinto, J. C.; Ribeiro, L. M.; Silva, J.; Viegas, D. X.: *Analysis of fire hazard in camping park areas*, http://hdl.handle.net/10316.2/34013, 2014.

[18] Amenyo, J. T.; Phelps, D.; Oladipo, O.; Sewovoe-Ekuoe, F.; Jadoonanan, S.; Jadoonanan, S.; Hossain, A.: *MedizDroids Project: Ultra-low cost, low-altitude, affordable and sustainable UAV multicopter drones for mosquito vector control in malaria disease management*, In IEEE Global Humanitarian Technology Conference (GHTC), pp. 590-596, 2014.

[19] Bendig, J.; Yu, K.; Aasen, H.; Bolten, A.; Bennertz, S.; Broscheit, J.; Gnyp, M. L.; Bareth, G.: *Combining UAV-based plant height from crop surface models, visible, and near infrared vegetation indices for biomass monitoring in barley*, International Journal of Applied Earth Observation and Geoinformation, vol. **39**, pp. 79-87, 2015.

[20] Bestaoui Sebbane, Y.: *Lighter than air robots: Guidance and control of autonomous airships*, Springer, 2012.

[21] Bestaoui Sebbane, Y.: *Smart autonomous aircraft: flight control and planning of UAV*, CRC Press, 2016.

[22] Breivik, O.; Allen, A. A.; Maisondieu, C.; Olagnon, M.: *Advances in search and rescue at sea*, Ocean Dynamics, vol. **63**, pp. 83-88, 2013.

[23] Bulgakov, A.; Evgenov, A.; Weller, C.: *Automation of 3D Building Model Generation Using Quadrotor*, Procedia Engineering, vol. **123**, pp. 101-109, 2015.

[24] Cassandras, C.; Ding, X.; Liu, X.: *An optimal control approach for the persistent monitoring problem*, IEEE Conference on Decision and Control and European Control Conference, pp. 2907-2912, 2011.

[25] Clarke, R.: *Understanding the drone epidemic*, Computer Law and Security Review, vol. **30**, pp. 230-246, 2014.

[26] Coifman, B.; McCord, M.; Mishalani, M.; Redmill, K.: *Surface transportation surveillance from unmanned aerial vehicles*, In Procedings of the 83rd Annual Meeting of the Transportation Research Board, 2004.

[27] Colorado, J.; Mondragon, I.; Rodriguez, J.; Castiblanco, C.: *Geo-mapping and visual stitching to support landmine detection using a low-cost UAV*, International Journal of Advanced Robotic Systems, DOI: 10.5772/61236, 2015.

[28] Cooper, J. L.; Goodrich, M. A.: *Integrating critical interface elements for intuitive single-display aviation control of UAVs*, International Society for Optics and Photonics Defense and Security Symposium, pp. 62260, 2006.

[29] Coppa, U.; Guarnieri, A.; Pirotti, F.; Vettore, A.: *Accuracy enhancement of unmanned helicopter positioning with low-cost system*, Applied Geomatics, vol. **1**, pp. 85-95, 2009.

[30] Coppa, U.; Guarnieri, A.; Camarda, M.; Vettore, A.: *Development of unmanned aerial vehicle at Padova University*, International Archives of Photogrammetry, Remote Sensing and Spatial Information Sciences, vol. b, Part 5, 2010.

[31] Cracknell, A. P.: *UAVs: regulations and law enforcement*, International Journal of Remote Sensing, vol. **38**, pp. 3054-3067, 2017.

[32] Cummings, M. L.: *Operator interaction with centralized versus decentralized UAV architectures*, In Handbook of Unmanned Aerial Vehicles, Springer Netherlands, pp. 977-992, 2015.

[33] Cunliffe, A. M.; Brazier, R. E.; Anderson, K.: *Ultra-fine grain landscape-scale quantification of dryland vegetation structure with drone-acquired structure-from-motion photogrammetry*, Remote Sensing of Environment, vol. **183**, pp. 129-143, 2016.

[34] Danilov, A. S.; Smirnov, U. D.; Pashkevich, M. A.: *The system of the ecological monitoring of environment which is based on the usage of UAV*, Russian Journal of Ecology, vol. **46**, pp. 14-19, 2015.

[35] De Biasio, M.; Arnold, T.; Leitner, R.; McGunnigle, G.; Meester, R.: *UAV-based environmental monitoring using multi-spectral imaging*, In SPIE Defense, Security, and Sensing, International Society for Optics and Photonics, 2010.

[36] DeGarmo, M.; Nelson, G. M.: *Prospective unmanned aerial vehicle operations in the future national airspace system*, In AIAA 4th Aviation Technology, Integration and Operations (ATIO) Forum, pp. 20-23, 2004.

[37] Donath, D.; Rauschert, A.; Schulte, A.: *Cognitive assistant system concept for multi-UAV guidance using human operator behaviour models*, Conference on Humans operating unmanned systems (HUMOUS), 2010.

[38] Eckerstorfer, M.; Buhler, Y.; Frauenfelder, R.; Malnes, E.: *Remote sensing of snow avalanches: Recent advances, potential, and limitations*, Cold Regions Science and Technology, vol. **121**, pp. 126-140, 2016.

[39] Egorova, T.; Gatsonis, A.; Demetriou, M.A.: *Estimation of gaseous plume concentration with an unmanned aerial vehicle*, AIAA J. of guidance, control and dynamics, vol. **39**, pp. 1314-1324, 2016.

[40] Ely, J.; Nguyen, T.; Wilson, J.; Brown, R.; Laughter, S.; Teets, E.; Richards, L.: *Establishing a disruptive new capability for NASA to fly UAV's into hazardous conditions*, In SPIE Defense Security (International Society for Optics and Photonics), 2015.

[41] Emelianov, S.; Bulgakow, A.; Sayfeddine, D.: *Aerial laser inspection of buildings facades using quadrotor*, Procedia Engineering, vol. **85**, pp. 140-146, 2014.

[42] Erdelj, M.; Natalizio, E.: *UAV-assisted disaster management: Applications and open issues*, In IEEE International Conference on Computing, Networking and Communications (ICNC), p. 1-5, 2016.

[43] Erdos, D.; Erdos, A.; Watkins, S. E.: *An experimental UAV system for search and rescue challenge*, IEEE Aerospace and Electronic Systems Magazine, vol. **28**, pp. 32-37, 2013.

[44] Eugster, H.; Nebiker, S.: *UAV-Based Augmented Monitoring-Real-Time Georeferencing and Integration of Video Imagery with Virtual Globes*, The International Archives of the Photogrammetry, Remote Sensing and Spatial Information Sciences, **37**, 2008.

[45] Federal Aviation Administration, Certification Process for Agricultural Aircraft Operations, Advisory Circular AC 137-1A, October 10, 2007.

[46] National Agricultural Aviation Association, Comments to the FAA. Docket ID: FAA-2014-0397, August 4, 2014.

[47] Federal Aviation Administration, Aerospace forecast, Fiscal years 2016-2036, March 24, 2016.

[48] Fairchild, C.; Harman, T. L.: *ROS robotics by example, bring life to your robot using ROS robotic applications*, PACKT publishing 2016.

[49] Fang, Z.; DeLaurentis, D.: *Dynamic planning of system of systems architecture evolution*, Procedia Computer Science, vol. **28**, pp. 449-456, 2014.

[50] Francis, M. S.: *Unmanned air systems: challenge and opportunity*, AIAA Journal of Aircraft, vol. **49**, pp. 1652-1665, 2012.

[51] Fregene, K.: *Unmanned aerial vehicles and control: Lockheed Martin advanced technology laboratories*, IEEE Control Systems, vol. **32**, pp. 32-34, 2012.

[52] Freeman, P. K.; Freeland, R. S.: *Agricultural UAVs in the US: potential, policy, and hype*, Remote Sensing Applications: Society and Environment, vol. **2**, pp. 35-43, 2014.

[53] Frew, E. W.; Argrow, B.; Houston, A.; Weiss, C.; Elston, J.: *An energy-aware airborne dynamic data-driven application system for persistent sampling and surveillance*, Procedia Computer Science, vol. **18**, pp. 2008-2017, 2013.

[54] Galinski, C.; Zbikowski, R.: *Some problems of micro air vehicles development*, Technical Sciences, vol. **55**, 2007.

[55] Gimenes, R.A.; Vismari, L.F.; Avelino, V.F.; Camargo, J.B.; de Almeida, J.H.; de Almeida, J.R.; Cugnasca, P.S.: *Guidelines for the integration of autonomous UAVs into the global ATM*, J. Intelligent Robot Systems, vol. **74**, pp. 465-478, 2014.

[56] Gomez-Candon, D.; De Castro, A. I.; Lopez-Granados, F.: *Assessing the accuracy of mosaics from unmanned aerial vehicle (UAV) imagery for precision agriculture purposes in wheat*, Precision Agriculture, vol. **15**, pp. 44-56, 2014.

[57] Goodrich, M. A.; Morse, B. S.; Gerhardt, D.; Cooper, J. L.; Quigley, M.; Adams, J. A.; Humphrey, C.: *Supporting wilderness search and rescue using a camera-equipped mini UAV*, Journal of Field Robotics, vol. **25**, pp. 89-110, 2008.

[58] Grimaccia, F.; Aghaei, M.; Mussetta, M.; Leva, S.; Quater, P. B.: *Planning for PV plant performance monitoring by means of unmanned aerial systems (UAS)*, International Journal of Energy and Environmental Engineering, vol. **6**, pp. 47-54, 2015.

[59] Ham, Y.; Han, K. K.; Lin, J. J.; Golparvar-Fard, M.: *Visual monitoring of civil infrastructure systems via camera-equipped Unmanned Aerial Vehicles (UAVs): a review of related works*, Visualization in Engineering, vol. **4**, pp. 1-9, 2016.

[60] Hausamann, D.; Zirnig, W.; Schreier, G.: *Monitoring of gas transmission pipelines-A customer driven civil UAV application*, In 3rd ODAS (ONERA-DLR Aerospace Symposium) Conference, 2003.

[61] Hrabar, S.: *3D path planning and stereo-based obstacle avoidance for rotorcraft UAVs*, In IEEE/RSJ International Conference on Intelligent Robots and Systems, pp. 807-814, 2008.

[62] Idachaba, F. E.: *Monitoring of Oil and Gas Pipelines by Use of VTOL-Type Unmanned Aerial Vehicle*, SPE Oil and Gas Facilities, vol. **5**, pp. 47-52, 2016.

[63] Ince, A. N.; Topuz, E.; Panayirci, E.; Isik, C.: *Principles of integrated maritime surveillance systems*, Springer, 2012.

[64] Jovanovic, M.; Starcevic, D.; Jovanovic, Z.: *Reusable Design of Data Visualization Software Architecture for Unmanned Aerial Vehicles*, AIAA Journal of Aerospace Information Systems, vol. **11**, pp. 359-371, 2014.

[65] Kanellakis, C.; Nikolakopoulos, G.: *Survey on Computer Vision for UAVs: Current Developments and Trends*, Journal of Intelligent and Robotic Systems, pp. 1-28, DOI 10.1007/s10846-017-0483-z, 2017.

[66] Karma, S.; Zorba, E.; Pallis, G. C.; Statheropoulos, G.; Balta, I.; Mikedi, K.; Statheropoulos, M.: *Use of unmanned vehicles in search and rescue operations in forest fires: Advantages and limitations observed in a field trial*, International Journal of Disaster Risk Reduction, vol. **13**, pp. 307-312, 2015.

[67] Krispel, U.; Schinko, C.; Ullrich, T.: *A survey of algorithmic shapes*, Remote Sensing, vol. **7**, pp. 12763-12792, 2015.

[68] Kruijff, G. J. M.; Tretyakov, V.; Linder, T.; Pirri, F.; Gianni, M.; Papadakis, P.; Priori, F.: *Rescue robots at earthquake-hit Mirandola, Italy: a field report*, In IEEE International Symposium on Safety, Security and Rescue Robotics (SSRR), pp. 1-8, 2012.

[69] Kruijff, G.; Colas, F.; Svoboda, T.; Van Diggelen, J.; Balmer, P.; Pirri, F.; Worst, R.: *Designing intelligent robots for human-robot teaming in urban search and rescue*, In AAAI Spring Symposium on Designing Intelligent Robots, 2012.

[70] Kruijff, G.; Janicek, M.; Keshavdas, S.; Larochelle, B.; Zender, H.; Smets, N.; Mioch, T.; Neerincx, M.; van Diggelen, F.: *Experience in System Design for Human-Robot Teaming in Urban Search and Rescue*, In Field and Service Robotics, Springer, pp. 111-125, 2014.

[71] Kumar, M.; Cohen, K.; Homchaudhuri, B.: *Cooperative control of multi-UAV for monitoring and fighting wildfires*, AIAA Journal of Aerospace, Computing, Information and Communication, vol. **8**, pp. 1-16, 2011.

[72] Kutnjak, H.; Leto, J.; Vranic, M.; Bosnjak, K.; Perculija, G.: *Potential of aerial robotics in crop production: high resolution NIR/VIS imagery obtained by automated unmanned aerial vehicle (UAV) in estimation of botanical composition of alfalfa-grass mixture*, In 50th Croatian and 10th International Symposium on Agriculture, pp. 349-353, 2015.

[73] Lee, S.; Morrison, J. R.: *Decision support scheduling for maritime search and rescue planning with a system of UAVs and fuel service stations*, In IEEE International Conference on Unmanned Aircraft Systems (ICUAS), pp. 1168-1177, 2015.

[74] Li, Z.; Bruggemann, T. S.; Ford, J. J.; Mejias, L.; Liu, Y.: *Toward automated power line corridor monitoring using advanced aircraft control and multisource feature fusion*, Journal of Field Robotics, vol. **29**, pp. 4-24, 2012.

[75] Lin, R.: *UAV Intelligent path planning for wilderness search and rescue*, PhD thesis, Brigham Young university, Utah, 2009.

[76] Luxhoj, J. T.: *A socio-technical model for analyzing safety risk of unmanned aircraft systems (UAS): An Application to Precision Agriculture*, Procedia Manufacturing, vol. **3**, pp. 928-935, 2015.

[77] Marta, A. C.; Gamboa, P. V.: *Long endurance electric uav for civilian surveillance missions*, In 29th Congress of the International Council of the Aeronautical Sciences, St. Petersburg, 2014.

[78] Mesas-Carrascosa, F. J.; Notario-Garcia, M. D.; de Larriva, J. E. M.; de la Orden, M. S.; Porras, A. G. F.: *Validation of measurements of land plot area using UAV imagery*, International Journal of Applied Earth Observation and Geoinformation, vol. **33**, pp. 270-279, 2014.

[79] Messina, E.; Jacoff, A.: *Performance standards for urban search and rescue robots*, In Defense and Security Symposium, International Society for Optics and Photonics, pp. 62301V-62301V, 2006.

[80] Miller, J.: *Robotic systems for inspection and surveillance of civil structures*, Doctoral dissertation, The University of Vermont, 2004.

[81] Mohamed, H. A.; Hansen, J. M.; Elhabiby, M. M.; El-Sheimy, N.; Sesay, A. B.: *Performance Characteristic Mems-Based IMUs for UAVs Navigation*, The International Archives of Photogrammetry, Remote Sensing and Spatial Information Sciences, vol. **40**, 2015.

[82] Morse, B. S.; Engh, C. H.; Goodrich, M. A.: *UAV video coverage quality maps and prioritized indexing for wilderness search and rescue*, In Proceedings of the 5th ACM/IEEE international conference on Human-robot interaction, pp. 227-234, 2010.

[83] Myung, H.; Wang, Y.; Kang, S. J.; Chen, X.: *Survey on robotics and automation technologies for civil infrastructure*, report of Dept. of Mechanical Engineering, University of Canterbury, Christchurch, New Zealand, pp. 15, 2014.

[84] Nebiker, S.; Annen, A., Scherrer, M.; Oesch, D.: *A light-weight multispectral sensor for micro UAV Opportunities for very high resolution airborne remote sensing*, International Arch. Photogrammetry and Remote Sensing Spatial Information, vol. **37**, pp. 1193-1198, 2008.

[85] Nebiker, S.; Eugster, H.; Flckiger, K.; Christen, M. : *Planning and Management of Real-time Geospatial UAS Missions within a Virtual Globe Environment*, In International Archives of the Photogrammetry, Remote Sensing and Spatial Information Sciences, Conference on Unmanned Aerial Vehicle in Geomatics, Zurich, Switzerland, Vol. **38**, 2011.

[86] Nedjati, A.; Vizvari, B.; Izbirak, G.: *Post-earthquake response by small UAV helicopters*, Natural Hazards, vol. **80**, pp. 1669-1688, 2016.

[87] Nehme, C. E.; Crandall, J. W.; Cummings, M. L.: *An operator function taxonomy for unmanned aerial vehicle missions*, In 12th international command and control research and technology symposium, pp. 1-9, 2007.

[88] Nex, F.; Remondino, F.: *UAV for 3D mapping applications: a review*, Applied Geomatics, vol. **6**, pp. 1-15, 2014.

[89] Nikolic, J.; Burri, M.; Rehder, J.; Leutenegger, S.; Huerzeler, C.; Siegwart, R.: *A UAV system for inspection of industrial facilities*, In IEEE Aerospace Conference, pp. 1-8, 2013.

[90] Nonami, K., Kendoul, F., Suzuki, S., Wang, W., Nakazawa, D.: *Autonomous flying robots: unmanned aerial vehicles and micro-aerial vehicles*, Springer, 2010.

[91] Ollero, A.; Alcazar, J.; Cuesta, F.; Lopez-Pichaco, F.; Nogales, C.: *Helicopter teleoperation for aerial monitoring in the COMETS multi-UAV system*, In 3rd IARP Workshop on Service, Assistive and Personal Robots, 2003.

[92] Ondracek, J.; Vanek, O.; Pechoucek, M.: *Monitoring oil pipeline infrastructures with multiple unmanned aerial vehicles*, In Advances in Practical Applications of Heterogeneous Multi-Agent Systems. Springer International Publishing, pp. 219-230, 2014.

[93] Ong, J. K.; Kerr, D.; Bouazza-Marouf, K.: *Design of a semi-autonomous modular robotic vehicle for gas pipeline inspection*, Proceedings of the Institution of Mechanical Engineers, Part I: Journal of Systems and Control Engineering, vol. **217**, pp. 109-122, 2003.

[94] Peng, L.; Silic, M.; Mohseni, K.: *A DDDAS Plume Monitoring System with Reduced Kalman Filter*, Procedia Computer Science, vol. **51**, pp. 2533-2542, 2015.

[95] Puri, A.: *A survey of unmanned aerial vehicles (UAV) for traffic surveillance*, Department of computer science and engineering, University of South Florida, Tampa, 2005.

[96] Quater, P. B.; Grimaccia, F.; Leva, S.; Mussetta, M.; Aghaei, M.: *Light unmanned aerial vehicles (UAVs) for cooperative inspection of PV plants*, IEEE Journal of Photovoltaics, vol. **4**, pp. 1107-1113, 2014.

[97] Quigley, M.; Goodrich, M. A.; Beard, R. W.: *Semi-autonomous human-UAV interfaces for fixed-wing mini-UAVs* In Proceedings of IEEE/RSJ International Conference on Intelligent Robots and Systems, Vol. **3**, pp. 2457-2462, 2004.

[98] Rao, B.; Gopi, A. G.; Maione, R.: *The societal impact of commercial drones*, Technology in Society, vol. **45**, pp. 83-90, 2016.

[99] Rathinam, S.; Kim, Z.; Sengupta, R.: *Vision-based following of structures using an unmanned aerial vehicle (UAV)*, report UCB-ITS-RR-2006-1, Institute of Transportation Studies, UC Berkeley, 2006.

[100] Rathinam, S.; Kim, Z.; Soghikian, A.; Sengupta, R.: *Vision-based monitoring of locally linear structures using an unmanned aerial vehicle*, Journal of Infrastructure Systems, vol. **14**, pp. 52-63, 2008.

[101] Remondino, F.; Barazzetti, L.; Nex, F.; Scaioni, M.; Sarazzi, D.: *UAV photogrammetry for mapping and 3d modeling: current status and future perspectives*, International Archives of the Photogrammetry, Remote Sensing and Spatial Information Sciences, vol. **38**, C22, 2011.

[102] Renault, A.: *A Model for Assessing UAV System Architectures*, Procedia Computer Science, vol. **61**, pp. 160-167, 2015.

[103] Ritter, B.: *Use of UAV for urban tree inventories*, PhD thesis, Clemson University, 2014.

[104] Ro, K.; Oh, J. S.; Dong, L.: *Lessons learned: Application of small uav for urban highway traffic monitoring*, In 45th AIAA Aerospace Sciences Meeting and Exhibit, pp. 596-606, 2007.

[105] Rodriguez-Fernandez, V.; Menendez, H. D.; Camacho, D.: *Automatic profile generation for UAV operators using a simulation-based training environment*, Progress in Artificial Intelligence, vol. **5**, pp. 3746, 2016.

[106] Roper, W. E.; Dutta, S.: *Remote sensing and GIS applications for pipeline security assessment*, ESRI User Conference Proceedings, Vol. **15**, 2005.

[107] Schon, S.; Band, R.; Pleban, J. S.; Creutzburg, R.; Fischer, A.: *Applications of multimedia technology on autonomous flying robots for university technology transfer projects*, In International Society for Optics and Photonics, IST/SPIE Electronic Imaging, pp. 86670Q-86670Q, 2013.

[108] Schuster, D.; Fincannon, T.; Jentsch, F.; Keebler, J.; William Evans, A.: *The role of spatial ability in the relationship between video game experience and route effectiveness among unmanned vehicle operators*, Report of University of Central Florida, Orlando, 2008.

[109] Shippen, A.: *RISER: 3D Contamination Mapping with a Nuclear Capable Drone*, International Workshop on the Use of Robotic Technologies at Nuclear Facilities, 2016.

[110] Shukla, A.; Karki, H.: *Application of robotics in onshore oil and gas industry: A review Part I*, Robotics and Autonomous Systems, vol. 75, pp. 490-507, 2016.

[111] Siebert, S.; Teizer, J.: *Mobile 3D mapping for surveying earthwork projects using an Unmanned Aerial Vehicle (UAV) system*, Automation in Construction, vol. **41**, pp. 1-14, 2014.

[112] Srinivasan, S.; Latchman, H.; Shea, J.; Wong, T.; McNair, J.: *Airborne traffic surveillance systems: video surveillance of highway traffic*, In Proceedings of the ACM 2nd international workshop on Video surveillance and sensor networks, pp. 131-135, 2004.

[113] Statheropoulos, M.; Agapiou, A.; Pallis, G. C.; Mikedi, K.; Karma, S.; Vamvakari, J.; Dandoukali, M.; Andritsos, F.; Thomas, C. P.: *Factors that affect rescue time in urban search and rescue (USAR) operations*, Natural Hazards, vol. **75**, pp. 57-69, 2015.

[114] Stevenson, J. D.; O'Young, S.; Rolland, L.: *Assessment of alternative manual control methods for small unmanned aerial vehicles*, Journal of Unmanned Vehicle Systems, vol. **3**, pp. 73-94, 2015.

[115] Theissing, N.; Kahn, G.; Schulte, A.: *Cognitive automation based guidance and operator assistance for semi-autonomous mission accomplishment of the UAV demonstrator SAGITTA*, Deutsche Gesellschaft fr Luft- und Raumfahrt-Lilienthal-Oberth eV, 2013.

[116] Teobaldelli, M.; Chirico, G. V.; Cona, F.; Rassi, F., Mandelli, A.; Medina, G. H.; Saracino, A.; Saulino, L.: *Forestry Decision Support System (FDSS) in Management of Short Rotation Woody Plantations*, Conference: AIIA Mid Term Conference New Frontiers of Biosystems and Agricultural Engineering for Feeding the Planet, Naples, Italy, 2015.

[117] Tomic, T.; Schmid, K.; Lutz, P.; Domel, A.; Kassecker, M.; Mair, E.; Burschka, D.: *Toward a fully autonomous UAV: Research platform for indoor and outdoor urban search and rescue*, IEEE robotics and automation magazine, vol. **19**, pp. 46-56, 2012.

[118] Tuna, G.; Nefzi, B.; Conte, G.: *Unmanned aerial vehicle-aided communications system for disaster recovery*, Journal of Network and Computer Applications, vol. **41**, pp. 27-36, 2014.

[119] Tyutyundzhiev, N.; Lovchinov, K.; Martnez-Moreno, F., Leloux, J.; Narvarte, L.: *Advanced PV modules inspection using multirotor UAV*, Aircraft engineering and aerospace technology, vol. **77**, pp. 352-360, 2005.

[120] Uhrmann, J.; Strenzke, R.; Schulte, A.: *Task-based Guidance of Multiple Detached Unmanned Sensor Platforms in Military Helicopter Operations*, In Proceedings of Cognitive Systems with Interactive Sensors, Crawley, UK, 2010.

[121] Vincenzi, D. A.; Terwilliger, B. A.; Ison, D. C.: *Unmanned Aerial System (UAS) Human-machine Interfaces: New Paradigms in Command and Control*, Procedia Manufacturing, vol. **3**, pp. 920-927, 2015.

[122] Vu, H.; Le, T. L.; Dinh, T. H.: *Selections of Suitable UAV Imagery Configurations for Regions Classification*, In Asian Conference on Intelligent Information and Database Systems, Springer Berlin Heidelberg, pp. 801-810, 2016.

[123] Wang, L.; Chen, F.; Yin, H.: *Detecting and tracking vehicles in traffic by unmanned aerial vehicles*, Automation in Construction, 2016.

[124] Wetjen, W.; Schilling, H.; Lenz, A.: *An advanced airborne hyperspectral remote sensing system for infrastructure monitoring and disaster control*, report of Fraunhofer Institute of Optronics, System Technologies and Image Exploitation IOSB, 2014.

[125] Zhang, J.; Jung, J.; Sohn, G.; Cohen, M.: *Thermal Infrared Inspection of Roof Insulation Using Unmanned Aerial Vehicles*, The International Archives of Photogrammetry, Remote Sensing and Spatial Information Sciences, vol. **40**, 381, 2015.

[126] Zheng, C.; Breton, A.; Iqbal, W.; Sadiq, I.; Elsayed, E.; Li, K.: *Driving-Behavior Monitoring Using an Unmanned Aircraft System (UAS)*, In International Conference on Digital Human Modeling and Applications in Health, Safety, Ergonomics and Risk Management, pp. 305-312, 2015.

CHAPTER 2

Mission Framework

2.1 INTRODUCTION

Mission planning can be considered from the point of view of artificial intelligence, control theory, formal methods and hybrid systems. An automated mission planning can enable a high level of autonomy for a variety of operating scenarios. The mission planning has to define a series of steps to define a flight plan. A strategic level regarding the choice of the way-points must be made before the operational decisions. These decisions are strongly related to factors such as UAV range and endurance, topography and communications and the mission requirements. The calculation of a flight plan involves the consideration of multiple elements. They can be classified as either continuous or discrete, and they can include non-linear aircraft structure and performance, atmospheric conditions, wind forecasts and operational constraints. Moreover, multiple flight phases can also be considered in flight planning [79]. Scheduling algorithms must be integrated into an embedded architecture to allow the system to adapt its behavior to its state and dynamic environments. The approach can reduce the uncertainty inherent in a dynamic environment through on-line re-planning and incorporation of tolerances in the planning process [57]. The motion plan is constrained by aircraft dynamics and environmental/operational constraints. In addition, the planned path must satisfy multiple, possibly conflicting objectives such as fuel efficiency and flight time. Integration in route optimization must take into account that the activation calculations plan is triggered by events.

Remark 15 ***Situation awareness*** *includes both monitoring and diagnosis. Plan generation and execution are grouped. A hierarchical planning approach is chosen because it enables a rapid and effective response to dynamic mission events.*

The embedded architecture must meet these two objectives by organizing the physical tasks of the mission and the tasks of reasoning. This reaction is conditioned by the inclusion of planning during execution in a control architecture.

Each controller is composed of a set of algorithms for planning, monitoring, diagnosing and executing. This architecture can be applied to complex problem solving, using hierarchical decomposition. As the mission problem is progressively broken down into smaller sub-problems, functional activities emerged. The control algorithms for multi-UAV missions typically have two levels, the higher-level task allocation and decision making algorithm that assigns UAVs to tasks based on their capabilities, and the UAV-level execution algorithms that include motion control algorithms. This approach provides a viable solution to real-time, closed-loop planning and execution problems:

1. The higher levels create plans with greatest temporal scope, but low level of detail in planned activities.

2. The lower levels' temporal scope decreases, but they have an increase in the detail of the planned activities.

At the lowest level of decomposition, the functional activities are operating on timescales of seconds. These activities are related to each other in a tree structure, with the lowest level nodes providing the output commands to the guidance, navigation and control system. In order to guarantee efficient execution and to maximize mission score, both algorithms have to work in close harmony [41]. The distinction between the strategic, tactical and operational decision making levels is widely recognized [47]. Forecasts that support such decisions are inherently different in nature. For example, strategic decisions require long-run forecasts at an aggregate level, while decisions at the highly dynamic operational level require short-term, very detailed forecasts. The differences in time granularity affect how these forecasts are generated. Since these forecasts are produced by different approaches and are based on different information sets, it is expected that they may not agree. This disagreement can lead to decisions that are not aligned.

2.2 AUTONOMY

It is important to develop a framework that provides standard definitions and metrics characterizing and measuring the autonomy level of a UAS.

Definition 16 ***Autonomy*** *is the quality of being self-governing. When applied to UAS, autonomy can be defined as UAS' own abilities of integrated sensing, perceiving, analyzing, communicating, planning, decision making and acting/executing, to achieve its goal as assigned by its human operator through a designed* ***human-robot interface*** *(HRI) or by another system that the UAS communicates with.*

Definition 17 *An* ***unmanned aerial system*** *is defined to be* ***autonomous relative to a given mission*** *(relational mission) when it accomplishes its assigned mission successfully, within a defined scope, with or without further interaction with human or other external systems. A UAS is fully autonomous* *if it accomplishes its assigned mission successfully without any intervention from a human or any other external system, while adapting to operational and environmental conditions.*

2.2.1 Levels of autonomy

The complex and dynamic nature of UAS greatly increases when the **levels of autonomy** (LOA) are considered and the assessment model must include this capability attribute. The dynamic interaction between the UAV and the operating environment of the UAV must be included in the assessment of the key performance attributes that involve safety.

Definition 18 ***Autonomy level*** *is defined as a set of progressive indices, typically numbers and/or names identifying a UAS capability of performing (mission complexity), the environments within which the missions are performed (environment complexity) and independence from any external system including any human element (external system independence).*

The following distinction between automatic, autonomous and intelligent systems can be made:

1. An **automatic system** will do exactly as programmed because it has no capabilities of reasoning, decision making or planning.
2. An **autonomous system** has the capability to make decisions and to plan its tasks and path in order to achieve its assigned mission.
3. An **intelligent system** has the capabilities of an autonomous system plus the ability to generate its own goals from inside by motivations and without any instruction or influence from outside.

Remark 19 *In addition to autonomy characterization and evaluation, two other aspects are important for comparing and evaluating UAS:* ***performance*** *and* ***dependability****. Autonomy is related to what the UAS can do, performance is related to how well the UAS meets mission requirements (accuracy, time, etc.), and dependability is related to how often the UAS accomplishes the mission without problems (success rate, failure rates, etc.).*

The following autonomy levels can be presented for **guidance, navigation and control** (GNC) [14, 52]:

1. **Level 0: Remote control**, all guidance functions are performed by external systems (mainly human pilot or operator). Sensing may be performed by the UAS; all data is processed and analyzed by an external system. Control commands are given by a remote external system (mainly human pilot).

2. **Level 1: Automatic flight control** concerns preprogrammed or uploaded flight plans (way-points, reference trajectories, etc.). All analyzing, planning and decision making is done by an external system. Most sensing and state estimation is made by the UAS and all perception and situation awareness by the human operator. Control commands are given by a remote external system.

3. **Level 2: External system independence navigation** (non-GPS) concerns pre-programmed or uploaded flight plans (way-points, reference trajectories, etc.). All analyzing, planning and decision making is done by an external system. All sensing and state estimation is made by the UAS (no external system such as GPS), all perception and situation awareness is made by the human operator. Control commands are given by a remote external system.

4. **Level 3: Fault/event adaptive UAS** concerns health diagnosis, limited adaptation, on-board conservative and low-level decisions, as well as execution of pre-programmed tasks. Most health and status sensing are made by the UAS, detection of hardware and software faults. Robust flight controller, reconfigurable or adaptive control compensates for most failures, mission and environment changes are recommended.

5. **Level 4: Real-time obstacle/event detection and path planning** concerns hazard avoidance, real-time path planning and re-planning, event driven decisions, and robust response to mission changes. Perception capabilities exist for obstacle and risks. Target and environment changes are detected, as well as real-time mapping and low-fidelity situational awareness are performed. Accurate and robust 3-D trajectory tracking capability is desired. Obstacles are a central issue for the flight path planning where a collision-free shortest path from an initial to a destination point has to be found.

6. **Level 5: Real-time cooperative navigation and path planning** concerns collision avoidance, cooperative path planning and execution to meet common goals, swarm or group optimization. There is relative navigation between UAS, cooperative perception, data sharing, collision detection, and shared low-fidelity situation awareness. Distributed or centralized flight control architectures, and coordinated maneuvers can be performed.

7. **Level 6: Dynamic mission planning** concerns reasoning, high-level decision making, mission-driven decisions, high adaptation to mission changes, tactical task allocation, and execution monitoring. High-level of perception to recognize and classify detected objects/events and to infer some of their attributes is used as well as situation awareness. Distributed or centralized flight control architectures, and coordinated maneuvers can be performed.

8. **Level 7: Real-time collaborative mission planning and execution** concerns evaluation and optimization of multi-UAV mission performance, and allocation of tactical tasks to each agent. Combination of capabilities of levels 5 and 6 in a highly complex, adversarial and uncertain environment, collaborative mid-fidelity situation awareness is preferred. Distributed or centralized flight control architectures can be used; coordinated maneuvers are made on-board.

9. **Level 8: Situational awareness and cognizance** concerns reasoning and higher level strategic decision making, strategic mission planning. Most of supervision is done by UAS, as well as choice of strategic goals. Cognizance is the conscious knowledge of complex environments and situations, inference of self/other intent, and anticipation of near-future events and consequences (high-fidelity situation awareness). The system is able to change or switch between different control strategies based on the understanding of the current situation/context and future consequences.

10. **Level 9: Swarm cognizance and group decision making** concerns distributed strategic group planning, selection of strategic goals, mission execution with no supervisory assistance, negotiating with team members and external system. Long track awareness of very complex environments and situations, inference and anticipation of other agent intents and strategies, and high-level team situation awareness exist. There is an ability to choose the appropriate control architecture based on the understanding of the current situation/context and future consequences.

11. **Level 10: Human-level decision making** concerns accomplishment of most missions without any intervention from an external system, and cognizance of all within the operation range. There exist human-like navigation capabilities for most missions, fast situation awareness that outperforms human situation awareness in extremely complex environments and situations is of interest. Same or better control performance for a piloted aircraft in the same situation and conditions is obtained.

Remark 20 *The highest level of autonomy might not be the most desirable operational level for several UAVs, missions or environments. Sometimes supervised autonomy or direct control over the UAV (assuring it will not lose its communications and sensors data link) will guarantee the best mission performance. For a UAV, a fully autonomous intelligent asset does not assure the best mission completion status in the case of an ever changing scenario or even a time sensitive target [29].*

UAV functional capability framework can be expanded by incorporating the concept of **functional level of autonomy** (F-LOA) with two configurations:

1. The lower F-LOA configuration contains sufficient information for the operator to generate solutions and make decisions to address perturbation events.

2. Alternatively, the higher F-LOA configuration presents information reflecting on the F-LOA of the UAV, allowing the operator to interpret solutions and decisions generated autonomously, and decide whether to veto from this decision.

F-LOA is used to describe how autonomous a specific functional sub-system of a UAV is, rather than the entire UAV entity described by levels of autonomy [15].

2.2.2 Decision making

One challenge facing coordination and deployment of UAVs is the amount of human involvement needed to carry out a successful mission. Currently, control and coordination of UAVs typically involves multiple operators to control a single agent. Operators are less involved in direct manual control of systems, but more involved in the higher levels of planning, decision making and remote operations.

Definition 21 ***Decision making*** *is the UAS ability to select a course of actions and choices among several alternative scenarios based on available analysis and information. The decisions reached are relevant to achieving assigned missions efficiently and safely. Decision making processes can differ in type and complexity ranging from low- to high-level decision making. Trajectory generation, path planning and mission planning are part of decision making processes.*

Another taxonomy is the human-automation collaboration taxonomy. It employs information processing flow model and concentrates on decision making. Dynamic decision making takes into account the state of the environment, situation awareness, decision and performance of actions [14].

2.2.2.1 Human impact

Definition 22 *The* ***human supervisory control*** *(HSC) is the process by which a human operator intermittently interacts with a computer, receiving feedback from and providing commands to a controlled process or task environment.*

HSC in UAV operation is hierarchical. The innermost loop represents the basic guidance and motion control, which is the most critical loop that must obey physical laws of nature such as aerodynamic and flight envelope constraints for UAVs. In this loop, operator actions are focused only on the short term and local control (keeping the aircraft in stable flight), and generally human control in this loop requires skill-based behaviors that rely on automation. The second loop, the navigation loop, represents the actions that some agent must execute to meet mission constraints such as routes to way-points, time on target, and avoidance of collisions and no-fly zones. The outermost loop represents the highest levels of control, that of mission and payload management. In this loop, sensors must be monitored and decisions made based on the incoming information to meet overall mission requirements. Finally, the system health and status monitoring loop represents the continual supervision that must occur, either by a human or automation or both, to ensure that all systems are operating within normal limits. The most exterior control loop line represents a highly intermittent loop in terms of the human, with the highest priority given to the innermost loop, health and status monitoring becomes a secondary task [42].

UAV scenarios present a number of challenges to designers of a dialogue system. In particular, the dynamic operating environment and the asynchronous, mixed-initiative nature of the dialogues require a flexible architecture to coordinate multiple asynchronous communicating processes [55]:

1. **Graphical user interface** (GUI) is a map display of the current operating environment which displays route plans, way-points and location of the UAV.

2. **Dialogue manager** is responsible for coordinating multi-modal inputs from the user, interpreting dialogue moves made by the operator and UAV, updating and maintaining the dialogue context, handling UAV reports and questions, and sending outputs to the operator.

3. **Control and report software** is responsible for translating commands and queries from the dialogue interface into commands and queries to the UAV, and vice versa for reports and queries received from the UAV.

Remark 23 ***Human-automation collaboration*** *can be beneficial due to the uncertainty inherent in supervisory control system. However, the inability of the human to understand the method by which the automation developed its solution, especially in time-pressured situations, can lead to automation bias. This automation bias can cause degradation in skills and performance and potentially loss of situational awareness. The human operator should have the ability to modify the way the automated planner works for collaborative decision making [18].*

2.2.2.2 Operational autonomy

It is important to distinguish between automation and autonomy. In an automatic system, actions are independent of context; the system responds to its sensor input in a fixed manner and does not engage in a more flexible behavior. In contrast, an autonomous system is designed to acquire knowledge about the context and use it to make more situation-dependent and effective decisions.

Definition 24 *The **operational autonomy** for UAVs is the decision making capacity of the UAV to attain operational objectives.*

As the **level of operational autonomy** (LOA) increases, the uncertainties faced by the UAV and the complexity of its decision making increase. The dependence on multiple human operators per UAV is perhaps the largest cost component of operating them. Operator attention and effort are subject to fatigue and error, and operators are subject to severe stresses. The difficulties in increasing the autonomy of UAVs arise from the propagation of several dynamic uncertainties into the performance of these systems: map errors, modeling errors, sensor errors, errors of actuation, wind gusts, electromagnetic (EM) and acoustic interference, etc. Arbitrary combinations of these uncertainties make it impossible to predict the performance of most systems as they are operated today.

Definition 25 *The **performance** can be quantified for a spectrum of mission profiles and can be assessed with three classes of metrics: **point performance capability**, **mission performance** boundaries for canonical mission profiles and the **likelihood that emergent missions can be accomplished**. Logic for determining which aircraft variant should be configured for each mission, should be examined based on the assessment of its performance flexibility and efficiency [70].*

Remark 26 *To become alternatives to current technologies in civilian applications, UAVs should be affordable and have the capacity for payload flexibility and multi-mission capability. To allow functional flexibility, UAVs can be required to perform different types of operations [16].*

In the near future, UAVs will be increasingly employed in collaborative interaction with humans and other robots, working together to achieve common goals. The design and implementation of interaction among collaborative UAVs as well as between UAVs and their human operators is critical to their successful employment. The key research areas for the integration in non-segregated airspace are as follows: detect and avoid; airspace and airport access; command, control and communications (C3); human factors; contingency; and security and autonomy.

UAV performance and wind An understanding of the nominal and potentially hazardous weather patterns that exist within the airspace is very important to the successful operation of UAVs. This is especially true at low altitude, where thermal effects and winds can be unpredictable and potentially catastrophic to UAV operations. Forecasting product, as well as real-time sensor data at a single location and computational fluid dynamics can be used in conjunction as they are complementary in essence. Each product differs based on the following: variables predicted, forecast horizon, spatial and temporal granularity, forecast grid, underlying prediction model, and model run-frequency. The real-time mobile atmospheric boundary-layer profiling may consist of two remote tropospheric profilers: a scanning Doppler lidar and a microwave temperature and humidity profiler combined with a mobile meteorological tower [28]. It is also important to develop an understanding of wind flows within urban areas, e.g., around buildings.

A common situation occurring amongst individual UAVs is in variability of the airspeeds. Between any two UAVs, it is expected that the cruising airspeed may differ by up to 10 percent of the nominal airspeed. Wind force is important because it has a significant non-linear effect on the guidance algorithm. The impact of wind effects on trajectory planning can be dealt with by iteratively solving a no wind case problem with a moving virtual target, or by generating candidates of smooth (continuous curvature) time-optimal paths between initial and final points in the plane. These approaches search only for a horizontal plan between the start and final configuration and do not take possible obstacles into account [7].

Health aware planning As coordination and communication amongst multiple UAVs become prohibitively complex, the application of swarm intelligence to teams of UAVs is foreseen. With swarm intelligent systems, control occurs at the micro-level of the agents, while the macro-level emergent behavior meets the application objective. However, the relationship between the low-level agent behavior and the macro-level emergent behavior is non-linear, not clearly understood, and difficult to predict. For a real-time swarm application such as a UAV swarm, additional tools and methods must be explored to improve control of the swarm. A framework is presented in [37], to incorporate a swarm application into a **dynamic data driven application system** (DDDAS). DDDAS entails the ability to incorporate additional data into an executing application. The data helps drive the decision making process, which in turn impacts the measurement of future real-time data, affecting future simulations. This synergistic feedback control loop between the simulations and an executing application can improve analytic and predictive capabilities of the real-time system.

The UAV health is characterized into four different categories [86]:

1. **Structural/actuator health**: Represents damage and functionality status of structural and actuator components of a UAV. Failures/faults such as wing damage, rudder damage, blade damage and mainframe damage are included in this category. These health components impact motion-related abilities, such as target pursuit.

2. **Sensor health**: Represents the functionality of the sensor-hardware such as imaging sensors, video recorders, sound detectors and infrared sensors.

3. **Communication**: Represents the functionality of the communication hardware of the UAV, such as its wireless modem. Health of this component affects the distance and reliability of UAV communication.

4. **Fuel**: Represents fuel or power consumption quantity of the UAV per transition which impacts all abilities.

Definition 27 *A **capability** of an agent is defined as a measure of how well an agent can perform a task, such as target pursuit, or steadily capturing images. A **task** is defined as sub-goal of the overall mission that can be completed by a single or multiple agents. Tasks have associated uncertainties, such as uncertain target dynamics in a tracking task or uncertain weather conditions in a precision drop-off task. Therefore, the success of completion of a task is a random variable that depends on the agent's capability.*

The generic **health aware planning problem** requires routing UAVs through different parts of the environment to accomplish different tasks such that the number of cumulative completed tasks is maximized. Furthermore, it is assumed that UAV health can degrade over time, and the health-aware planner needs to determine when to optimally return UAVs to base for repair. The environment can be visualized by imagining a graph whose vertices are task zones and edges depict zones between which the UAV can transition. Avionics capabilities can be assessed using the diagnostic offered by those subsystems. However, estimating the motion-related capabilities of the UAV is difficult, because they must be inferred through the UAV's dynamic response. To counter this problem, online generated dynamic health models are used to characterize UAV capabilities after failures. It is necessary to transfer the knowledge contained in the estimated dynamic model of a UAV after a failure to the planning layer as a change in agent capabilities.

A taxonomy for **multi-robot task allocation** (MRTA) problems is based on the main characteristics of UAVs, tasks, and time, as follows:

1. **Single-task robots** (ST) versus **multi-task robots** (MT): Single-task robots can do at most one task at a time, while multi-task robots can work on multiple tasks simultaneously.

2. **Single-robot tasks** (SR) versus **multi-robot tasks** (MR): Single-robot tasks require exactly one robot in order to be completed, while multiple robots are needed to complete a multi-robot task.

3. **Instantaneous assignments** (IA) versus **time-extended assignments** (TA): In instantaneous assignments, tasks are allocated as they arrive, while in time-extended assignments, tasks are scheduled over a planning horizon.

4. **Time window** (TW) versus **synchronization and precedence** (SP) constraints:

(a) Hard temporal constraints versus soft temporal constraints. Hard temporal constraints require that no temporal constraint is violated, while soft temporal constraints allow some violations with a penalty.

(b) Deterministic versus stochastic models. In deterministic models, the output of the model is completely determined by the initial conditions, while stochastic models assume a model of the uncertainty is available.

Synchronization and precedence constraints can be used to model different types of temporal relationships between tasks. Precedence constraints typically are in the form that the start time of a task cannot occur earlier than the end time of any of its predecessors (end to start). Other precedence models include start to start, start to end, and end to end constraints. The nature of the temporal constraints in multi-UAV task allocation problems is very broad; for example, in search and rescue domains, tasks are discovered over time and have to be done as quickly as possible. In dynamic environments, UAVs might arrive late to some tasks and might miss some. On the other hand, some surveillance tasks require not arriving late to tasks. In some cases tasks need to be executed in a specific order, such as in urban disaster scenarios in which police must clear blockades from roads before ambulances can travel to carry injured people. Some tasks may need to be done concurrently, as in surveillance where UAVs have to track people while avoiding obstacles [68].

2.2.3 Fundamentals

2.2.3.1 Graph theory basics

Graph theory is extensively used in this book, and some definitions are introduced in this section [93].

Definition 28 *A **graph** $G = (V, E)$ consists of V, a non-empty set of vertices (or nodes) and a set of edges. Each edge has either one or two vertices, associated with it, called its endpoints. An edge is said to connect its endpoints.*

Definition 29 ***Triangle inequality**: If for the set of vertices $a, b, c \in V$, it is true that $t(a, c) \leq t(a, b) + t(b, c)$ where t is the cost function, then t satisfies the triangle inequality.*

Definition 30 ***Adjacent vertices**: Two vertices u and v in an undirected graph G are called adjacent (or neighbors) in G if u and v are endpoints of an edge e of G. Such an edge e is called incident with the vertices u and v and e is said to connect u and v.*

Definition 31 *The **degree of a vertex** in an undirected graph is the number of edges incident with it, except that a loop at a vertex contributes twice to the degree of that vertex.*

Definition 32 *A* ***directed graph*** *(or digraph)* (V, E) *consists of a non-empty set of vertices* V *and a set of directed edges (or arcs)* E*. Each directed edge is associated with an ordered pair of vertices. The directed edge associated with the ordered pair* (u, v) *is said to start at* u *and end at* v*. A directed graph* G *consists of a vertex set* $V(G)$ *and an edge set* $E(G) \subseteq V(G) \times V(G)$*. For an edge* $e = (u, v) \in E(G)$*,* u *is called the head vertex of* e *and* v *is called the tail vertex.*

If $(u, v) \in E(G)$ for all $(v, u) \in E(G)$, then the graph is undirected. The graph is simple if there are no edges of the form (u, u) for $u \in V(G)$. Two standard ways exist to represent a graph without multiple edges $G = (V, E)$: as a collection of adjacency lists or as an adjacency matrix. Either way applies to both directed and undirected graphs. For the adjacency-matrix representation of a graph $G = (V, E)$, the vertices are numbered $1, 2, ..., |V|$ where $|V|$ represents the size of V.

Definition 33 *The* ***adjacency matrix*** *representation* $\mathbf{A}$ *of a graph* $G = (V, E)$ *consists of a* $|V| \times |V|$ *matrix* $A = (\mathbf{A}_{ij})$ *such that*

$$\mathbf{A}_{ij} = \left\{ \begin{array}{cc} 1 & if\ (i, j) \in E \\ 0 & otherwise \end{array} \right\} \tag{2.1}$$

If G *is directed,* $\mathbf{A}_{ij} = true$ *if and only if* $< v_i, v_j >$ *is in* E*. There are at most* $|V|^2$ *edges in* E*.*

Remark 34 *The adjacency matrix* $\mathbf{A}$ *of an undirected graph is its own transpose* $A = A^T$*. An adjacency matrix can also represent a weighted graph. When the graph is dense, an adjacency matrix representation is preferred. When the graph is sparse, the adjacency list is preferred.*

Definition 35 *The* ***indegree*** *of a vertex* u *is the number of edges that have* u *as their head vertex. The indegree matrix of a graph* G*, denoted* $D(G)$*, is a diagonal matrix of size* $|V(G) \times V(G)|$ *defined as follows:*

$$D(G)_{ii} = indegree(u_i) \qquad u_i \in V(G) \tag{2.2}$$

Definition 36 *The* ***outdegree*** *of a vertex* u *for a graph* G *is defined analogously, as the number of edges having* u *as the trail vertex.*

Definition 37 *The* ***Laplacian*** $\mathbf{L}$ *of the graph* G *with the incidence matrix* $\mathbf{A} \in \mathbb{R}^{n \times m}$ *is the* $n \times n$ *matrix:*

$$\mathbf{L} = \mathbf{A}.\mathbf{A}^T \tag{2.3}$$

Definition 38 *The* ***Fiedler value****, also called* ***algebraic connectivity****, is a spectrum measure to evaluate the connectivity of a graph. It is a global connectivity measure associated with the entire graph. The Fiedler value is the second smallest eigenvalue of the Laplacian matrix. The larger the Fiedler value, the better the overall connectivity. The algebraic distance is a local connectivity measure.*

Definition 39 *A* ***weighted adjacency matrix*** *is a matrix representation of a weighted graph. Let v_i and v_j be numbered vertices for $1 < i, j < n$. Let w_{ij} be the weight of a directed edge if there exists an edge $e = (v_i, v_j)$, and assume that all edge weights are positive; $w_{ij} > 0$. The matrix $\mathbf{M}(i,j)$ is an adjacency matrix where $\mathbf{M}(i,j) = w_{ij}$ if there exists an edge $e = (v_i, v_j)$. For each vertex i in the graph, $\mathbf{M}(i,i) = 0$, and where no path exists $\mathbf{M}(i,j) = \infty$.*

Definition 40 *The edge (vertex)* ***connectivity*** *of a graph G is the smallest number of edge (vertex) deletions sufficient to disconnect G. Communication graph connectivity is generally referred to as the vertex connectivity of the defined communication graph.*

The algebraic connectivity is considered to be a measure of how well-connected a graph is. It has a direct connection to the number of connected components. A graph with small algebraic connectivity is easier to bisect than one with large algebraic connectivity.

Definition 41 *A graph $G = (V, E)$ is* ***strongly connected*** *if and only if there exists a path from every vertex $v \in V$ to every other vertex $u \in V$.*

Definition 42 *Given a* ***directed graph*** *$G, P = (u_1, ..., u_k)$, is a directed path in G if for every $1 \leq i < k$, there exists edge $(u_i, u_{i+1}) \in E(G)$.*

Definition 43 *A* ***tree*** *is a directed graph with no cycles, in which all nodes have several outgoing edges to each of which corresponds another node, called a child. A child node can in turn have other children. All nodes, excluding the root, have one and only one incoming edge, which comes from the parent. The root has no parent.*

Definition 44 *Given a directed graph G, T is called a* ***rooted directed spanning tree*** *of G if T is a subgraph of G, it has no cycles and there exists a directed path from at least one vertex, the root to every other vertex in T.*

2.2.3.2 Temporal logic

Symbolic planning is mainly based on linear temporal logic and automaton. Thus, some basic notions are introduced in this section. Linear temporal logic is a rich specification language that can express many desired properties, including safety, reachability, invariance, response and/or a combination of these. Linear temporal logic formulas are made of temporal operators, Boolean operators and atomic propositions connected in any sensible way [44]. Standard linear temporal logic is built upon a finite set of atomic proposition, logical operator such as negation, disjunction and the temporal modal operators [58]. Some introductory definitions are introduced below.

Examples of temporal operators include $\mathbf{X}$ (next time), $\mathbf{F}$ (eventually or in the future), $\diamondsuit$ (always or globally), $\mathbf{U}$ (until). The Boolean operators are the $\neg$ (negation), $\wedge$ (conjunction), $\vee$ (disjunction), $\Longrightarrow$ (implication), $\Longleftrightarrow$

(equivalence). The atomic propositions are properties of interest about a system, such as the set of obstacles and targets. The semantics of linear temporal logic formulas are given over labeled transition graphs.

Definition 45 *Any* ***atomic proposition*** π *is a linear temporal logic formula : if* ϕ *and* ψ *are linear temporal logic formulas so are* $\neg\phi, \phi \vee \psi, \phi \wedge \psi, \phi \mathbf{U} \psi$ *with* $\neg$ *negation,* $\vee$ *conjunction,* $\wedge$ *disjunction are standard Boolean operators,* $\mathbf{U}$ *being a temporal operator.*

Definition 46 *A* ***linear temporal logic formula*** ϕ *over the atomic propositions* Π *is defined inductively as follows:*

$$\phi ::= \mathbf{True}| \quad \alpha| \quad \phi \vee \phi| \quad \phi \wedge \phi| \quad \neg\phi| \quad \boldsymbol{X}\phi| \quad \phi \mathbf{U} \phi \tag{2.4}$$

where **True** *is a predicate true in each state of a system,* $\alpha \in \Pi$ *is an atomic proposition, and* $\mathbf{X}$ *and* $\mathbf{U}$ *are temporal operators.*

A linear temporal logic formula over Π is interpreted over $\omega-$words, i.e., infinite sequence in 2^{Π}. Let $\omega = \omega_0 \omega_1 \omega_2 \dots$ be such a word.

Definition 47 *The* ***satisfaction of an LTL*** *formula* ϕ *by* ω *at position* i *written* $\omega_i \Diamond \phi$ *is defined recursively as follows:*

1. *for any atomic proposition* $\pi \in \Pi, \omega_i \pi$ *if and only* $\pi \in \omega_i$
2. $\omega_i o \phi$ *if and only if* $\pi \in \omega_i$
3. $\omega_i \phi \mathbf{U} \psi$ *if and only if* $\omega_i \phi$ *or* $\omega_i \psi$

This description allows to represent the relative position between robots. Such information is necessary for optimizing the UAV motion [85]. Linear temporal logic formulas are interpreted over infinite words, generated by the transition system **True**. For example, $\mathbf{X}\alpha$ states that at the next state of a word, proposition α is true and $\alpha_1 \mathbf{U} \alpha_2$ states that there is a future moment when proposition α_2 is true and proposition α_1 is true at least until α_2 is true. From these temporal operators, two other temporal operators can be constructed:

1. **Eventually** (i.e., future) **F**: defined as $\mathbf{F}\phi = \mathbf{True}\mathbf{U}\phi$ states that ϕ eventually becomes true in the word.
2. **Always** (i.e., globally) **G** defined as $\mathbf{G}\phi = \neg\mathbf{F}\phi$ states that ϕ is true at all positions of the word.

By combining the temporal and Boolean operators, an LTL formula can be built up from a set Π of atomic propositions, the logic connectives and the temporal modal operators. An LTL formula is interpreted on infinite strings $\sigma_0 \sigma_1 \sigma_2 \dots$ where $\sigma_i \in 2^{\Pi}$ for all $i \geq 0$. Given propositional formulas p_1 and p_2 examples of LTL formulas include $p_1 \mathbf{U} p_2$, states that p_1 has to remain true until p_2 becomes true. A safety formula $\square p_1$ simply asserts that property p_1 remains invariantly true throughout an execution. A reachability formula $\diamond p_1$ states that property p_1 becomes true at least once in an execution (i.e., there exists a reachable state that satisfies p_1). Transition systems and automata are also important notions in linear temporal logic. A transition system is obtained from the dual graph of the partition induced by the regions of interest, if the vertices are labeled according to their being part of obstacles or of targets, and if the edges are seen as transitions that a UAV can take. A UAV can take a transition from a region r_i to a region r_f if a feedback controller $u_{ri,rf}$ can be designed to fly the UAV in finite time from region r_i to a region r_f through the separating facet, irrespective of the initial position of the robot in r_i. State x_i has a self-transition if a feedback controller $u_{ri,ri}$ can be designed to keep the UAV in finite time in this state. The computation of such controllers can be performed by polyhedral operations. Specifically, this procedure produces an automaton exhaustively describing all possible solutions to the problem. This corresponds to the set of all discrete solutions.

Higher degrees in autonomy imply higher-level tasks involving decision making in route guidance, traditionally optimal way-point navigation. For UAVs, high-level tasks as well as safe behaviors can be formulated with linear temporal logic such as this example: perform persistent surveillance in region A until target is found; then, report data to region B, never fly in region C, and finally return to base [6]. Guidance and control algorithms are then required to plan and execute UAV motions to satisfy these specifications.

2.2.3.3 *Sensor coverage*

When the sensing coverage of a sensor node is assumed to follow the **binary detection model**, an event occurring within the sensing radius of a sensor node is assumed to be detected with probability 1 while any event outside the unit circle is assumed to be undetected. The recognition possibility of an event is equal to 1 within the sensing range, otherwise the probability is 0:

$$P(S_i) = \left\{ \begin{array}{cc} 1 & \text{if } d(S_i, P) < r \\ 0 & \text{otherwise} \end{array} \right\} \tag{2.5}$$

where $d(S_i, P)$ is the distance between the point P and the sensor S_i, $P(S_i)$ is the probability of the event within the sensing range of the ith sensor. There are two kinds of binary sensing models:

1. In the perfect binary sensing model, each node is able to recognize in case the target falls within its sensing range R.

2. In the imperfect or frail binary sensing model, the target is always diagnosed within the inner disk of radius R_{in}.

In [92], two completing hypotheses are summarized as follows:

$$\left\{ \begin{array}{c} H_0 : X = 0 \\ H_1 : X = 1 \end{array} \right\} \tag{2.6}$$

where H_0 is the null hypothesis representing the event has not occured yet and H_1 the alternative hypothesis representing the event has already happened. Then $P(X;H_0)$ and $P(X;H_1)$ are respectively the probability distribution function of the decision that the event occurred under H_0 or H_1. Generally, the purpose is to minimize the probability of missed detection $P(H_0;H_1)$, holding the probability of false alarms $P(H_1;H_0)$.

Problem formulation Let the workspace $W \subset \mathbb{R}^2$ be a compact set, the minimal and maximal velocities are such that $0 \leq v_{min} < v_{max}$ (0 can be acceptable for a rotary-wing UAV not for an airplane, due to the stall velocity), the state of the UAV (position of center of mass, magnitude and direction of its velocity vector) is $\zeta = (x, y, v, \psi) \in D = W \times [v_{min}, v_{max}] \times S$, $x(\zeta)$ is the projection of $\zeta \in D$ on the set W. The UAV kinematic model at a constant altitude can be given by:

$$\begin{aligned} \dot{x}(t) &= v(t)\cos\psi(t) \\ \dot{y}(t) &= v(t)\sin\psi(t) \\ \dot{v}(t) &= u_1(t) \\ \dot{\psi}(t) &= \frac{u_2}{v(t)} \end{aligned} \tag{2.7}$$

where the tangential and lateral accelerations respectively u_1, u_2 are the control inputs. The set of admissible control input values is the compact domain:

$$U = \left\{ (u_1, u_2) \in \mathbb{R}^2; \frac{u_1^2}{a^2} + \rho^2 u_2^2 \leq 1 \right\} \tag{2.8}$$

where the constants a, ρ are prespecified. The set W is partitioned into convex polytopic sub-regions called cells. The intersection of two cells is empty or a single vertex or a finite-length segment that lies on the boundaries of both cells. Let the number of cells be $N^c \in \mathbb{N}$ and $R^i \subset W$ the sub-region associated with the ith cell, $i = 1, \ldots, N^c$. An undirected graph $G = (V, E)$ is associated with this partition such that each vertex of G is uniquely associated with a cell and each edge of G is associated with a pair of geometrically adjacent cells. Two cells are considered geometrically adjacent if their intersection is a finite-length segment.

Definition 48 *The* ***G-trace*** *of the trajectory $\zeta(t;\zeta_0;u), t \in [0, t_f]$ is the path $tr(\zeta, G) = (v_0, v_1, v_p) \in L_G$ of minimal length such that:*

1. *$x(\zeta(t;\zeta_0;u)) \in cell(v_0)$*
2. *There exists a positive and strictly increasing sequence $\{0, t_1, \ldots, t_f\}$ and $x(\zeta(t;\zeta_0;u)) \in cell(v_k), t^i n\, [t_{k-1}, t_k]$ for each $k = 1, \ldots, P$.*

Linear temporal logic (LTL) can express specifications on the behavior of a system over time.

Problem 49 *Given an LTL formula ϕ and $\zeta_0 \in D$, determine the collection of paths $L_{\Gamma_\phi} \subset L_{\Gamma_{\zeta_0}}$ such that every path in L_{Γ_ϕ} satisfies the formula ϕ.*

Route planning algorithm Route planning involves the search for a collision-free path, taking into account the flight capabilities of the UAV. The solution proposed in [20] relies on traversability analysis and assignment of transition costs to successions of edges in the graph G. It uses the notion of lifted graph: Certain reachability properties of the UAV model are associated with edge transitions in the graph. A transition cost function is defined to compute the cost of paths. The graph and the associated edge transition cost computations define a finite state transition system that represents the UAV dynamic model. **Dijkstra's algorithm** is used to find the shortest path on this graph. Typically, the basic high-quality aerial photography oriented flight route refers to one with minimum altitude, no ground collision, sufficient trajectory smoothness and sufficient terrain coverage:

$$F_{cost} = \omega_1 C_{coverage} + \omega_2 C_{altitude} + \omega_3 C_{collision} + \omega_4 C_{smooth} \tag{2.9}$$

where $C_{coverage} = S_{terrain}/S_{coverage}$ penalizes the route that cannot cover the entire terrain, $C_{altitude}$ penalizes the route with the high flight altitude integrating the altitude error between the route and the ground during the entire flight, $C_{collision} = L_{under}/FR$ penalizes a flight route which could collide with the ground, $C_{smooth} = \sum_{i=1}^{Dim} \theta_i/Dim$ penalizes routes that are not sufficiently smooth, the parameters $\omega_1, \dots, \omega_4$ are user-specified penalty parameters. $S_{terrain}$ is the surface of the whole terrain, $S_{coverage}$ is the area covered by the UAV flight over the terrain, L_{under} is the total length of a flight route crashing into the ground, FR is the flight route, $\theta_i \in [0, \pi/2]$ is the intersection angle between the ith and $(i+1)^{th}$ grid points $(i = 1, \dots, Dim-1)$. In order to compute $S_{coverage}$, L_{under}, N_{f_e} sample points are uniformly generated along each of the $(Dim + 1)$ line segments, yielding $(Dim + 2 + (Dim + 1)N_{f_e})$ uniformly distributed sample points along the entire flight route: $L_{under} = n.FR/(Dim + 2 + (Dim + 1)N_{f_e})$ [59].

2.3 HOMOGENEOUS UAV TEAM

In many application scenarios, the use of several autonomous UAVs working together can speed up the sensing or exploration process by splitting the task into different parts that can be carried out simultaneously. In addition, the use of teams can increase accuracy by fusing information that has been obtained from different sources. In some other cases, collaboration enables teams of UAVs, when working together, to complete a task that would be impossible with only one UAV. The swarm intelligence based search strategy uses a completely distributed approach where every physical agent operates autonomously and collaborates with surrounding neighbors to explore the environment and compile sensing data. In this strategy the UAVs operate following two phases. During the first stage, the UAVs perform an exploration behavior. While they are doing this, they are also sensing the environment seeking data values above a fixed alarm threshold. Also during this stage, the UAVs start broadcasting the data sensed using their sensors by means of the different communications channels available to them so that other agents of the system can receive these data (at least those that are close enough). If during this state, the UAV detects a value that is higher than the alarm threshold value, the UAV changes its state to a search state. In any other case, the UAV continues with the exploration behavior until it receives or senses a data value above the alarm threshold [88]. Currently, operating multiple small UAVs is prohibited.

2.3.1 Modeling

A group of N UAVs are modeled as rigid bodies in $\mathbb{R}^3$. The configuration of the ith UAV is represented by its position $p_{B_i} \in \mathbb{R}^3$ and rotation matrix $\mathbf{R}_{B_i} \in SO(3)$ with respect to a common inertial frame. The rotation may be also described by the Euler angles: yaw ψ , pitch θ and roll ϕ angles. With a special focus on the rotary-wing UAV platform, the UAV is only able to track a smooth reference trajectory $(p_i(t); \psi_i(t))$ in the 4-dimensional space $\mathbb{R}^3 \times \mathbb{S}^1$. This is the case, for example, for helicopter and quadrotor UAVs, as well as for any other UAV whose position and yaw angle $p_{Bi}, \psi_{Bi}(t)$ are flat outputs, i.e., algebraically defining with their derivatives, the state and control of the UAV [32]. Due to their small size and limited airspeed envelope, followers have to maintain their position at a comparatively small spot in the wake of the leading vehicle, requiring both more accurate relative localization and more accurate position control.

2.3.1.1 SISO systems

A group of N UAVs is each described by an input-affine non-linear differential equation of the form:

$$\begin{aligned} \dot{x}_k &= f_k(x_k) + g_k(x_k)u_k \\ y_k &= h_k(x_k) \end{aligned} \tag{2.10}$$

where the state is $x_k \in X_k \subset \mathbb{R}^{n_k}$, the input $u_k \in \mathbb{R}$ and the output of UAV k is $y_k \in \mathbb{R}$, f_k, g_k are smooth vector fields defined in $X_k, \forall k \in \{1, \dots, N\}$. Hence, each UAV is a **single input single output** (SISO) system. To synthesize a cooperative behavior of the group, the output trajectories of all UAVs shall agree upon and converge to some common output trajectory and show a desired behavior.

$$y_k(t) - y_j(t) \to 0 \quad \text{as } t \to \infty \tag{2.11}$$

A hierarchical control scheme consisting of two levels is presented in [77]: On the UAV level, each UAV is equipped with a local controller that achieves asymptotic tracking of a given reference trajectory. On the network level, a synchronization mechanism achieves agreement of the reference signals.

Definition 50 ***Synchronization constraints*** *specify temporal constraints among tasks.*

The local tracking control problems and the synchronization problem on the network level are decoupled. The main challenge is to establish feedback from the UAVs to the network level such that the group can react cooperatively on disturbances acting on individual UAVs.

UAV level Each UAV is equipped with a controller that achieves asymptotic output tracking for some reference output $y_k^*(t)$. If this reference signal $y_k^*(t)$ is generated by a linear dynamical reference model of the form:

$$\begin{aligned} \dot{z}_k &= \mathbf{A} z_k + \mathbf{B} w_k \\ y_k^* &= \mathbf{C} z_k \end{aligned} \tag{2.12}$$

where $z_k \in \mathbb{R}^n, y_k \in \mathbb{R}, w_k \in \mathbb{R}, k \in N$. The asymptotic output tracking control problem turns into an asymptotic model matching problem.

Network level The coordination problem on the network level reduces to an output synchronization problem of the linear reference models (2.12). These models are identical for all UAVs and each UAV has access to the full state of its model since it is part of the controller. Hence, the synchronization problem on the network level can be solved via static diffusive couplings of the form:

$$w_k = \mathbf{K} \sum_{j=1}^{N} a_{kj} \left(z_j - z_k \right) \tag{2.13}$$

where $\mathbf{K}$ is a coupling gain matrix, and a_{kj} are the entries of the adjacency matrix of the communication graph on the network level. If the pair $(\mathbf{A}, \mathbf{B})$ is controllable and stabilizable, then there exists a matrix $\mathbf{K}$ such that (2.13) solves the synchronization problem.

2.3.1.2 MIMO systems

Each UAV is a **multiple input multiple output** (MIMO) system. Modeling UAVs as kinematic yaw-agents is a common assumption in the multi-robot literature.

$$\dot{\psi}_i = w_i \tag{2.14}$$

$w_i \in \mathbb{R}$ is the yaw-rate input. The position is steered at the first-order level or kinematics level as:

$$\dot{p}_i = u_i \tag{2.15}$$

where, u_i is a linear velocity.

At the second-order level or dynamics level, the position and velocity are given by the following differential equations:

$$\dot{p}_i = v_i \qquad \mathbf{M}_i \dot{v}_i = u_i - \mathbf{B}_i v_i^2 \tag{2.16}$$

where, u_i is a force, v_i is a velocity and $\mathbf{M}_i$ is the symmetric positive definite inertia matrix of UAV i and $\mathbf{B}_i \in \mathbb{R}^3$ is a positive definite matrix representing an artificial damping added to asymptotically stabilize the behavior of the UAV and also take into account typical physical phenomena such as wind/atmosphere drag. Due to their higher complexity, dynamic agents are less commonly adopted. Nevertheless dynamic agents provide a better approximation of the actual UAV dynamics, and therefore are more appropriate whenever the dynamical properties of the UAVs are more stressed, as in the case of intermittent interactions. The inputs are generated by task controller, obstacle controller and topological controller. The topological controller generates an input in order to implement a desired mutual interaction, e.g., to reach a prescribed set of precise mutual configurations or to approximately keep a given separation distance among the UAV. The collection of all the mutual interactions can be described by an **interaction-topology graph** $G = (V; E)$, where the vertices V represent the UAVs and the weighted edges $E \subset V \times V$ represent the intensity of the interaction between two UAVs. A non-zero weight models the presence of interaction, while a zero weight means no interaction and is equivalent to the absence of that edge. Three possible classes of interaction graphs can be considered:

1. **Constant topology**, when the task requires that the interaction pairs are always the same;
2. **Unconstrained topology**, when the topology can freely change over time, allowing the group to even disconnect into several sub-groups;
3. **Connected topology**, when the interaction graph can still freely change but with the constraint of remaining connected at all times, i.e., to ensure the group cohesion.

Assuming an interaction-topology graph is chosen, a formation is then commonly described by assigning the desired relative behavior between every UAV pair $(i; j) \in E$.

2.3.2 Planning

In many cases there may be a human operator who directs the semi-autonomous system, but increasingly the system must work on its own for long periods without such interventions. There is a clear need for such systems, particularly:

1. When deployed in remote or dangerous environments where direct and local human control is infeasible; or
2. When the complexity or speed of the environmental interactions is too high for a human to handle.

Agent-based modeling and simulation represent a system as autonomous agents that interact amongst one another as well as with the environment. Agent behavior is designed to capture local interactions, through which, after repeated interaction, the collective characteristics of a system emerge over time. Modeling a system at the agent-level is considered a bottom-up approach, in contrast to conventional top-down models, like equations [64]. One of the problems in UAV swarm control and navigation is **mission scheduling**. The autonomous and decentralized task allocation problem in dynamic environments is considered in [69]. The focus is the overlapping coalition in that agents are allowed to be members of multiple coalitions. A market-based decentralized coalition formation algorithm in dynamic environment with limited communication ranges is proposed. The agents in a virtual market negotiate within their neighborhood, defined as agents within their communication range, using local information to maximize the group utility. The network topology between agents is a connected graph such that there exists a path between every pair of vertices. Agents communicate with each other in a synchronous manner, according to the scheduled time table. The virtual market follows the designed market mechanism, which has different phases such as the advertisement preparation, consensus on a project manager followed by the application and finally the task allocation.

UAV team mission planner In UAV domain, the mission planning issue focused not only on trajectory planning but also on task planning. The mission can be defined as the following data structures:
Mission = {Waypoint, Task, UAV}, where Waypoint = $\{x, y, z\}$,
Task = {Task position, Task status, Goal} ,
UAV = {UAV status, Sensor, Task},
UAV status = {Unstarted, Working, Active, Out-of-service},
Task status = {Unallocated, Allocated, Done, Failed}.

For a set of UAVs and a set of tasks, the allocation of tasks to UAVs is based on the time to go from the current position of the UAVs to the position of tasks and the priority of these tasks. So the multiple criteria function for task allocation is described as the following [83]:

$$J = min_{i=1,\dots,M} \left(\sum_{j=1}^{N} (t_{ij}, p_j) \right) \tag{2.17}$$

Where N and M are the number of tasks and UAVs, respectively, p_j is the priority factor of the task j, and the parameter t_{ij} is the travel time from the position of the UAV i to the position of the task j.

UAV swarm mission planner The problem of dynamic mission planning for a UAV swarm is investigated. A centralized-distributed hybrid control framework is proposed for mission assignment and scheduling. Dynamic data driven application system principles are applied to the framework so that it can adapt to the changing nature of the environment and the missions. In the framework, a central controller is responsible for assigning tasks in a mission to different UAVs based on its latest knowledge about the swarm. Upon receiving the task assignment, the UAV tries to schedule the new task into its local task queue in order to minimize total task completion cost. Since the central controller does not have real time information about the swarm, it incorporates the status information sent from the swarm and utilizes this information to update its swarm model for future task assignments. To evaluate the proposed framework, a prototype simulation framework is implemented in [90]. The swarm is capable of carrying out multiple missions, as well as accepting new missions on the fly.

Problem 51 *Given a swarm S, either with or without existing missions, and a new mission MS, the problem is how to schedule the mission onto the swarm so that the total mission cost is minimized.*

A mission consists of multiple tasks. A task can be finished by any UAV. The scheduling process produces a mapping, or an assignment, M, from the set of tasks T to the set of UAVs, S. The mapping specifies which task is assigned to which UAV. The mission cost is the sum of all its tasks' costs. The cost of a task can be different for different types of tasks. Generally, it consists of the traveling cost and the task completion cost. To solve this problem, a hybrid approach can be employed which combines global task assignment and local task scheduling. Two types of agents are used. One is the **swarm control agent** (SCA), and the other is the UAV agent. A swarm operation environment includes a single SCA and multiple UAVs. The SCA is the interface between the human operator and the swarm, it usually runs in the ground station and its design objective is to decompose a newly arrived mission into multiple tasks, assign those tasks to UAVs, and monitor the status of the swarm and missions. On the other hand, a UAV agent represents an operational UAV in the swarm which is capable of completing different tasks. Agents communicate with each other through messages. These messages either require the recipient to take an action, or contain the latest information about the sender.

Let $S = \{v_1, v_2, ..., v_n\}$ denote a multi-UAV system, and $n = |S|$ denote the number of UAVs in the team. The number of UAVs is a constant number during run-time. A UAV has the following properties:

1. Speed, currently constant speed is assumed. But different UAVs can have different speeds;

2. Maximum capacity, a UAV's capacity is power source to sustain the UAV's flight, such as fuel or battery.

During a swarm's mission execution, for any given time t, each UAV has a state, defined as $X_i(t)$. A state contains the following information:

1. Current position of the UAV. All UAVs fly on the same altitude, so the current position contains x and y coordinates;
2. Current heading of the UAV;
3. Residual capacity;
4. Measured capacity consumption rate, this value, along with the UAV's max capacity, determines how long the UAV can stay in the operation area before having to return to base. This measurement changes overtime;
5. A list of tasks assigned to this UAV could be empty. Here all UAVs are homogeneous, that is, a UAV can take over any task.

A team's state consists of two parts. One is the collection of its UAVs' states, the other is a collection of mission states for all missions scheduled on the team [90]. A mission consists of one or more tasks specifying the detailed steps of the mission. If each task is considered as a node, and the dependency relationship between two tasks as a directed edge, then a mission can be represented by a **directed acyclic graph**. After a mission is scheduled on a team, there is a state associated with it for any given time t. A mission's state is a 3-tuple $MS_t = \{C, P, W\}$ where C is the set of completed tasks, P is the set of tasks that are in progress, and W is the set of tasks that are waiting to be finished. A mission is considered completed when $P = W = \emptyset$, and $C = T$. Since tasks in a mission have dependencies among each other, they cannot be all assigned at the same time. So the swarm control agent needs to decide which tasks are ready to be assigned.

Definition 52 *A **ready task** is the one that does not have any predecessor tasks or all of its predecessors have been finished. As a task is completed, other tasks that are dependent on it may become ready.*

The swarm control agent monitors the status of each mission and assigns tasks to UAVs as they become ready. For each ready task, the swarm control agent selects the assignment target UAV by the following steps:

1. Calculate, for each UAV, the cost of finishing the task. The swarm control agent will use its latest knowledge about the swarm to estimate the cost of finishing it. This cost is different for different UAVs, because UAVs are on different locations and may or may not have existing tasks. There are two situations:
   (a) The UAV does not have scheduled tasks. In this situation, the cost is the traveling cost from UAV's current location to task location plus the task's completion cost. The capacity used for comparison is the UAV's current residual capacity;

   (b) There are scheduled tasks on the UAV already. In this case, the estimated final location and estimated final residual capacity values are used to calculate the task cost.

2. Sort the calculated costs and choose the UAV with the smallest cost value as the candidate for the task.

The task assignment gives a task to a UAV, but it does not specify how this task should be scheduled on the UAV. The swarm control agent does not specify whether the new task should be executed immediately or after all existing tasks are completed. The UAV controls this decision. There are several possible scheduling policies:

1. **First come first serve** (FCFS): the new task will be executed after all earlier tasks are completed;

2. **Insertion based policy**: if there are no scheduled tasks on the UAV, the new task will start execution immediately. Otherwise, the UAV will schedule the new task.

3. **Routing problem**: the scheduling problem can also be modeled as a routing problem.

4. **Adaptive policy**: this policy simply uses all the other policies to calculate task costs and pick the policy that produces the minimum cost each time the UAV needs to perform a task scheduling.

The pseudo-code given in algorithm 1 is to create a schedule which is driven to use the available resources in the following manners, which in general leads to a time-efficient characteristic [53]:

1. Balanced: task assignment towards earliest available UAV, which indicates its relative idleness compared to other UAVs.

2. Safe

   (a) no multiple UAVs are allowed to occupy a position simultaneously,

   (b) task execution following precedence.

3. Early

   (a) recharge station which eventually delivers a recharged UAV to its next destination early is chosen,

   (b) short makespan for the whole task completion.

The input to this algorithm is sequence of tasks (sequence), list of UAVs (uavs), while the output is schedule of tasks on UAVs (schedule).

UAV swarms have high redundancy and are comprised of relatively simple UAVs compared to the complexity of the task at hand. Furthermore, they

Algorithm 1 Earliest available time algorithm

1. Position availability
2. **For** each task in sequence **do**
   (a) An abstraction of task is broken down into possible flights and other actions which involves start and end position.
   (b) Task is checked for existing preceding task. If preceding tasks exist, the last completion time-stamp for the task is kept.
3. **For** each uav in uavs **do** (UAV ready time is saved, task-occupancy of each UAV is checked)
   (a) UAV must have enough battery level to at least fly to the nearest recharge station.
   (b) If UAV does not have enough battery to go to recharge station after executing a task, then UAV needs to go to the nearest recharge station.
   (c) Overall availability check: incorporating this task availability and UAV availability check, the start timestamp of the respective task is calculated.

use distributed control with the use of local rules and local communication. These characteristics of UAV swarms make them highly robust, scalable and flexible. Three main advantages exist scalability, flexibility and robustness:

1. **Scalability** is defined as the ability of a swarm UAV system to operate with larger or smaller numbers of individuals without impacting performance considerably. The use of local sensing and communication is the main reason for scalability in swarm UAV systems.

2. **Flexibility** refers to the ability of a swarm UAV system to adapt to new, different, or changing requirements of the environment. When the problem changes, the system has to be flexible enough to promptly respond to the new problem. In swarms, flexibility is promoted by redundancy, simplicity of the behaviors and mechanisms such as task allocation and stochasticity.

3. **Robustness** can be defined as the degree to which a system can continue to function in the presence of partial failures or other abnormal conditions. Swarm UAV systems are more reliable due to high redundancy. Any individual is expendable as others can compensate for their loss. Due to their simple and minimalist design, the individual UAVs are less prone to failures. Because of decentralized control with either no leaders or intermittent-replaceable leaders, losing an individual UAV or even a group of them will not drastically affect the overall operation of the swarm. Furthermore, the distributed sensing makes the system less susceptible to noise.

2.3.3 Cooperative path following

In the study of cooperative control of multiple UAVs, tracking of each UAV requires the estimation of its attitude and position as well as its neighbors in real time so that the task can be completed through a collaboration strategy. Whether rigid-body motion can be predicted fast enough and whether integral accuracy maintained over long-duration times are critical issues for successful multi-UAV control. The traditional method involves modeling the dynamics using differential equations and solving those equations using numerical methods. In recent years, developments have occurred in modeling rigid-body motion using the theory of differential geometry, discretizing the **Lagrange-d'Alembert** equation directly, and then using the variational principle to derive the equations of the rigid-body motion. These methods ensure that geometrical characteristics of the system in the discretization process are retained, and related physical quantities are conserved in the iterative calculations. The methods include the **discrete Moser-Veselov** (DMV) algorithm and the **Lie-group variational integrator** [60].

2.3.3.1 *Formation control*

In gradient-based formation control, stabilizable formations are identified by employing rigidity graph theory, in which the vertices of a graph represent the agents and the edges stand for the inter-agent distance constraints to define the shape of the formation. Rigid formations are associated with a potential function determined by the agent's relative positions. The potential has a minimum at the desired distances between the agent's; thus, its gradient leads to the formation controller that stabilizes rigid formations locally. For a formation of n agents associated with the neighbor relationship graph $\mathbb{G}$, the agents can be modeled in the simplest way as $\dot{p}_i = u_i$ where $u_i \in \mathbb{R}^m, i = 1, \ldots, n$ is the control input. For each edge E_k, a potential function V_k can be constructed with its minimum at the desired distance $\|z_k^*\|$ so that the gradient of such functions can be used to control inter-agent distances distributively:

$$V\left(\bar{\mathbf{B}}^T p\right) = V(z) = \sum_{k=1}^{|E|} V_k(z_k) \tag{2.18}$$

where the elements of the incidence matrix $\mathbf{B}$ are defined as:

$$B_{ik} = \left\{ \begin{array}{rl} +1 & \text{if } i = E_k^{tail} \\ -1 & \text{if } i = E_k^{head} \\ 0 & \text{otherwise} \end{array} \right\} \tag{2.19}$$

where E_k^{tail}, E_k^{head} denote the tail and head nodes, respectively, of the node E_k. Applying the gradient descent control to each agent gives:

$$u_i = -\nabla_{p_i} \sum_{k=1}^{|E|} V_k(z_k) \tag{2.20}$$

If the following family of quadratic potential functions:

$$V_k(\|z_k\|) = \frac{1}{2l}\left(\|z_k\|^l - d_k^l\right) \quad l \in \mathbb{N} \tag{2.21}$$

with the following gradient along z_k:

$$\nabla_{z_k} V_k(\|z_k\|) = z_k \|z_k\|^{l-2} \left(\left(\|z_k\|^l - d_k^l\right)\right) \tag{2.22}$$

The factor l plays an important role for the design of the desired steady-state motion of the formation. The closed-loop system dynamics can be written as:

$$\begin{aligned} \dot{p} &= -\bar{\mathbf{B}} D_z D_{\tilde{z}} e \\ \dot{z} &= -\bar{\mathbf{B}}^T \dot{p} \\ \dot{e} &= l D_z D_{\tilde{z}}^T \dot{z} \end{aligned} \tag{2.23}$$

where $e_k = \|z_k\|^l - d_k^l$, $\tilde{\in}\mathbb{R}^{|E|}$ is the stacked column vector consisting of all the $\|z_k\|^{l-2}$. The matrices $D_z, D_{\tilde{z}}$ are the block diagonal matrices of $z, \tilde{z}$ [35].

2.3.3.2 *Cooperative mission*

The cooperative missions considered in [91] require that the UAVs follow collision-free paths with the same arrival time at their final destinations (time-critical operations). In the adopted setup, the UAVs are assigned desired nominal paths and speed profiles along them. The paths are then appropriately parametrized, and the UAVs are requested to execute cooperative path following, rather than open-loop trajectory-tracking maneuvers. This strategy yields robust performance in the face of external disturbances by allowing the UAVs to negotiate their speeds along the paths in response to information exchanged over the supporting communications network. To enforce the temporal constraints that must be met in real time to coordinate the entire fleet of UAVs, the speed profile of each UAV is adjusted based on coordination information exchanged among the UAVs over a time-varying communications network.

Problem 53 ***Time-critical cooperative path-following problem:*** *Given a fleet of n UAVs supported by an* ***inter-vehicle communications network*** *and a set of desired 3-D time trajectories $p_{d,i}(t_d)$, design feedback control laws such that*

1. *All closed-loop signals are bounded;*
2. *For each vehicle $i, i \in \{1, \dots, n\}$, the path-following generalized error vector converges to a neighborhood of the origin; and*
3. *For each pair of UAVs i and j; $i, j \in \{1, \dots, n\}$ the coordination error converges to a neighborhood of the origin, guaranteeing (quasi-) simultaneous time of arrival and ensuring collision-free maneuvers.*

Under an appropriate set of assumptions, the path-following and coordination control laws are able to ensure stability and ultimate boundedness of the path-following and time-critical cooperation problems when treated separately.

In [9], the proposed mission design involves the assignment of UAVs, targets and sensors to the mission. Its main purpose is not the generation of a flight plan, but the correct assignment and configuration of sensors along a route, where the route is drawn to cover in the best way the targets assigned to the mission; each target may be a point, a linear object (river, road), or an area. In a multi-UAV scenario this goal is more complex, because the mission optimization has to be achieved combining in the best way the plan characteristics and sensor usage of all UAVs without conflicts. The UAV reasoning capabilities are based on the availability of several kinds of knowledge and information:

1. Physical and operative characteristics about UAVs and sensors; as the max speed and max fuel capacity of a UAV, and the max resolution or **field of view** (FOV) of a sensor.

2. Models for estimating parameters that are not explicitly given in the mission plan, but are needed for checking its validity; as the estimation of the fuel consumption given the length of the route and the kind of UAV, and the coverage of a target given the altitude of the UAV and the field of view of the sensor.

3. Mission constraints which restrict the contents of the mission plan by also taking into account the mission requirements; typical examples are checking that each leg of the route belongs to exactly one task, and that the estimated duration of the mission falls within the specified time envelope.

UAVs involved in a cooperative mission often operate in a common workspace. Self-collisions must be taken into account during cooperative planning such that all planned reference trajectories are feasible. Mitigating the risk of collision is especially crucial for airborne systems, where collision is likely to result in catastrophic loss of the platform and possible damage to the environment. If team planning does not explicitly consider safety constraints, the optimized team decisions may not be executed as planned. An additional layer of real-time reactive approaches may be frequently activated. This may result in wasted computational power and communication effort during the team planning process, and may also induce uncertainty in the overall system behavior. Additional constraints often appear in real-world scenarios, and neglecting these constraints can limit the applicability of multi-UAV coordination strategies. Decentralized data fusion and decentralized decision making may require the team to maintain a certain communication topology. A planned trajectory that disregards team network integrity could potentially result in disconnected communication links that limit further information sharing and decentralized planning. There may exist further spatial constraints such as obstacles and prohibited regions of entry on the field, or temporal-coverage constraints where UAVs must pass by a ground station or a user to report their observations at arbitrary time intervals. A common approach to planning for information gathering tasks in continuous domains is to formulate the problem as the optimization of an information-theoretic objective function over a finite future action horizon. Integrating constraints into this optimization problem is challenging for several reasons. Collision and network connectivity constraints are continuous along the UAV's action horizon. Constraints may also vary with time due to moving obstacles. Furthermore, constraints may not be fixed during planning. For example, when considering UAV-UAV collisions, constraints are coupled during optimization. Finding a feasible trajectory for one UAV depends on the trajectories chosen by other UAVs in the team. In [34], these continuous constraints are converted to discrete form while preserving continuity guarantees, and analytical gradients derived. Constraints are integrated into a unified optimization framework using an augmented Lagrangian formulation and solved using asynchronous decentralized gradient-descent. Planning is then performed using a multi-step receding horizon approach. Because the constraints are represented discretely and have closed-form gradients, decision making can be performed efficiently and in an online manner while ensuring continuous constraint satisfaction along the UAV trajectories.

2.3.4 Communication

2.3.4.1 Basics

Agents moving in an environment need to communicate to achieve coordination for a common objective.

1. Using **explicit communication** methods, UAVs can broadcast or send a signal to the other agents. This forces the introduction of a mechanism of declaration of intentions, for which either each agent communicates its position at fixed time instants, or a supervisor exists that is able to know the topology of the network of agents.

2. Using **implicit communication**, the agents do not need to broadcast their position, because they can feel the presence of other agents, or objects in general, avoiding collision. Moreover, implicit communication is also used to catch radio (or similar) signals that involve perturbation of the environment via sound waves embedding messages. Implicit communication does not necessarily substitute direct communication, but can be used as a redundant and faster way of communication in case of immediate response to an environment perturbation [12]. UAVs can communicate in an implicit way, i.e., using their sensors to catch perturbations of the environment due to other agents. Implicit communication may be used in case of presence of noise and environmental hostilities, which prevent direct communication, as well as in case of being subject to faults and failures affecting the sender or the receiver.

Agents communicate and coordinate to achieve a common goal. Moreover, they move within an environment that changes dynamically and detect each other's presence not necessarily via direct communication but rather by observations of the environmental changes. World variables are, hence, used to represent the changes of the environment as perceived by the agents, achieving implicit communication.

All the constituents of the UAV communication networks pose challenging issues that need resolution. Unlike many other wireless networks, the topology of UAV networks remains fluid with the number of nodes and links changing and also the relative positions of the nodes altering. UAVs may move with varying speeds depending on the application, causing the links to be established intermittently. Such a behavior could pose certain challenges such as:

1. Some aspects of the architectural design would not be intuitive. The fluid topology, the vanishing nodes and links would all challenge the designer to go beyond the normal ad-hoc mesh networks.

2. The routing protocol cannot be a simple implementation of a proactive or a reactive scheme. The inter-UAV backbone has to repeatedly reorganize itself when UAVs fail. In some cases, the network may get partitioned. The challenge would then be to route the packet from a source to a destination while optimizing the chosen metric.

3. Users' sessions should be maintained by transferring them seamlessly from an out-of-service UAV to an active UAV.

4. Energy of UAVs should be conserved for increasing the life of the network. Nodes could die out for many reasons and may be replaced by new ones. The UAVs could move randomly or in organized swarms not only in 2-D but also in 3-D with rapid change in position. Energy constraints are much greater in small UAV networks. It is important to characterize a network to understand its nature, constraints and possibilities.

The following questions could be asked [38]:

1. How fast does the topology change with time?

2. How frequently does the network get partitioned as the nodes fail or move away?

3. How can the network life be increased? What type of architecture would be more suitable?

4. Does it require self-organizing, self-healing capabilities?

5. Which protocols can be run at different layers? Does it support addition and removal of nodes dynamically?

6. Are the links intermittent and what is their quality?

While there is an uninterrupted line-of-sight between the operator and the UAV, operation does not pose a major challenge. Under line-of-sight conditions, the wireless link may experience few and short interruptions. As the UAV starts to move away and objects obstruct the link, commanding the UAV starts to become gradually challenging. Losses over the link may occur frequently and the control commands get lost in transit. This results in a stuttering operation of the UAV. **Micro-autonomy** allows making decisions under uncertainty tracking the state of a remotely controlled system. UAV has a physical component that can be controlled over a packet switched link. A wireless link over which, due to the nature of the channel, packet losses will occur, can be used [73]. It can decide which short-term action should be executed by the physical component when packet losses occur. This compensation mechanism enables the system to potentially correct or neutralize the effects of packet losses. This mechanism is not intended to provide long-term autonomy for the system but rather compensate for link interruptions that last from tens to hundreds of milliseconds.

2.3.4.2 Information architectures

A fundamental task in formation control is to maintain some prescribed geometric shape to keep the vehicles in an optimal sensing configuration [81]. A graph theoretic concept called **rigidity** is proposed to describe information architectures to maintain formation shape. A formation is rigid if it can move as a cohesive whole, with inter-vehicle distances preserved throughout the motion. It is possible to control a particular inter-vehicle distance in two ways:

1. Assign the responsibility to all UAVs to actively maintain the distance: Symmetric information architecture modeled by an undirected graph.
2. Assign the responsibility to only one UAV to maintain the distance, whereas the others do nothing to maintain that particular distance: Asymmetric information architecture modeled by a directed graph.

A formation is denoted $F(G, p)$ where G is the information architecture graph and $p : V \rightarrow \mathbb{R}^{2|V|}$ is a mapping that assigns a position in the plane to each vertex. The vertices represent UAVs and the edges represent information flow amongst the UAVs. An edge is present between two vertices whenever the distance between the two UAVs is actively maintained. If a vertex j is connected by an edge to a vertex i, j is a neighbor of i. The distance is maintained using control laws to govern the motion of each vehicle. The control law for each UAV requires the relative position of its neighbors in an arbitrary local coordinate basis. Consider the rigidity function $f : \mathbb{R}^{d|V|} \rightarrow \mathbb{R}^{|E|}$ defined by

$$f(p) = [\ldots, ||p_i - p_j||^2, \ldots] \tag{2.24}$$

where the kth entry of f corresponds to the squared distance between vertices i and j when they are connected by an edge. Assuming that the formation moves but $f(p)$ stays constant (the edges in E correspond to links where distances are preserved) and expanding about the constant value in a Taylor series while ignoring higher-order terms (for a formation undergoing smooth motion), the following relation is obtained:

$$\mathbf{J}_f(p)\delta p = 0 \Rightarrow \mathbf{J}_f(p)\dot{p} = 0 \tag{2.25}$$

where δp is an infinitesimal perturbation of the formation and the rigidity matrix $\mathbf{J}_f$ is the Jacobian of f.

When the formation is rigid, the only permissible smooth motions are translation and rotation of the whole formation. In the plane, this accounts for three linearly independent vectors, and so the kernel of $\mathbf{J}_f$ has a dimension of three.

Theorem 54 *A formation $F(G, p)$ is **rigid** if and only if rank $|\mathbf{J}_f| = 2|V| - 3$, which is the maximum rank $\mathbf{J}_f$ can have.*

For generic configurations, information about formation rigidity is contained in the graph, which allows for a purely combinatorial characterization of rigidity.

Theorem 55 *A graph $G(V, E)$ in the plane is a **rigid graph** if and only if there exists a sub-set E" of edges that the induced sub-graph $G'(V', E')$ satisfies the following:*

1. $|E'| = 2|V| - 3$;
2. *Any sub-graph $G"(V", E")$ of G' with at least two vertices satisfies* $|E"| \leq 2|V"| - 3$.

Definition 56 *A graph is called **minimally rigid** if it is rigid and there exists no rigid graph with the same number of vertices and a smaller number of edges.*

A graph is minimally rigid if removing any edge results in a loss of rigidity. The minimum number of required edges is linear in the number of UAVs. In contrast, in an all-to-all information architecture, i.e., every inter-UAV distance is actively maintained, the number of required edges is quadratic in the number.

An asymmetric information architecture is modeled by a directed graph in which a direction is assigned to every edge with an outward arrow from the UAV responsible for controlling the inter-UAV distance. Because only one UAV is responsible for controlling a particular inter-UAV distance, the overall information complexity is reduced by half. This structure minimizes the number of information links and may be necessary when there is limitation on sensor range. In this case, rigidity is necessary, but not sufficient, to preserve all inter-UAV distance in a formation with a directed graph.

Constraint consistence is needed to generalize the notion of rigidity to directed graphs. To distinguish this notion from rigidity, it is called **consistence**. A graph is called **minimally persistent** if it is both minimally rigid and constraint consistent. Constraint consistence rules out certain information flow patterns that make it impossible to maintain formation shape.

Theorem 57 *Any directed graph in the plane with no vertex having more than two outgoing edges is constraint consistent.*

Theorem 58 *A directed graph in the plane is minimally persistent if and only if it is minimally rigid and constraint consistent.*

A special case of a persistent information architecture is a **leader-follower** structure in which one vehicle can move freely in the plane and the remaining vehicles maintain a shape around the leader.

Remark 59 *There is an important distinction to be made between decentralized design and decentralized implementation.* ***Rigidity*** *and* ***persistence*** *are inherently centralized properties of the formation in that it is impossible for a single UAV to guarantee rigidity of the whole formation. Therefore, rigid and persistent information architectures provide a basis for the design of control laws that have decentralized implementation. Once the architecture is established, the implementation of the design is decentralized in that the UAVs can operate using only local information.*

2.3.5 Task assignment

2.3.5.1 Intra-path constraints

Some approaches to coordination in independent task domains exclusively reason about instantaneous allocation, where each agent is assigned only a single task at a time. Other approaches use time-extended allocation, where agents are assigned a set of tasks to accomplish over a period of time. Reasoning about time-extended allocation can improve performance, as agents can discover synergies and dependencies between tasks. Agents must determine the order in which to perform a number of tasks. The presence of intra-path constraints makes it necessary for agents to recruit assistance in order to determine which paths to take through an environment and form accurate fitness estimates for tasks. One domain where agent's schedules are coupled due to intra-path constraints is **precedence-constrained disaster response**.

Definition 60 *Precedence constraints specify partial ordering relationships between pairs of tasks.*

In [49], one time-extended tiered auction system uses two heuristics methods to allocate groups of tasks that share preconditions. The shape of the mission objective function is related to the difficulty of creating good task assignment. The task assignment problem can be written in the following mixed integer linear or non-linear program:

$$max_{x,\tau} \sum_{i=1}^{N_a} \sum_{j=1}^{N_t} F_{i,j}(x,\tau) x_{ij} \tag{2.26}$$

subject to:

$$G(x,\tau) \leq d; x \in \{0,1\}^{N_a \times N_t} \tag{2.27}$$

where $x \in \{0,1\}^{N_a \times N_t}$ is a set of $N_a \times N_t$ binary decision variables x_{ij} which are used to indicate whether or not task j is assigned to agent i; $\tau \in \mathbb{R}^{+N_a \times N_t}$ is the set of real-positive decision variables τ_{ij} indicating when agent i will service its assigned task j, $F_{i,j}$ is the score function for agent i servicing task j given the overall assignment and $G = (g_1, \dots, g_{N_c})$ with $d = (d_1, \dots, d_{N_c})$ defines a set of N_c possibly non-linear constraints of the form $g_k(x,\tau) \leq d_k$ that captures transition dynamics, resource limitations, tasking assignment constraints, cooperative constraints, etc. This programming problem is NP-hard and difficult to solve. Local information algorithms may produce arbitrarily bad allocations or, in the case of many algorithms, may not even converge. An algorithm called **bid warped consensus-based bundle** is proposed in [48].

2.3.5.2 Urban environments

The UAV mission scenario is recognition and mapping of buildings using multiple UAV in urban environments [13]. The communication link is limited to visual line-of-sight, therefore the link connecting the UAVs and the ground station can be obscured by any obstacle placed between them. The proposed approach in [89] decomposes the task assignment problem into:

1. NP_T **points of interest** (buildings) distribution among exploring agents using a heuristic function. The initial distribution is based on the position of the ground station with respect to the center of the area to cover and the number of points NP_i assigned to each agent. The distribution tries to equalize the number NP_i for all the agents NA_{ex}:

$$NP_i = floor\left(\frac{NP_T}{NA_{ex}} + \delta_i\right); \forall i = 1, \dots, NA_{ex} \tag{2.28}$$

where $\delta_i = \left\{ \begin{array}{ll} 1 & \text{if } i \leq S \\ 0 & \text{if } i > S \end{array} \right\}$ and

$$S = \left[\frac{NP_T}{NP_{ex}} - floor\left(\frac{NP_T}{NA_{ex}}\right)\right] .NA_{ex} \tag{2.29}$$

In order to improve this distribution, an iterative process based on this heuristic function where the heuristic value in percentage for the UAV i is

$$H_i = 100 \frac{H_i^p}{\sum_{i=1}^{NA_{ex}} H_i^p} \tag{2.30}$$

where the raw heuristic value

$$H_i^p = A_i + M_1 F_i + M_2 NP_i + M_3 P_i + M_4 MD_i \tag{2.31}$$

with A_i the area enclosed by the convex hull of each set of points, F_i the distance from the ground base to the furthest point from each set, P_i the length of the perimeter of the convex hull, MD_i the furthest distance of two consecutive points in the hull and M_j the corresponding multipliers.

2. **Visiting sequence generation** for each set of nodes: a **vehicle routing problem** is developed to assign the UAVs to set of points of interest and sequence them. The first route involves only the convex hull of the set of points, that is the best sequence in which these points can be visited, as any changes to this will result in a self-intersecting route. After this initial sequence is obtained, the process of adding the remaining points starts. The closest point to any road already in the sequence is added to this road. This process continues until all points are in the sequence. The final step is to perform a heuristic improvement to the obtained path, checking whether any of the points is closer to any road rather than to the lines connecting the points before and after itself in the sequence.

3. **Path generation**, once the visiting sequence is generated, is conducted using the **visibility graph**. The visibility graph is the graph of nodes from different obstacles that can see each other. Path generation considers only the obstacles between the consecutive points at each step, instead of dealing with the whole set of obstacles. Each building in the scenario may become an obstacle. At each step, path planning counts only obstacles between these two points. The connection of consecutive points is performed using the visibility approach.

4. **Communication link maintenance**: communication is available between any agents visible to each other within a certain communication range. Although two UAVs are within the communication range, the visibility between them can be blocked by a building. In order to ensure the link, each communication relay will monitor two exploring agents. The distribution of the exploring agents among the communication relay vehicles is designed to minimize the average distance between the monitored UAVs, the communication link being checked at discrete time steps.

2.4 HETEROGENEOUS UAVS TEAM

Heterogeneous UAVs have varying capabilities that affect their task performance. The performance of a UAV at a task relates to the composition of the team. Team performance can be computed as the sum of individual UAV capabilities, e.g., the amount of resources it possesses. There is synergy among the UAVs in the team, where team performance at a particular task depends not only on the individual UAV's capabilities, but also on the composition of the team itself. Specific UAVs may have or acquire a high task-based relationship that allows them to perform better as a team than other UAVs with equivalent individual capabilities but a low task-based relationship.

2.4.1 Consensus algorithm

A motivating scenario is the **urban search and rescue** (USAR) domain. When a disaster occurs, researchers from around the world arrive with their urban search-and-rescue UAVs. Due to safety and space constraints, only a subset of these UAVs may be able to be deployed to the site. Selecting an effective team is an ad-hoc problem, where the agent capabilities and synergy are initially unknown. Some of these UAVs may have been designed to work well with other UAVs developed by the same group, and in some cases, UAVs from different sources may have synergy in a team. Thus, it is necessary to model and learn the synergy of these UAVs and select the best team of UAVs to be deployed [56].

A network is considered whose communication topology is represented by an undirected graph G which is composed by a set of nodes $V = \{v_1, \dots, v_n\}$ linked by edges such that an edge e_{ij} means that the node v_i is able to communicate with node v_j. For each UAV $a_i \in N_j$, there exists an arc from node i to node j associated to UAVs a_i and a_j respectively. Given a graph G, a **consensus algorithm** is an interaction rule that specifies how each node v_j updates its estimates of the generic information $s \in S$ shared among neighbors based on any received value v_{ij}, i.e., it specifies the function $\zeta : S \times S \to S$ which is used to compute:

$$s_i^+ = \zeta(s_i, s_j) \quad \text{for } i, j = 1, \dots, n \tag{2.32}$$

If the iteration of each node converges toward a common value, a consensus is reached. Typical consensus algorithms assume that exchanged data are represented by real numbers and are typically combined according to a weighted average rule. A more general class of consensus algorithm is introduced in [63] to allow UAVs to share locally collected information and eventually converge to a unique network decision. Nodes are UAVs monitoring a common neighbor, and are supposed to communicate in G in order to reach an agreement on the reputation of the observed UAV. A solution is proposed where UAVs share any information that is directly measured or reconstructed by inspecting its neighborhood through logical consensus. After having established an agreement for the value of the encoder map for a generic agent, consensus algorithms will use the same decision rule and hence decide for the same classification vector.

To model task-based relationships, in a connected weighted graph structure, where the vertices represent the agents, and the edges represent the task-based relationships, the level of synergy of a set of agents can be defined as a function of the shortest path between agents. A non-binary metric of team performance can be based on a Gaussian model of the individual agent capabilities, to capture the inherent variability in team performance in a dynamic world. This formulation of team performance captures many interesting characteristics, such as the effects of including new agents into the team.

2.4.2 Task assignment

2.4.2.1 Dynamic resource allocation

The resource allocation system proposed in this paragraph offers dynamically reconfigurable UAV teams, which are temporarily formed to perform a mission or part of it. Teams get dismantled once the mission is serviced so that the team members can join various other teams attending different missions. The majority of current UAV resource allocation systems tend to be **problem-specific**, focusing on, for example, wide-area search, sense and react, or data gathering missions. Each of these missions has a set of UAVs associated with it that can only perform that particular mission. Problem specific planning solutions generally do not accommodate the operational constraints imposed by multiple mission scenarios. The resource allocator is required to map on the following principal performance attributes: **effectiveness, efficiency and robustness**. The resource allocation problem is modeled as a connected network of nodes in which each node represents a position and a characteristic related to that position, whether it is a task, base, or initial position. Nodes model the three categories: bases, initial positions and tasks. The resources travel among the connected nodes, receiving benefits from tasks achieved at certain nodes and expending cost due to traveling, maintenance, and sensor operation. Nodes have different requirements and constraints and will affect the resources in different ways. The main objective is to identify a mission plan that meets all requirements at minimum overall cost. Because each platform and sensor combination offers different effectiveness in performing a certain task, the importance of the effectiveness is dependent on the value (gain) of a certain task. A detailed mathematical formulation of this problem can be found in [50].

2.4.2.2 Auction-based approach

Auction-based approaches have been developed to perform distributed task allocation amongst teams of heterogeneous UAVs. These approaches assume UAVs can fail at any time and that communication may be unreliable.

Auction-based approaches assume bidders will bid only on tasks they are capable of carrying out. **Auction-based approaches** typically assume all UAVs have the necessary capabilities to assign tasks, where task allocation is assumed itself a task which is delegated to a more capable UAV. Descriptions of possible tasks are created in advance and describe the units of work required to complete the mission.

Definition 61 *The minimum requirements determine the* ***suitability*** *or set of capabilities a UAV must possess in order to carry out the task. For UAVs that meet these minimum requirements, the suitability expression defines the degree to which a UAV is suited to carry out that task. The minimum requirements and suitability expressions are heuristic in nature, and are determined by a human expert in advance of the mission.*

The minimum requirements expression for a mapping task may, for example, describe that a task requires a UAV with the capability to map an area. The suitability expression of this task could assign greater suitability to a UAV possessing a laser range-finder versus a sonar. Describing tasks in this manner forms the basis on which UAVs can reason about the best available team member to carry out a specific task. This knowledge also serves to indicate when the current team structure is less than adequate, and describes the needs of desirable new team members. To facilitate efficient task allocation and assign a general responsibility of duties, roles are defined in terms of the tasks a UAV filling the role is normally expected to be able to perform.

Remark 62 *A role can be thought of as a characterization of the type of work a UAV filling the role would normally encounter. Thus, the tasks expected of a role determine the capabilities required of a UAV filling that role. Since task requirements are formulated such that a UAV's suitability to carry out a task can be calculated, it follows that a UAV's suitability to fill a role is the aggregate of its suitability to complete each task normally expected of the role. A task can be assigned to a team member occupying a role that is normally expected to carry out that task, eliminating the need for further reasoning. The desired team identifies the required roles and the quantity of each in order to make an effective team. This description then forms the goal that the framework team maintenance operations aims to achieve. The desired team composition is highly domain- and equipment-dependent and is determined by a human in advance of operation [37].*

2.4.2.3 Deadlock problem

An integrated approach based on graph theory can be presented for solving the deadlock problem in the cooperative task assignment of multiple heterogeneous UAVs, which is concerned with the cooperative decision making and control. Because of heterogeneity, one task cannot be performed by arbitrary UAVs in the heterogeneous group. A UAV that performs multiple tasks on targets needs to change its path, waiting for others if another vehicle that executes a former or simultaneous task has not finished or arrived. This creates risks of deadlock. Two or more UAVs may fall into a situation of infinite waiting due to shared resources and precedence constraints among various tasks. A **task-precedence graph** of solutions is constructed and analyzed for detecting deadlocks. Transposing operations are used to unlock solutions involved in deadlocks. In addition, the topological sort of tasks is used in the path elongation of vehicles. Thus, deadlock-free solutions are obtained, and the path coordination is done. The problem of multiple consecutive task assignment of heterogeneous UAVs is studied in a centralized form so as to present the method systematically. Multiple heterogeneous UAVs are allocated to perform a set of predefined consecutive tasks on known ground targets in [22]. The task assignment problem is described as a **combinatorial optimization problem**. Each assignment that allocates multiple vehicles to perform multiple tasks on multiple targets is a candidate solution. An integrated graph-based method for solving deadlock is presented. Each feasible assignment of tasks is a feasible solution of the combinatorial optimization problem. The non-deadlock condition is a prerequisite for the subsequent process. An initial assignment is first processed into two groups according to two types of task relation. Then a **task precedence graph** (TPG) is constructed for analyzing solutions and detecting deadlocks. If an initial solution is found whose task precedence graph has strongly connected components through a graph algorithm (i.e., the graph is not acyclic), it must encode **deadlocks** and will be modified by transposing operations until unlocked. Finally, the topological sort of each feasible solution is used in path elongation of the assignment.

2.5 UAV–UGV TEAMS

By incorporating different types of robots in a mission, accomplishment of more complex tasks with higher precision will be possible, for the complementary skills provided by each type to overcome the specific limitations of the others. For instance, UAVs are by far faster than **unmanned ground vehicles** (UGVs) and also have a wider view. As a consequence, they are able to provide a rough map. However, they are not able to detect small ground objects, precisely. On the other hand, UGVs are closer to ground objects and could be deployed for long time missions. Nevertheless, using heterogeneous types of agents in a multi-agent system makes the problem more challenging. This complexity is due to the different dynamics and constraints of heterogeneous agents. UAVs and UGVs collaboratively play central roles in information gathering. Specifically, UAVs offer the searching capability in a wider area and a global view of the crowd. However, the accuracy of the information on target localization is restricted by various factors, such as the speed and altitude limits, on-board sensor resolution, and environmental disturbances. UGV observation range is often smaller than that of UAVs and may be obstructed by nearby obstacles. The problem of ground target tracking by aerial agents can be considered as the interaction among the ground target and the aerial vehicle [2].

A **dynamic data-driven adaptive multi-scale simulation** (DDDAMS) based planning and control framework is proposed for effective and efficient surveillance and crowd control via UAVs and UGVs in [30]. Surveillance and crowd control is a complex task, as it involves various topics including surveillance region selection and vehicle assignment, vehicle searching and path generation, sensory data collection and processing, target detection and classification, object movement tracking, vehicle formation control and coordination. The framework is mainly composed of integrated planner, integrated controller, and decision module for DDDAMS. The integrated planner, which is designed in an **agent-based simulation** (ABS) environment, devises best control strategies for each function of crowd detection, crowd tracking, and UAV/UGV motion planning. The integrated controller then controls real UAVs/UGVs for surveillance tasks via sensory data collection and processing, control command generation based on strategies provided by the decision planner for crowd detection, tracking, and motion planning, and control command transmission via radio to the real system. The decision module for DDDAMS enhances computational efficiency of the proposed framework via dynamic switching of fidelity of simulation and information gathering based on the proposed fidelity selection and assignment algorithms.

2.5.1 Coordination framework

The proposed framework allows the coordination of teams of UAVs and UGVs according to a mission-oriented paradigm. Coordination is dynamically enabled by specific executive parameters and system conditions. In particular, the challenging issues related to the ability of properly registering and fusing data obtained by both aerial and terrestrial robots have presented a plan database component to manage and execute multi-robot plans; finally, different merging strategies for information from multiple UAVs and UGVs in a cooperative search scenario have demonstrated that those strategies enable cooperative search better than uncooperative search even with pre-defined and fixed path mobility of UAVs[46].

The definition of the missions, consisting of a sequence of tasks, is provided through a **mission-oriented application** description on the top of the structure. The **core framework** block takes charge of the mission and transfers related information to the group of UAVs involved. Information is customized for both UAVs and UGVs using dedicated adaptation modules. The bottom of the general framework contains a set of platform-dependent blocks that are expressly designed to exchange specific information with each type of UAVs or UGVs. On top of the stack there is the **application layer** composed of two sub-layers:

1. The missions sub-layer contains the definition of the whole mission (e.g., Survivor Search and Rescue Area exploration);

2. The tasks sub-layer provides the sequence of individual tasks of the mission (e.g., Search Target, Reach Target, etc.).

The **object-to-object layer** manages the cooperative execution of each task that is handled through the exchange of specific messages between UAVs. The **Internet working layer** allows communications and collaborative coordination between UAVs. It hosts the **application-level air-ground protocol** that is essential to both create/maintain a network of UAVs and to allocate tasks among the UAVs to accomplish an assigned mission. In particular, this protocol transports messages coming from the upper layer and distributes them among the UAVs in either uni-cast or multi-cast mode. Finally, the **access layer** provides communication technology support to the information exchange in accordance with the networking interfaces supported by the UAVs. In addition to the UAVs, the architecture requires the presence of a remote controller node. The definition of the mission is, in any case, decided outside the group of UAVs-UGVs and the remote controller entity represents the headquarter mission interface. Once a mission is defined, the controller has the task of transferring this information to the team of UAVs-UGVs to decide who will perform the related tasks and how they will be performed. The **controller node** also contains a **mission controller block**. This module provides first a user interface that allows the definition of a mission as a set of task to be executed by the UAVs, and the monitoring of the mission execution. The mission description can be transferred to all UAVs via the air-ground protocol support [45].

2.5.2 Relative localization approach

A heterogeneous robotic system with both UAVs and UGVs can enhance the capabilities in a 3-D sense and improve accessibility. Localization methods available for mobile agents can be categorized into major groups:

1. **Absolute localization** where the robots are empowered with absolute positioning capabilities. The most direct method is the use of GPS or differential GPS sensors where the accuracy is limited to the GPS accuracies.

2. **Relative localization** has been developed as a solution for effective and accurate execution of multi-robot collaborative missions where robots mutually track neighbors in their individual body-fixed coordinate frames. Moreover, relative localization has been identified as a feasible method of establishing the localization for less capable robots in a heterogeneous multi-robot system. In the relative localization a robot tracks its neighbors with respect to a body fixed coordinate system on the robot using inter-robot relative measurements, typically the range and bearing information. Most relative observation-based approaches employ filter-based solutions or geometry-/model-based solutions for localization. The common filters have been the **extended Kalman filter**

(EKF), **particle filter** and **unscented Kalman filter** (UKF). The geometry-/model-based approaches compute inter-robot transformation using relative measurements and the pose of the observed robot.

3. **Simultaneous localization and mapping** (SLAM) uses a common reference frame defined at the very beginning of the mission, and cooperative localization where through sharing sensor information and inter-robot observations, members in a multi-robot system refine their pose with respect to a common reference frame. Different variants of Kalman filters are applied to solve a SLAM problem where non-linear relative measurements have been algebraically transformed into the linear format for sensor fusion. This pseudo-linear-based approach is introduced in [59] to overcome instability and bias problems associated with the measurement linearization in EKF.

In a cooperative UAV-UGV mission, the motion of the ground vehicle is controlled independently and a priori unknown. The UAV is meant to assist the ground vehicle by acting like an extended (mobile) sensor device that provides additional information through complementary sensors on-board the UAV and the different field of view. The UAV can provide information that might not be available to the UGV at all or provides information earlier than the UGV would be able to detect with its own sensors. Relative localization between the vehicles is required. Visibility-based relative localization can be realized in different ways, e.g., by detecting the UAV hovering above the UGV with a camera mounted on the ground robot. The region of visibility and the higher mobility of the UAV can be used to gather additional information. Gathering as much information as possible by covering the area can be realized with a dynamic coverage strategy and a circular motion approach [32].

2.5.3 Logistics service stations

The integration of UAVs and UGVs, as automated **logistics service stations** (LSS), can enable the UAVs to provide a persistent service to multiple customers. The UAVs provide service to the customers in turn, handling the tasks to replacement UAVs in a cooperative fashion as required. Logistics service stations replenish consumables, such as a UAV's battery, and enable the persistent operations. The communication network distributes the system state information and control/planning commands across the field to the networked resources. Ensuring that essential resources are provided to the UAVs before they are required while maximizing the system objectives is a task conducted by the task-planning algorithms. The control algorithms direct the UAV actions in real time. For automated logistics service stations for UAVs, the location of stations on an $n \times n$ grid is determined by solving the **p-median problem**. The p-median problem is that of locating p facilities to minimize the demand weighted average distance between demand nodes and the nearest of the selected facilities. Automatic battery replacement systems can be developed and tested to support nearly uninterrupted UAV flights. Networked cooperative systems also include a cooperative planner and learning algorithm to control UAVs with fuel limitations and stochastic risks. A system focused on the interaction between UGVs and UAVs to extend the endurance of the UAVs. The UGVs act as docking stations and host the UAVs during the mission. The synchronization and coordination is managed by a **ground control station** (GCS). Localizing a stationary target in a GPS denied environment was addressed using a team of UAVs equipped with bearing-only sensors. The UAVs used cooperative localization to localize themselves and the target. The approaches for task planning are categorized into Markov decision processes, integer programming, and game theory. For Markov decision process and **mixed integer linear programming** (MILP) approaches, there have been some efforts to incorporate logistics service stations into the planning process. With Markov decision process approaches, centralized real time algorithms enabled persistent operation in a stochastic environment, directing UAVs to conduct multiple flights in the planning horizon and including visits to logistics service stations, were conducted. Uncertainties such as UAV health and fuel levels were addressed. They combined approximate dynamic programming and reinforcement learning to address challenges associated with Markov decision process formulations. MILP approaches with fuel awareness have been conducted but persistence is not pursued. The decision variables are inspired by the classical MILP formulation for the **vehicle routing problem**. A **receding horizon task assignment** (RHTA) approach was developed to enable persistence with a single site for the logistics service stations. Policies and reduced MILP models were used to manage the health of UAVs. MILP-based methods for persistent operations of a system of UAVs with logistics service stations distributed across a field of operations were conducted. Their task was to provide uninterrupted security escort service to customers over a finite horizon [80].

2.5.3.1 Continuous approximation model

A mathematical model of a logistic network is developed in [17] employing UAV-UGV as the potential modes of supplies transportation, a sufficient number of these robots are available at all times to serve the demand points. The objective is to minimize the expected overall cost which includes the cost of locating facilities. An arborescent network is considered where each distribution center can serve multiple customer demand points. The demand at each distribution center is assumed to follow a Poisson process as it is generated by the demand originating from customer demand points in its influence area assumed to be circular; the demand per unit time for customer demand points in the cluster C_i is assumed to be an independent and identically distributed **Poisson process** with rate θ_i, the cluster C_i groups customers with similar demand. Each distribution center is assumed to be located at the center of the influence area. If the entire area R is covered with influence areas of size

$A_{r_i}(x)$, then the total number of distribution centers $N_{r_i}(x)$ needed at each cluster C_i is given by:

$$N_{r_i}(x) = \int \left(\frac{R}{A_{r_i}(x)} \right) + 1 \tag{2.33}$$

Three cost functions can be considered: the total facility cost, the total transportation cost and the average inventory holding cost for the distribution centers. There are a number of significant factors that contribute to the unit cost calculation of UAVs such as climbing, hovering, descending, turning, acceleration and deceleration, rotational, and constant speed cost. To solve this problem, a two-phase algorithm can be used: in the first phase, a grid cover-couple approach is used and in the second phase, a distribution center influence area using continuous approximation approach can be proposed.

2.5.3.2 *Interoperable framework*

The inherent modular and distributed nature of the multi-agent system offers scalability, fault tolerance and parallelism. Behavior-based multi-agent architectures improve the rapid online decision making ability of a robot by decomposing a system into sub-systems with task achieving behaviors. Multi-agent architectures for service-oriented UAV applications are generally platform oriented, leading to varied architectures for the same applications. Algorithms for service-oriented UAV applications are generalized and it is possible to use these algorithms on various platforms irrespective of the details of their implementations. **Coupled layered architecture for robotic autonomy** (CLARAty) is a framework for heterogeneous robot platforms with generic and reusable robotic components. CLARAty provides a framework for generalized algorithms applied to rover platforms irrespective of the implementation details. Agent design patterns defining common features allow introduction of new hardware and software components without modifying the architecture. The **hybrid deliberative/reactive architecture** (HDRA) is a distributed architecture for unmanned aircraft systems. In HDRA, the essential generic functionalities of a UAV are isolated for effective integration of low-level (navigational sub-system, low-level control with motion planning) and high-level (mission planning and execution) functionalities [4].

2.6 MISSION ANALYSIS

The contextual autonomy capability can be considered as:

1. Mission complexity:
   (a) commanding structure
   (b) types of tasks, knowledge required
   (c) dynamic planning, analysis

   (d) situation awareness

2. Human independence:

   (a) interaction time percentage, planning time percentage
   (b) UAV communication initiation
   (c) interaction levels
   (d) workload/skill levels

3. Environmental complexity

   (a) static;
   (b) dynamic: object frequency/density/types;
   (c) urban, rural;
   (d) operational: weather, landforms, mapping, threats, etc.

2.6.1 Methodology

UAV missions in general can be divided into planning, management, and re-planning segments to identify functions that will be assumed by human operators. For example, UAV mission segments can comprise the following tasks:

1. **Mission planning** implies the use of a scheduling mechanism to plan health and status reports, threat area and no-fly zone information to designate the area of deployment, and a decision support mechanism to designate loiter locations;

2. **Mission management** implies the use of indicators to monitor the health, status and progress of a UAV;

3. **Mission re-planning** implies the use of path planning to re-designate deployment areas.

Given these tasks, UAV missions require human operators to supervise mission planning, and to monitor the health and status of UAVs.

2.6.1.1 Photography with UAV

UAV imagery differs from conventional aerial photography in a number of important ways [84]:

1. Flight and camera configurations are often arbitrary.

2. Flying height is low in relation to the height of objects on the ground, resulting in significant perspective distortions.

3. There is a big variation in the amount of image overlap and the rotation angles between images.

Hence, UAV photography has characteristics of both traditional aerial photography and terrestrial photography, and should be applicable to both types of imagery. To manage a complete video delivery in autonomy, mission preparation to the delivery of a finished film should

1. Analyze a film application,
2. Recall the legal aspects of shooting,
3. Achieve artistic plans with UAV,
4. Make a video frame with UAV,
5. Propose shooting movements adapted to the order, the situation/location shooting,
6. Produce images in accordance with customer demand,
7. Identify and understand the influence of the conditions of shooting images (wind, light, turbulence, trajectory), and
8. Select relevant framing axes following situations.

Photogrammetry with UAV The temporal resolution of UAV imagery is superior to imagery collected by satellite and conventional aerial photography platforms which are restricted due to limitations in the availability of aircraft, weather, and satellite orbits. Due to the small footprint of UAV imagery, automated techniques must be developed to geometrically rectify and mosaic the imagery such that larger areas can be monitored. The technique for geometric correction and mosaicking of UAV photography using **feature matching** and **structure from motion** (SfM) photogrammetric techniques can be used. Currently, the combination of a small UAV equipped with a high-definition camera and a free SfM-photogrammetry software can produce virtual 3-D objects that are suitable for visualization or 3-D printing. SfM photogrammetry requires generally expensive survey equipment (e.g., real time kinematic global positioning system (RTK-GPS) or total station) to provide accurate and precise real-world map coordinates of objects in the scene. **Direct georeferencing** (DG) refers to a method where the photogrammetric solution is determined with precise and accurate knowledge of the camera position and orientation at the time of image acquisition. This methodology is routinely used in the case of full-sized airborne photography and LiDAR surveys (without the camera calibration requirement) [11]. Images are processed to create 3-D point clouds, initially in an arbitrary model space. The point clouds are transformed into a real-world coordinate system using either a direct geo-referencing technique that uses estimated camera positions or via a **ground control point** (GCP) technique that uses automatically identified ground control points within the point cloud. The point cloud is then used to generate a **digital terrain model** (DTM) required for rectification of the images. Subsequent geo-referenced images are then joined together to form a mosaic of the study area. The impacts can be split into two categories [1]:

1. The first category includes the main environmental concerns that are known for conventional aircraft, such as **emissions/pollution**, **noise**, and **third party risk** (TPR). Emissions may be of less concern as the UAVs are smaller than conventional aircraft. TPR concerns the safety of the people on the ground who are involuntarily exposed to an aircraft accident. By definition, people who are on board of an aircraft (aircrew and passengers) and people who work within an airfield are considered as first party and second party, respectively.

2. The second category relates to concerns that may gradually become important such as **light pollution** in case of night operations, **privacy** and **effects on the biotope**. The impact on society is still not clear and requires further analysis to study the real impact.

The most important objectives are

1. Learn automatic flight techniques to collect data and geo-referenced images;
2. Make photo recorded with UAV;
3. Understand and process the images collected for analysis;
4. Recall the legal aspects of the visible spectrum;
5. Make a practical case of photogrammetry with UAV on website;
6. Prepare mission;
7. Select and calibrate the camera;
8. Conduct surveys with UAV from precise specifications;
9. Set automatic flight plan based on measurement needs;
10. Perform automatic flight;
11. Control measurement and emergency situations.

A typical feature, for instance, of mission plans for UAS photogrammetry is the large forward and cross overlap to compensate for aircraft instability. One feature of mission management is to accommodate flight plans to the actual wind conditions of the mission area at the time of mission execution. Mission planning and real-time mission management sub-systems are the key to a competitive exploitation of UAS for photogrammetry and remote sensing. An autopilot usually includes its navigation system. An **orientation system** (OS), depending on the requirements, usually includes a mapping-grade or geodetic-grade set of sensors. The cm-level positioning can be achieved and both direct and indirect sensor orientation are possible. Automatic image matching is also applicable to irregular UAS blocks, as well as the structure from motion approach. **Digital surface models** (DSM) and **orthophotos** are the two main mapping products of UAS [19].

Coastal erosion monitoring with UAV For coastal areas, the complex geographical attributes, the vast coasts, as well as the increasing interest for activities in even remote off-shore locations, have raised challenges for the marine environment protection and for a sustainable management. A detailed example is given for coastal erosion monitoring in [61]. In order to develop the most efficient method for UAV-based digital elevation modeling, the overall method is divided into four distinct categories:

1. **Mission planning** involves combining all aspects of the project and ensuring that every goal will be met and planned for. These aspects include obtaining usable resolution images, adequate image coverage, sufficient images for data processing, and precise GPS coordinates. It utilizes principles of photogrammetry such as Nadir, angle of incidence, ground resolution, etc. In **digital elevation modeling** (DEM) generation, Z-value accuracy can be the most important factor and is generally two to three times less accurate than the X- and Y- values obtained by the same GPS. This margin of error must be accounted for when designing a mission. The quality and accuracy of the GPS equipment is a priority because, while the structure from motion software is accurate and will correctly place points according to the scale assigned, the model still relies upon ground control points generated by the user and the GPS equipment.

2. **Site preparation** is based upon methods in both **remote sensing** and **DEM**. These methods include the use of ground control points and preliminary surveys. It also entails planning flight paths that are appropriate to the site and conditions. A layout is developed for ground control points, checking the accuracy of the satellite image used for mission planning, ensuring no obstacles are in the path of the UAV. Environmental conditions are evaluated to determine if a flight can be safely undertaken. Site preparation is perhaps the most important step because the DEM and orthophotos depend on the proper placement of the ground control points. The minimum requirements for ground control point placement must be achieved.

3. **Flight operations** may rely on methods developed and practiced by the hobbyist community such as open-source software.

4. **Data processing** relies upon resource-intensive computing and highly automated scripts in order to take the images obtained during flight operations and translate the data into a 3-D digital elevation model.

All of these categories are components of a larger method and cannot be singled out if accurate results are desired. Each category engages a particular aspect of photogrammetry, UAV operations, and erosion monitoring. An algorithmic approach that allows to tackle in a common framework the problems of area decomposition, partition and coverage for multiple heterogeneous UAS is presented in [5]. The approach combines computational geometry techniques and graph search algorithms in a multi-UAS context. In particular, the following requirements are considered:

1. The area should be exactly decomposed regardless of its complexity or the existence of no-fly areas, without cells being outside or partially inside the areas of interest. Then, each sub-area should be the exact sum of its decomposition into cells.

2. Area partition should be done according to the relative capabilities of the involved UAS in order to balance their workloads. Hence, each UAS will be responsible for a given percentage of the whole area, and measured in number of cells of the above-mentioned area decomposition process.

3. Every configuration space of each UAS cannot be disjointed or intersected by another configuration space or obstacle. Resulting partitioned areas have to prevent overlapping coverage paths.

4. The shape of the sub-areas should be generated growing uniformly from the locations as the probability to find the objects of interest is decreasing uniformly from these locations. Moreover, in order to reduce the number of turns and to minimize the time for coverage while providing smoother trajectories, each area has to be as symmetric as possible.

Infrastructure monitoring with UAV To perform quality inspections with a UAV requires specific technical and organizational skills. The most important objectives are:

1. Learn flying techniques to collect data and images in the visible spectrum: Know the different types of works, construction and operation of structures, the methodologies adapted to different types of structures or networks;

2. Conduct technical inspection missions with UAV: Make a practical case of technical inspection with UAV; prepare mission with UAV; calibrate and adjust the sensors; program the hovering UAV; and conduct surveys with UAV from precise specifications;

3. Understand and process the collected data: Know the data processing capabilities data processing and analysis, control measurement and emergency situations; and conduct surveys.

Imagery collection in disaster management Documentation of infrastructure response to a disaster is an important objective; however, the task can be hazardous, and collecting detailed information of an affected area while keeping costs down is typically challenging. Especially for large-scale events, accommodations and provisions may not be available to survey teams for hundreds of kilometers, making field data collection even more difficult. When UAVs acquire photogrammetry data with appropriate imagery data, the capabilities of UAVs for disaster research and management can be further realized [2]:

1. High-resolution images can be analyzed and used to produce hazard maps, dense surface models, detailed building renderings, comprehensive elevation models, and other disaster area characteristics.
2. These data can then be analyzed using remote sensing methods or visual interpretation to coordinate rescue efforts, record building responses to the disaster, detect building failures, investigate access issues, and verify experimental disaster modeling.
3. The data can also be gathered before a disaster in order to document immediate pre-event conditions of critical facilities and infrastructure, monitor susceptible environmental concerns, and document historical conditions and sites.

Data collection in precision agriculture The ideal survey path must balance endurance and sample density. Such a flight pattern may focus on a point of interest while remaining out of less-interest regions. The following procedure may be followed [67]:

1. Define boundary vertices;
2. Transform the boundary into the augmented space by considering the dimensions of the UAV;
3. Define a planning grid;
4. Define a uniform survey path from the planning grid.

Environmental modeling Environmental modeling is at the intersection of **information fusion** (IF) and **dynamic data driven applications systems**. A **wide area motion imagery** (WAMI) application is used in [8]. Hurricane modeling for weather and climate prediction and emergency response are also potential applications. In 2016, for Hurricane Matthew, the plan of action was the following: pre-landfall mission planning, crew deployment, live news coverage, news flights, infrastructure inspections, redeployment, and post-mission debrief with customers.

3-D mapping is essential in surveying, geographic information system for applications that include mapping streets, tunnels, and civil infrastructure; mines and industrial sites; buildings; cultural heritage sites; and natural terrain, caves, and forests. Laser scanning technology is employed to capture precise range measurements for generating 3-D point cloud models. Scanning from a low-altitude airborne platform affords access to some sites not otherwise reachable and provides the ability to measure surfaces that cannot be viewed from ground level. A **simultaneous localization and mapping** (SLAM) solution is the preferred way to address this problem as it avoids the need for additional positioning infrastructure or a priori knowledge of the environment. The sensor repeatedly acquires measurements within a relatively wide 3-D field of view at an appropriate rate to facilitate reliable incremental motion estimation. For accurate modeling, the scanner trajectory must be represented as a continuous-time function, as it is not possible to rely on a UAV remaining sufficiently motionless at discrete poses while scanning with such a sensor configuration. The continuous-time SLAM solution presented in [51] is based on a view-based approach focused on estimating the trajectory of the sensor payload. There are two main algorithmic components:

1. A non-rigid trajectory estimation and map registration algorithm. Non-rigid registration is used both for online incremental motion estimation (**laser odometry**) as well as global optimization of the overall trajectory.

2. A place recognition solution. For long-duration data-sets in which accumulated drift error may be significant, a coarse registration step is necessary to provide a better initial trajectory estimate for the global optimization. Multiple data-sets can also be automatically merged using place recognition followed by global non-rigid registration.

Non-rigid registration can be expressed as a non-linear optimization problem that takes a prior trajectory as input and computes corrections to that trajectory to minimize errors between the measurements and motion constraints. A trajectory is defined as a function $T(\tau)$ that specifies a 6 **degrees of freedom** (DoF) transformation for any time value τ in the domain. In practice, the trajectory transforms a point p measured at time τ from the sensor frame S to the world frame W as:

$$p_w = T_w^s(\tau) \oplus p_s = r_w^s(\tau) \oplus p_s + t_w^s(\tau) \tag{2.34}$$

where p is a point in the specified frame, t, r are respectively the translational and rotational components of this 6 DoF transformation and $\oplus$ is the transformation composition operator. The trajectory is decomposed into a baseline trajectory $t_0(\tau), r_0(\tau)$ and a small correction $\delta t(\tau), \delta r(\tau)$:

$$T(\tau) = (t_0(\tau) + \delta t(\tau); r_0(\tau) + \delta r(\tau)) \tag{2.35}$$

Trajectories are stored as samples and a spline interpolates transformations for times between the samples. The primary constraint types included in the non-rigid registration formulation for this application are as follows: minimizing correspondence errors in the laser data; minimizing deviations of the trajectory with respect to inertial measurements; minimizing deviations of the velocity with respect to the prior trajectory; and ensuring smoothness and continuity in the trajectory.

2.6.1.2 *Emergency response with UAV*

Emergency response is classified by needs for:

1. Damage assessment,
2. Surveillance,
3. Medical aid supplies, and
4. Search and rescue tools.

Needs can be entered into the mission coordination system as requests either by operators or by authorized emergency responders in the field via a mobile device application. Each request is associated with a geographic location, which can be generated automatically by the mobile device or entered manually by the device operator. The most common means of assessing the situation are by:

1. Drive through,
2. Aerial flyover, and
3. Site visit by foot.

Rescue robotics is a domain in which robots have the potential to make the difference, with their ability of working in environments forbidden to humans. A fleet of autonomous vehicles should be available to serve the entered set of requests. To facilitate a broad range of capabilities, the fleet may consist of vehicles of three different types:

1. Fixed-wing aircraft with medium payload and high speed to fly reconnaissance sorties,
2. Ground vehicles to set up local deployment stations, with large payload capability and longevity, and
3. Rotary-wing UAV to deliver supplies and tools, with a small payload and high agility.

The emergency response system aims to serve aid requests from the field in a time-optimal manner. More formally, the objective of the system is to minimize the summation of the service time for each request, weighted by the request priorities. This problem decomposes into two parts:

1. Determining optimal deployment locations for the ground vehicles that will serve as depots for deliveries, and
2. Solving a vehicle routing problem to determine trajectories for the delivery rotor-craft along the locations of aid requests, which includes determining the cargo of each UAV in each sortie.

Control over the UAVs can be changed from the autonomous way-point following to manual control with a mobile device. By allowing an emergency responder to seize control over the UAV, it can be guided in and landed based on human sensory information and situational awareness. During the operation, the UAVs communicate information back to the mission command and control center [65].

2.6.2 Mission specificity

To define a wide array of UAV missions, a generic ontology for UAV missions has to be developed. Ontology can be viewed as content theory that focuses on properties and relationships among objects from a specific domain, acting as a **body of knowledge** that is based on a vocabulary used to describe the domain [39]. An ontology is shared knowledge.

Definition 63 ***Ontology*** *is a tuple* $\langle S, A \rangle$ *where* S *is the vocabulary (or signature) of the ontology and* A *is the set of ontological axioms specifying the intended domain vocabulary. The signature* S *is broken down into three sets:*

1. *The set of concepts (C)*
2. *The set of relations (R)*
3. *the set of instances (I)*

Ontology engineering for UAVs is at a crossroads, since experts from different fields bring their own language, practice and methodology to create and store knowledge in order to create ontologies. Any UAV mission can be depicted as a combination of points and paths. A database contains the flight profile for each point/path segment such as height, speed, and loiter times. Moreover, missions can be repeated at intervals and feature a priority value [74]. The formalization of a multi-UAV mission can be specified using high-level concepts. The motivation to develop a new language that could aggregate mission information derives from the development of a framework for coordination methodologies among agents representing heterogeneous vehicles. In order to specify the mission to be performed, a **mission description language** (MDL) expresses several kinds of missions, their description and specific characteristics. The language should be easily readable, by both machine and humans and independent of the system being used; also, several concepts must be included, representing the possible multiple phases of a mission, the targets, requirements, etc. The language should be flexible enough to describe a number of typical entities and their higher-level characteristics that allow for a rapid mission deployment [78]. The description of teams and missions can be divided into two categories:

1. The **static component** includes:
   (a) The **scenario description language** (SDL) describes the scenario the UAVs will operate in, including a description of facilities that can be used, global environment restrictions and some global control structures.
   (b) The **team description language** (TDL) describes the teams, including specific team restrictions, and a description of all vehicles that compose a team, their characteristics and the sensors/cargo they transport.
2. The **dynamic component** also includes the **disturbance description language** (DDL), which presents a description of disturbances that exist in the mission environment. UAVs must be increasingly equipped with more autonomy. The routes the UAV should take may depend on parameters of the task, additional actions might be necessary at certain points of the route and unforeseen situations must be handled. The operational state, including the battery lifetime, and environment conditions, such as wind and weather conditions, should be checked constantly to safely abort the mission if necessary.

Real-time health condition diagnosis and adaptation to failed condition are important control-level trends. Monitoring information and results may benefit decision making for mission execution. Establishing internal relations between monitoring data and system performance is essential to confirm plan mission execution. Moreover, a mission-execution decision making problem focuses on the best execution option in a set of possible missions, and mission control is expected to operate safely and robustly under external and internal disturbances and to be able to accommodate fault conditions without significant degradation of their performance. A centralized-distributed hybrid control framework is proposed for UAV mission assignment and scheduling in [36].

2.6.2.1 High–level language

A higher-level and platform-independent approach is needed for specifying missions. An approach to describe UAV missions with a graphical **domain specific language** (DSL) is based on an analysis of use cases and grouping of concepts, with an extendable model [75]. The conceptual model of the developed language can be based on identifying numerous missions whose purposes vary significantly. A common set of concepts can be identified and used as basis for the language.

1. **Routing elements** are the central concept to describe UAV missions. They are used to move the UAV from some location to another one. Examples for routing elements are taking off, touching down or flying to a certain position or area.

2. **Actions** can be used to give a mission a purpose while routing elements provide means for building the structure of a mission. A broad range of activities can be carried out at various points during a mission. Examples include taking a picture, using a laser scanner, or scanning for a signal such as wireless local area network (LAN).

3. **Branches** are available. They allow for specifying alternative mission flows depending on run time conditions. This concept is adapted from general programming language theory and allows expressing a broad range of UAV missions.

4. **Processing actions** are introduced, e.g., to recognize an image or interpret a laser scan. For conditional branching, input data is needed as well as some kind of processing of this data.

5. **Filters** concept is introduced for enhancing routing elements with individual movement strategies. This is based on the observation that UAV movements frequently need to respect certain constraints, e.g., maintaining a specified velocity or avoiding obstacles on-line. Thus, filters may influence routing elements.

6. **Parallel blocks** are provided to support activities that are meant to run simultaneously to the main flow. This can be employed, e.g., to monitor sensor outputs, record a video or take a picture every minute. In conjunction with conditional branches, parallel blocks are able to influence the program flow.

Influence diagrams A multi-criteria decision making framework is presented in [76]. It reasons over a library of coalition formation algorithms for selecting the most appropriate subset of algorithms to apply to a wide spectrum of missions. The framework is based on influence diagrams in order to handle uncertainties in dynamic real-world environments. A number of coalition formation algorithms have been proposed in order to address the combinatorial optimization problem, each tailored for specific mission conditions. For example, group-rational agents can be used as a heuristic to generate both overlapping and disjoint coalitions by minimizing the overall system cost. Real-world tasks involve inter-task constraints, intra-task constraints and spatial constraints that require UAVs to plan and schedule tasks. For example, a two stage distributed formation algorithm can concentrate on minimizing coalition sizes and minimizing task completion time. The proposed decision making module in [76] has as components:

1. **Taxonomy table** stores the taxonomy features and the respective domain values that facilitate classification of the coalition algorithms. The taxonomy dimensions are partitioned into four relation-based categories: agent, task, domain and algorithm. Let F be a set that contains the N taxonomy features where each feature has its respective non-empty domain set. Let Dom be a collective set containing all the respective domain sets of N taxonomy features. All this information is captured by:

$$\forall F_i \in F, \exists D_i \in Dom | 1 \leq i \leq N, D_i \neq \emptyset \tag{2.36}$$

   where D_i is the domain value set of feature F_i. A feature $F_i \in F$ can be instantiated with any particular value of its domain value set, D_i.

2. **Utility calculation** determines the feature-value pair utility scores that are essential for creating the influence diagram's utility table. Each coalition formation algorithm can be linked to a subset of related feature-value pairs that govern the algorithm's applicability. Therefore, the algorithms and the feature-value pairs can be visualized as **hubs** and **authorities**, respectively. Let V be a set of size d containing all possible feature-value pairs derived from the set F of taxonomy features and their corresponding domain sets Dom, such that:

$$V = \{(F_x, d_i) | F_x \in F, d_i \in D_x, D_x \in Dom\} \tag{2.37}$$

   The size of the feature-value pair set V is defined as $d = \sum_{x=1}^{N} |D_x|$ where $|D_x|$ represents the domain size of $F_x \in F$.

3. **Feature extraction** determines the most important features that discriminate the algorithms, thus reducing the problem dimensionality. The selection algorithm leverages a $p \times N$ matrix $\mathbb{U}$, where p is the number of algorithms and N is the number of taxonomy features. An element u_{ij} is the base utility score for the specific feature-value pair, with feature j associated with algorithm i. Each eigenvector of this matrix $u_{ij}, i = 1, \ldots, p; j = 1, \ldots, N$ accounts for some variance in the original data-set and is expressed as a linear combination of the N taxonomy features. The kth eigenvector p_{c_k} is defined by

$$p_{c_k} = z_{k_1} F_1 + z_{k_2} F_2 + \cdots + z_{k_N} F_N = FZ \tag{2.38}$$

   where F is the row feature vector of size N with $F_i \in F$ representing the ith taxonomy feature. Z is a matrix of size $N \times N$ containing the weight coefficients of all the N eigenvectors. The relative sizes of the coefficients z_{k_i} indicate the relative contributions of the corresponding feature $F_i \in F$.

4. **Influence diagram** builds the system's influence diagram/decision network dynamically at run-time based on the extracted prominent features. Once the feature extraction algorithm identifies the most prominent features, the influence diagram is built dynamically at run-time. An

influence diagram augments a Bayesian network by introducing decision variables and a utility function that characterizes the decision maker's preferences. This decision problem is solved by determining the optimal strategy that maximizes the expected utility score for the framework.

The curse of dimensionality is addressed by extracting prominent features that discriminate the coalition algorithms.

Decision diagrams A phased mission system consists of consecutive non-overlapping stages, or phases, where various tasks are performed in each phase. The phases are usually executed in sequential order, so that the success of a given phase often depends on the success of its preceding phase. Analyses of phased missions require determining the failure probability of each phase as well as the entire mission. Decision diagrams are rooted **directed acyclic graphs** (DAGs) that can be used to represent large switching functions. Modeling different operational modes other than just the binary case of failure or normal operation are critical in analyzing the disaster-tolerance of phased mission systems. To address these issues, the application of **multiple-valued logic decision diagrams** can be proposed, applying graph algorithms that compute the probability of mission success while also modeling mission component degradation or failure [62].

Definition 64 *The* ***decision table*** *is the ordered five-tuple* $DT =< U, C, D, V, f >$, f *is the information function,* U *is the set of objects,* V *is the set of all possible values of features, the elements of the set* C *are conditional features while the elements of the set* D *are decision features.*

UAVs are constructed in a component based fashion with an agent-based decision maker and a continuous control system. The agent based decision maker is viewed as the replacement for a human pilot or operator who would, otherwise, interact with the control system. Since autonomous systems are frequently safety or mission critical this verification gap is a significant concern. One of the most crucial aspect of verifying complex decision making algorithms for autonomous systems, is to identify that the controlling agent never deliberately makes a choice it believes to be unsafe [23].

A methodology that supports the demands of designing autonomous decision making systems is needed. Autonomous systems can operate in dynamic domains requiring the application of strategic and tactical decision making. Effective performance in such domains requires capabilities such as the balancing of **reactivity** with **proactivity**. The autonomous system must not only be **goal directed**, but must also be able to switch focus when the environment changes in an important way, or when it discovers that one of the assumptions underlying its current tactical approach is invalid. It may also need to coordinate its activities with peers who are working towards the same goal. These capabilities, namely **autonomy**, **reactivity**, **proactivity** and **social ability**, are characteristic of agent-based systems. A mission expresses what the autonomous system needs to achieve as well as other information such as potential risks [30]. Under certain circumstances, even a simple UAV reconnaissance scenario may require sophisticated decision making capabilities. In the simplest case, all it has to do is fly to the location, photograph the target, and return to base. However, unexpected events may require that it reconsiders its current plan on how to achieve the mission objective, or may even mean that the objective must be abandoned.

A **cognitive model** (CM) of the decision problem consists of a directed graph that comprises factors, and connections between the factors. It may be represented by a matrix of inter-relationships. In the direct graph, the nodes designate factors and the arrows denote the corresponding relations. A cognitive model involves such phenomena as concepts, analytic hierarchy, objects, denotations, association, synonyms, antonyms, connection, and influences. It is developed using the following sequence of procedures:

1. Gather information and identify the problem.
2. Select an appropriate method for solving the problem.
3. Formulate questionnaires based on fuzzy scales.
4. Find groups of experts.
5. Send questions to the experts.
6. Collect answers from the experts.
7. Build the computational cognitive model.
8. Solve the direct problem and estimate the different scenarios of actions.
9. Solve the inverse problem and find the optimal decision.
10. Formulate the decision.

Direct problem-solving evaluates the temporal dynamics of the output factors that depend on different input impulse combinations. In a cognitive model, the solution is optimized using the inverse cognitive problem-solving approach. The inverse problem-solving method may use soft computing techniques to determine the optimal input combination that fulfills specified conditions for attaining the desired values for the output. The input factors are the control variables and the output factors are the goal variables. In inverse modeling, the decision making participants search for the values of input values that result in the required output values [72].

2.6.2.2 Model checking of missions

Mission optimization constitutes a **multi-objective problem** because UAV missions are characterized by multiple properties. The problem of UAV **scheduling/routing** has also been modeled as a **minimum cost network flow problem**; a **traveling salesman problem**; and a **dynamic programming problem**. When integrated, planning and simulation software should ideally provide information advantage. But these systems are currently challenged by the size and complexity of the UAV domain. Mission correctness is contingent on multiple factors including UAV flight trajectories, interoperability and conflict resolution capabilities; and a dynamic environment with variable terrain, unpredictable weather and possible moving targets. In this context, mission plans may contain errors that compromise mission correctness. Mission developers should be supported by analyzing UAV mission plans with software verification methods and in particular probabilistic model checking that can detect mission-critical errors before real-world execution. A method for domain-specific model checking, called cascading verification described in [94], is applied to the probabilistic verification of complex UAV mission plans. Model checking is a formal verification method whereby a model checker systematically explores the state space of a system model to verify that each state satisfies a set of desired behavioral properties. In particular, model builders use a high-level **DSL** to encode system specifications that can be analyzed with model checking. A compiler uses automated reasoning to verify the consistency between each specification and domain knowledge. If consistency is deduced, then explicit and inferred domain knowledge is used by the compiler to synthesize a **discrete-time Markov chain** (DTMC) model and **probabilistic computation tree logic** (PCTL) properties from template code. Thus, verification cascades through several stages of reasoning and analysis. **DSM** is a model-driven software development process that uses domain-specific languages to encode system aspects. The UAV domain exhibits complexity at different levels of granularity. UAVs incorporate sophisticated payloads, multiple sensors and increasing computational power. These capabilities could, in time, enable UAVs to execute complex, multi-task missions with reduced human supervision.

Some definitions useful for the sequel are introduced as follows:

Definition 65 *A* ***discrete time Markov Chain*** *(MC) is a tuple* $\mathcal{M} = \langle S, P, s_{init}, \Pi, L\rangle$ *where* S *is a countable set of states,* $P : S \times S \to [0,1]$ *is the transition probability function such that for any state* $s \in S, \sum_{s' \in S} P(s,s') = 1$, $s_{init} \in S$ *is the initial state,* Π *is a set of atomic propositions and* $L : S \to 2^{\Pi}$ *is a labeling function.*

An observable first-order discrete Markov chain is encoded as the matrix of state transition properties. Its rows sum to one, but the columns do not necessarily do so. A state S_i in a **Markov chain** is said to be absorbing if $a_{ii} = 1$. Otherwise such a state is said to be transient.

Definition 66 *A* ***Markov decision process*** *(MDP) is defined in terms of the tuple $\langle S, A, T, R \rangle$, where:*

1. *S is a finite set of environmental states.*
2. *A is a finite set of actions.*
3. *$T : S \times A \to S$ is the state transition function. Each transition is associated with a transition probability $T(s, a, s')$, the probability of ending in state s', given that the agent starts in s and executes action a.*
4. *$R : S \times A \to S$ is the immediate reward function, received after taking a specific action from a specific state.*

Definition 67 *A particular* ***finite Markov decision process*** *is defined by its state and action sets and by the one-step dynamics of the environment. Given any state s and action a, the probability of each possible state, s', is*

$$P^a_{ss'} = Prob\left\{s_{t+1} = s'|s_t = s, a_t = a\right\} \tag{2.39}$$

where $P^a_{ss'}$ represents transition probabilities and t denotes a finite time step.

In the Markov decision process, the value of $P^a_{ss'}$ does not depend on the past state transition history. The agent receives a reward r every time it carries out the one-step action. Given any current state s and action a, together with any next state s', the expected value of the next reward is

$$R^a_{ss'} = E\left[r_{t+1}|s_r = s, a_t = a, s_{t+1} = s'\right] \tag{2.40}$$

where $P^a_{ss'}$ and $R^a_{ss'}$ completely specify the dynamics of the finite MDP. In the finite MDP, the agent follows the policy Π. The policy Π is a mapping from each state s and action a to the probability $\Pi(s, a)$ of taking action a when in state s. In the stochastic planning calculation, based on the MDP, the policy Π is decided so as to maximize the value function $V^\Pi(s)$. The $V^\Pi(s)$ denotes the expected return when starting in S and following Π thereafter. The definition of $V^\Pi(s)$ is

$$V^\Pi(s) = E_\Pi\left[\sum_{k=0}^{\infty} \gamma^k r_{t+k+1}|s_t = s\right] \tag{2.41}$$

where E_Π denotes the expected value given when the agent follows the policy Π and γ is the discount rate $0 < \gamma < 1$. If the values of $P^a_{ss'}$ and $R^a_{ss'}$ are known, dynamic programming is used to calculate the best policy Π that maximizes the value function $V^\Pi(s)$. When the values of $P^a_{ss'}$ and $R^a_{ss'}$ are unknown, a method such as on-line reinforcement learning is useful in obtaining the best policy Π in the learning environment [26]. After the planning calculation has finished, a greedy policy that selects action value a that maximizes $V^\Pi(s)$ is optimal.

Definition 68 *A* ***path*** *through an MDP is a sequence of states, i.e.,*

$$\omega = q_0 \overset{(a_0,\sigma_{a_0}^{q_0})(q_1)}{\longrightarrow} q_1 \longrightarrow \dots q_i \overset{(a_i,\sigma_{a_i}^{q_i})(q_{i+1})}{\longrightarrow} \longrightarrow \dots \tag{2.42}$$

where each transition is induced by a choice of action at the current step $i \geq 0$. *The* i^{th} *state of a path* ω *is denoted by* $\omega(i)$ *and the set of all finite and infinite paths by* $Path^{fin}$ *and* $Path$*, respectively.*

A control policy defines a choice of actions at each state of an MDP. Control policies are also known as **schedules**.

2.6.3 Human-UAV Team

Human-robot interface (HRI) takes into account the asymmetric, two-way nature of the awareness relationship, as well as the fact that teams of people may be working with teams of robots with different needs [27]. Treating humans' understanding of UAVs and other team members separately from UAVs' knowledge of humans and other UAVs results in a definition tailored to humans' and UAVs' distinct needs. Similarly, a definition that concentrates on locations, identities, activities, status, and surroundings is not dependent upon a particular UAV implementation. The HRI awareness definition, however, does not meet the third criterion: that of specificity to the UAV domain. Major advances are still needed in sensors and sensor interpretation to facilitate lower level situation awareness activities so that the operator will have a higher level situation awareness; and that ways must be found to facilitate appropriate communication to support productive team processes. Human-robot interaction should not be thought of in terms of how to control the UAV, but rather as how a team of experts can exploit the UAV as an active information source [66].

Definition 69 ***Situation awareness*** *(SA) is the perception of the elements in the environment within a volume of time and space, the comprehension of their meaning, and the projection of their status in the near future.*

People directly controlling the UAV, whether directing the flight controls or the sensor controls, are in the **operator role**; pilots of manned aircraft in the vicinity are peers to UAV operators, and air traffic controllers are in a supervisory role since they direct the activities of both pilots and UAV controllers. The operators of the UAV ground control stations will have the greatest need for awareness about the UAV, but the air traffic controllers responsible for directing UAVs in their airspace and pilots of manned aircraft in the UAV's vicinity also have awareness needs regarding UAVs. The physical separation of the crew from the aircraft presents challenges to the effective design of the UAV control station. Numerous human factors issues such as system time delays, poor crew coordination, high workload, and reduced situational awareness may have negative effects on mission performance. When on-board an aircraft, a pilot and crew receive a rich supply of multi-sensory information instantaneously regarding their surrounding environment. UAV operators, however, may be limited to a time-delayed, reduced stream of sensory feedback delivered almost exclusively through the visual channel [10].

2.6.3.1 Human–UAV Interaction

Autonomy is a capability that enables the joint human-machine system to accomplish a given mission. Although significant achievements have been made in autonomy, most autonomous systems still rely on human intervention to some degree due to their incapability to handle complex, uncertain, and dynamically changing tasks. The true value of autonomy is to extend and complement human capability. The collaboration between humans and autonomous systems will bring out the best capabilities. To fully realize the potential of collaborative robots, the correctness of the software with respect to safety and functional requirements needs to be verified. **HRI** systems present a substantial challenge for software verification, the process used to gain confidence in the correctness of an implementation, i.e., the UAV's code, with respect to the requirements. The UAV responds to an environment that is multi-faceted and highly unpredictable. Formal methods can achieve full coverage of a highly abstracted model of the interactions, but are limited in the level of detail that can practically be modeled. Robotic code is typically characterized by a high level of concurrency between the communicating modules that control and monitor the robot sensors and actuators, and its decision making. **Coverage-driven verification** (CDV) can be used for the high-level control code of robotic assistants, in simulation-based testing. CDV is a systematic approach that promotes achieving coverage closure efficiently, i.e., generation of effective tests to explore a system under test, efficient coverage data collection, and consequently efficient verification of the system under test with respect to the requirements [3].

Interaction awareness For each human m of all M humans, and for each UAV n of all N UAVs working together on a synchronous task, human-UAV interaction awareness consists of four parts:

1. **Human-UAV interaction awareness**: the understanding that m has of: N identities, current 3-D spatial relationships between N and other objects (points on Earth, other aircraft, terrain, and targets, if applicable), predicted future 3-D spatial relationships, weather near N, health of N, other (non-health related) status of N, the logic used by N when acting on M commands, N missions, their progress towards completing their missions, and the trust m has for each of these items.

2. **Human-human interaction awareness**: the understanding that m has of the locations, identities, and activities of M.

3. **UAV-human interaction awareness**: the knowledge that n has of M commands necessary to direct their activities, any conflicts among commands given to n by M, and any constraints that may require a modified course of action or command non-compliance.

4. **UAV-UAV interaction awareness**: the knowledge that n has of the commands by N; the tactical plans of N; any health conditions present in N; any weather conditions present near N; and any other coordination necessary to dynamically re-allocate tasks among N if needed.

Since the general case of the human-UAV interaction awareness decomposition assumes the possibility of multiple UAVs, complete human-UAV awareness means that the interface enables each human to understand the state of each of the UAVs; and not just the state of one UAV. Sensor operators use imagery from a gimbal-mounted camera to conduct a wide variety of activities. Video imagery quality can be compromised by narrow camera field of view, data-link degradations, poor environmental conditions, bandwidth limitations, or a highly cluttered visual scene. If imagery interpretation could be enhanced and made more robust under a wide variety of situations, UAV mission effectiveness will increase substantially. Current modeling efforts are focused on developing and validating a fine-grained cognitive process model of the UAV operator. The model is implemented in the **adaptive control of thought rationale** (ACT-R) cognitive modeling architecture. The question is to assess how accurately the model represents the information processing activities of expert pilots as they are flying basic maneuvers with a UAV simulation. Pilots give attention to performance instruments more often than control instruments. The distribution of operator attention across instruments is influenced by the goals and requirements of the maneuver [71]. ACT-R provides theoretically motivated constraints on the representation, processing, learning, and forgetting of knowledge, which helps guide model development.

Stochastic shared autonomy Stochastic shared autonomy systems capture three important components: the operator, the autonomous controller and the cognitive model of the human operator. This representation may include a Markov model representing the fully autonomous system, a Markov model representing the fully operated system and a Markov model representing the evolution of the human's cognitive states under requests from autonomous controller to human or other external events. Switching from the autonomous controller to the operator can occur only at a particular set of the human's cognitive states, influenced by requests from the autonomous controller to the operator. This problem can be formulated at solving a multi-objective **MDP** with temporal logical constraints. One objective is to optimize the probability of satisfying the given temporal logic formula and another objective is to minimize the human's effort over an infinite horizon, measured by a given cost function. The stochastic system controlled by the human operator and the autonomous controller gives rise to two different MDPs with the same set of states S, the same set of atomic propositions and the same labeling function: $L \rightarrow 2^{AP}$, but possibly different set of actions and transition probability functions:

1. **Autonomous controller**: $M_A = \langle S, \Sigma_A, T_A, AP, L\rangle$ where $T_A : S \times \Sigma_A \times S \rightarrow [0,1]$ is the transition probability function under autonomous controller, S, Σ are finite state and action sets, AP is a finite set of atomic proposition and L is a labelling function which assigns to each state $s \in S$ a set of atomic propositions $L(s) \subset 2^{AP}$ that are valid at the state s.

2. **Human operator**: $M_H = \langle S, \Sigma_H, T_H, AP, L\rangle$ where $T_H : S \times \Sigma_H \times S \rightarrow [0,1]$ is the transition probability function under human operator.

The evolution of the modeled cognitive state can be also modeled as a MDP.

Definition 70 *The* ***operator's cognition*** *in the shared autonomy system is modeled as a MDP:*

$$M_c = \left\langle H, E, D_0^H, T_C, Cost, \gamma, H_s \right\rangle \tag{2.43}$$

where H represents a finite set of cognitive states, E is a finite set of events that trigger changes in cognitive state, $D_0^H : H \rightarrow [0,1]$ is the initial distribution, $T_C : H \times E \times H \rightarrow [0,1]$ is the transition probability function, $Cost : H \times E \times H \rightarrow \mathbb{R}$ is the cost function: $Cost(h, e, h')$ is the cost of human effort for the transition from h to h' under event e. $\gamma \in (0,1)$ is the discount factor. $H_s \subset H$ is a subset of states at which the operator can take control.

Given two MDPs, M_A for the controller and M_H for the operator and a cognitive model for the operator M_c, a shared autonomy stochastic system can be constructed as another Markov decision process as: $M_{SA} = \langle S, \Sigma, T, D_0, AP, L, Cost, \gamma\rangle$.

The problem can be formulated as follows:

Problem 71 *Given a stochastic system under shared autonomy control between an operator and an autonomous controller, modeled as MDP M_H and M_A, a model of human's cognition M_c and a linear temporal logic specification ϕ, compute a policy that is Pareto optimal with respect to two objectives:*

1. *Maximizing the discounted probability of satisfying the linear temporal logic specification ϕ.*

2. *Minimizing the discounted total cost of human effort over an infinite horizon.*

Details on the proposed solution can be found in [33].

2.6.3.2 Operator versus UAV

Many missions require multiple UAVs, with possibly diverse capabilities, to collaborate over long time periods. A team of UAVs can simultaneously collect information from multiple locations and exploit the information derived from multiple points to build models that can be used to take decisions. Team members can exchange sensor information, collaborate to track and identify targets and perform detection and monitoring activities among other tasks. Furthermore, the multi-UAV approach leads to redundant solutions offering greater fault tolerance and flexibility. Robustness and flexibility constitute the main advantages of multiple-robot systems with respect to single-robot ones. The degree of autonomy of the multi-UAV system should be tuned according to the specificities of the mission under consideration. In general the use of semi-autonomous groups of UAVs, supervised or partially controlled by one or more human operators, is currently the only viable solution in order to deal with the complexity and unpredictability of real-world scenarios. In addition to that, the human presence is also often mandatory for taking the responsibility of critical decisions in highly risky situations.

Ratio operator per UAV The appropriate conditions and requirements under which a single pilot would be allowed to control multiple UAVs simultaneously must be investigated [21]. To enable a single human operator to control and coordinate a group of UAVs, **decision support** must be provided to the pilot. The pilot communicates with the UAV team remotely and controls the UAV team to execute a mission. Pilot-controlled mode and autonomous mode can be allowed. In both of these modes, the UAV team is controlled in a leader-follower manner, and the leader UAV is assigned by the pilot. The followers are positioning themselves with respect to the other UAVs. In the autonomous mode, the leader UAV executes the mission without intervention of the pilot. At any time, the pilot is allowed to take over and directly control the leader UAV. Hence, the pilot can interrupt the mission and release control of the UAV. The UAV team should automatically resume the execution of the given mission [24].

For a notional system architecture to be required for single operator control of multiple UAS, operators need to interact with an overall mission and payload manager while relegating navigation and motion control tasks to automation. The challenge in achieving effective management of multiple UAVs is to determine if automation can be used to reduce workload, and what kind of decision support will be needed by operators given the high-workload environment. An increasing number of tasks have to be automated to achieve single operator control of multiple UAVs. If system reliability decreases in the control of multiple UAVs, trust declines with increasing numbers of UAVs but improves when humans are actively involved in planning and executing decisions. Guiding multiple UAVs by just one pilot means a large number and variety of supervision and monitoring tasks, interrupted by time-critical re-planning and re-configuration tasks as reaction to unexpected events. To support pilots, especially in critical workload situations, a cognitive associate system requires the machine awareness of the actual task the operator is working on, and the ability to detect critical workload situations to initiate assistant system interventions. An approach of building human operator behavior models is described in [25] to determine both the current task of the operator, and derivations in task accomplishment.

The number of UAVs that a single operator can control is not just a function of the level of decision support automation, but also the operational demands. To achieve the goal of one person controlling many UAVs, operators should only monitor the piloting/maneuvering of the UAV, not do it themselves. The multiple vehicle supervisory control represents an attention allocation problem that requires an operator to determine how and when to allocate limited cognitive resources to multiple, often competing tasks. The most efficient and effective ratio operator/UAV is sought. At present, the estimates of this ratio vary widely around unity, where unity is a ratio of 1 operator/1 UAV. Anyway, regulations allow currently only this configuration, however, they are likely to evolve rapidly. Current operational systems require many human supporters to launch, control, direct, recover, and maintain even one single UAV. Despite these difficulties, design aspirations are for ratios of 1 operator/4 UAVs [40].

Human management is crucial, as automated planners do not always perform well in the presence of unknown variables and possibly inaccurate prior information. When supervising multiple UAVs in dynamic environments, where events are often unanticipated, automated planners have difficulty responding to unforeseen problems. As UAVs fly, new tasks emerge and mission needs actualization. The automated planner should adapt to assist in real-time decision making and be capable of dynamic mission re-planning. Providing the human operator with the ability to modify the way the automated planner works for collaborative decision making, the human operator can choose and rank criteria to adjust the weights of variables in the objective function, this is a **multivariate resource allocation task. Collaborative human/automation planning** reduces overall workload. To evaluate the ability of a UAV to generate some measure of self-confidence to an operator, the methods by which the UAV perceives the environment can be investigated. In addition, representative algorithms that use this collection of sensor data can also be evaluated to determine the scenarios where false perception of the UAV's environment is present, or an inability to properly use the sensor data due to mission constraints is possible. Competency boundary estimation for autonomous sensors involves mainly output quality and mission impact. While the first dimension provides clear quantitative metrics, the second is more contextual, i.e., a poorly performing sensor may not affect a mission if high degrees of accuracy are not needed. Output quality (i.e., can the algorithm generate feasible solutions?), and mission impact (i.e., if the solution is viable, how could variation in output affect the mission?) are representations of the **competency boundaries** of the optimization/verification block [45].

Integration of autonomous functions and operator assistance Automation functions are tied into work processes as tools actively performing well-defined sub-tasks leaving the human operator as a high-end decision component supervising the process. This hierarchical composition of the human component and the automation can be the source of various automation-induced drawbacks with respect to the overall performance in mission accomplishment. The problem regards a work system comprising a single operator working in a ground or even airborne control station supervising several UAVs. The setup of such a manned-unmanned teaming scenario can be analyzed in a top-down manner using the approach of the work system as a human factors engineering framework [54]. Conventional automation, such as autopilot or flight management systems, does not assist the operator in performing tasks like decision making with the aim to achieve the top-level mission goals.

In contrast to automation aiming at autonomous performance in the sense of taking over human responsibilities, cooperative automation in the sense of human-machine co-agency shall be enabled: to collaborate with the human operator, to negotiate the allocation of tasks adapted to the needs of the current situation, and to jointly supervise the performance of sub-tasks under the consideration of the overall work objective by semi-autonomous systems, if applicable [54]. This system is described by two embedded work systems:

1. The superior work system is constituted by the work objective of performing the mission. Its overall work objective will be issued by an external authority also providing suitable information and resources to the work system. It consists of an operating element formed by a human pilot and a cooperative assistant system, providing work objectives to the UAV guidance sub-work system in teamwork between human and machine.

2. The second system is constituted by the work objective of guiding UAVs performing the mission. Its work objective will be given by the superior work system. It consists of the human UAV operator supported by a cooperative assistant system, forming the operating element of this sub-work system.

Intelligent adaptive interfaces (IAIs) are emerging technologies for enhancing performance in multiple UAV control. To guide interface design, the framework integrates a user-centered design approach with the concept of proactive use of **adaptive intelligent agents** (AIA) aiming at maximizing overall system performance. The framework uses a multiple-agent hierarchical structure to allocate tasks between operators and agents for optimizing operator/agent interaction, as a means of reducing operator workload, and increasing situation awareness and operational effectiveness [43]. The interaction of operators with various levels of automation to enhance overall human-machine performance is domain-dependent.

Definition 72 *An* ***adaptive intelligent agent*** *is a software program that exhibits adaptation, autonomy, and cooperation in accomplishing tasks for the user.*

Since AIAs aim to help users by automatically taking over some tasks and adapting themselves to changes in users and the intelligent adaptive interfaces, they should be more human-like and resemble human behavior for effective interactions. An example of an agent paradigm resembling cognitive and adaptation processes of human behavior is the **belief desire intention** (BDI) approach. The key data structure in a BDI model is a plan library, which is a set of plans for an agent, specifying courses of actions that an agent may take to achieve its intentions. A plan library of an agent represents its procedural knowledge about producing system states. An interaction model should represent at least three properties of the control and controlled systems as well as the operator, which are as follows:

1. What changes to the system does the operator want to make?
2. Why should the changes be made with respect to system goals and current state?
3. How can the needed changes to the system be made (i.e., the operator activities undertaken to achieve the desired state)?

Additionally, if there are concurrent activities, the model should present their nature and choices available to the operator, given the current system state. To be useful to the design, an effective interaction model must be both descriptive and prescriptive. It describes what an operator actually does and specifies what the operator should do next, as a decision making aid.

2.7 CONCLUSION

In the first part of this chapter, levels of autonomy are considered. Then, different kinds of teams are considered: human-UAV teams, homogeneous and heterogeneous UAV teams and finally UAVs-UGVs teams. In the last part of the chapter, mission analysis methodology is introduced because the framework of operations is comparable even if different missions are performed.

Bibliography

[1] Aalmoes, R.; Cheung, Y. S.; Sunil, E.; Hoekstra, J. M.; Bussink, F.: *A conceptual third party risk model for personal and unmanned aerial vehicles*, In IEEE International Conference on Unmanned Aircraft Systems (ICUAS), pp. 1301-1309, 2015.

[2] Aghaeeyan, A.; Abdollahi, F.; Talebi, H. A.: *UAV-UGVs cooperation: With a moving center based trajectory*, Robotics and Autonomous Systems, vol. **63**, pp. 1-9, 2015.

[3] Araiza-Illan, D.; Western, D.; Pipe, A.; Eder, K.: *Coverage-Driven Verification-An approach to verify code for robots that directly interact with humans*, arXiv preprint arXiv:1509.04852, 2015.

[4] Arokiasami, W. A.; Vadakkepat, P.; Tan, K. C.; Srinivasan, D.: *Interoperable multi-agent framework for unmanned aerial/ground vehicles: towards robot autonomy*, Complex and Intelligent Systems, vol. **2**, pp. 45-59, 2016.

[5] Balampanis, F.; Maza, I.; Ollero, A.: *Area Partition for Coastal Regions with Multiple UAS*, Journal of Intelligent and Robotic Systems, pp. 1-16, DOI 10.1007/s10846-017-0559-9, 2017.

[6] Belta, C; Bicchi, A.; Egerstedt, M.; Frazzoli, E.; Klavins, E.: Pappas, G. J.: *Symbolic planning and control of robot motion*, IEEE Robotics and Automation Magazine, vol. **14**, pp. 61-70, 2007.

[7] Bestaoui Sebbane, Y.: *Smart Autonomous Aircraft: flight control and planning of UAV*, CRC Press, 2016.

[8] Blasch, E.; Seetharaman, G.; Reinhardt, K.: *Dynamic data driven applications system concept for information fusion*, Procedia Computer Science, vol. **18**, pp. 1999-2007, 2013.

[9] Boccalatte, M.; Brogi, F.; Catalfamo, F.; Maddaluno, S.; Martino, M.; Mellano, V.; Prin, P.; Solitro, F; Torasso, P.; Torta, G.: *A multi-UAS cooperative mission over non-segregated civil areas*, Journal of Intelligent and Robotic Systems, vol. **70**, pp. 275-291, 2013.

[10] Calhoun, G. L.; Draper, M. H.; Abernathy, M. F.; Patzek, M.; Delgado, F.: *Synthetic vision system for improving unmanned aerial vehicle operator situation awareness*, In Proceedings of SPIE Enhanced and Synthetic Vision, vol. **5802**, pp. 219-230, 2005.

[11] Carbonneau, P. E.; Dietrich, J. T.: *Cost-effective non-metric photogrammetry from consumer-grade sUAS: implications for direct georeferencing of structure from motion photogrammetry*, Earth Surface Processes and Landforms, vol. **42**, pp. 473-486, 2017.

[12] Capiluppi, M., Segala, R.: *World automata: a compositional approach to model implicit communication in hierarchical hybrid systems*, arXiv preprint arXiv:1308.5335.

[13] Cheng, M. X.; Ling, Y.; Sadler, B. M.: *Network connectivity assessment and improvement through relay node deployment*, Theoretical Computer Science, vol. **660**, pp. 86-101, 2017.

[14] Chen, T.; Gonzalez, F.; Campbell, D.; Coppin, G.: *Management of multiple heterogeneous UAVs using capability and autonomy visualisation: Theory, experiment and result* In International Conference on Unmanned Aircraft Systems (ICUAS), pp. 200-210, 2014.

[15] Chen, T. B.; Campbell, D. A.; Coppin, G.; Gonzalez, F.: *Management of heterogeneous UAVs through a capability framework of UAVs functional autonomy*, 15th Australian International Aerospace Congress (AIAC15), 2013.

[16] Chowdhury, S.; Maldonado, V.; Tong, W.; Messac, A.: *New modular product-platform-planning approach to design macroscale reconfigurable unmanned aerial vehicles*, AIAA Journal of Aircraft, vol. **53**, pp. 309-322, 2016.

[17] Chowdhury, S.; Emelogu, A.; Marufuzzaman, M.; Nurre, S. G.; Bian, L.: *Drones for Disaster Response and Relief Operations: A Continuous Approximation Model*, International Journal of Production Economics, vol. **188**, pp. 167-184, 2017.

[18] Clare, A. S.; Cummings, M. L.; How, J. P.; Whitten, A. K.; Toupet, O.: *Operator object function guidance for a real-time unmanned vehicle scheduling algorithm*, AIAA Journal of Aerospace Computing, Information, and Communication, vol. **9**, pp. 161-173, 2012.

[19] Colomina, I.; Molina, P.: *Unmanned aerial systems for photogrammetry and remote sensing: A review*, ISPRS Journal of Photogrammetry and Remote Sensing, vol. **92**, pp. 79-97, 2014.

[20] Cowlagi, R. V.; Zhang, Z.: *Route guidance for satisfying temporal logic specifications on aircraft motion*, AIAA Journal of Guidance, Control and Dynamics, vol. **40**, pp. 390-401, 2017.

[21] Cummings, M. L., Bruni, S., Mercier, S.; Mitchell, P. J.: *Automation architecture for single operator, multiple UAV command and control*, The International C2 Journal, vol. **1**, pp. 1-24, 2007.

[22] Deng, Q.; Yu, J.; Mei, Y.: *Deadlock-Free Consecutive Task Assignment of Multiple Heterogeneous Unmanned Aerial Vehicles*, AIAA Journal of Aircraft, vol. **51**, pp. 596-605, 2014.

[23] Dennis, L. A.; Fisher, M.; Lincoln, N. K.; Lisitsa, A.; Veres, S. M.: *Practical verification of decision-making in agent-based autonomous systems*; Automated Software Engineering, vol. **23**, pp. 305-359, 2016.

[24] Ding, X. C.; Powers, M.; Egerstedt, M.; Young, S. Y. R.; Balch, T.: *Executive decision support, Single-Agent Control of Multiple UAVs*, IEEE Robotics and Automation Magazine, vol. **16**, pp. 73-81, 2009.

[25] Donath, D.; Schulte, A.: *Behavior Based Task and High Workload Determination of Pilots Guiding Multiple UAVs*, Procedia Manufacturing, vol. **3**, pp. 990-997, 2015.

[26] Doshi-Velez, F., Pineau, J., Roy, N. : *Reinforcement learning with limited reinforcement: Using Bayes risk for active learning in POMDP* Artificial Intelligence, vol. **187**, pp. 115-132, 2012.

[27] Drury, J. L.; Riek, L.; Rackliffe, N.: *A decomposition of UAV-related situation awareness*, In Proceedings of the 1st ACM SIGCHI/SIGART Conference on Human-Robot Interaction, pp. 88-94, 2006.

[28] D'Souza, S.; Ishihara, A.; Nikaido, B.; Hasseeb, H.: *Feasibility of varying geo-fence around an unmanned aircraft operation based on vehicle performance and wind*, In IEEE/AIAA 35*th* Digital Avionics Systems Conference (DASC), pp. 1-10, 2016.

[29] Durst, P. J.; Gray, W.: *Levels of Autonomy and Autonomous System Performance Assessment for Intelligent Unmanned Systems*, (No. ERDC/GSL-SR-14-1), Engineer research and development center, Vicksburg MS Geotechnical and structures LAB, 2014.

[30] Evertsz, R.; Thangarajah, J.; Yadav, N.; Ly, T.: *A framework for modelling tactical decision-making in autonomous systems*, Journal of Systems and Software, vol. 110, pp. 222-238, 2015.

[31] Flushing, E. F.; Gambardella, L. M.; Di Caro, G. A.: *Strategic control of proximity relationships in heterogeneous search and rescue teams*, arXiv preprint arXiv:1312.4601, 2013.

[32] Franchi, A.; Secchi, C.; Ryll, M.; Bulthoff, H.H.; Giordano P.R.: *Shared control: Balancing autonomy and human assistance with a group of quadrotor UAVs*, IEEE Robotics and Automation Magazine, vol. **19**, pp. 57-68, 2012.

[33] Fu, J.; Topcu, U.: *Pareto efficiency in synthesizing shared autonomy policies with temporal logic constraints*, IEEE International Conference on Robotics and Automation, pp. 361-368, 2015.

[34] Gan, S. K.; Fitch, R.; Sukkarieh, S.: *Online decentralized information gathering with spatial-temporal constraints*, Autonomous Robots, vol. **37**, pp. 1-25, 2014.

[35] Garcia de Marina, H.; Jayawardhana, B.; Cao, M.: *Distributed rotational and translational maneuvering of rigid formations and their applications*, IEEE Transactions on Robotics, vol. **32**, pp. 684-697, 2016.

[36] Geng, J.; Lv, C.; Zhou, D.; Wang, Z.: *A mission execution decision making methodology based on mission-health interrelationship analysis*, Computers and Industrial Engineering, vol. **95**, pp. 97-110, 2016.

[37] Gunn, T.; Anderson, J.: *Dynamic heterogeneous team formation for robotic urban search and rescue*, Journal of Computer and System Sciences, vol. **81**, pp. 553-567, 2015.

[38] Gupta, L.; Jain, R.; Vaszkun, G.: *Survey of important issues in UAV communication networks*, IEEE Communications Surveys and Tutorials, vol. **18**, pp. 1123-1152, 2015.

[39] Haidegger, T.; Barreto, M.; Gonalves, P.; Habib, M. K.; Ragavan, S. K. V.; Li, H.; Vaccarella, A.; Perrone, R.; Prestes, E.: *Applied ontologies and standards for service robots*, Robotics and Autonomous Systems, vol. **61**, pp. 1215-1223, 2013.

[40] Hancock, P. A., Mouloua, M.; Gilson, R.; Szalma, J.; Oron-Gilad, T.: *Provocation: Is the UAV control ratio the right question?*, Ergonomics in Design, vol. **15**, pp. 7-12, 2007.

[41] Hayat, S.; Yanmaz, E.; Muzaffar, R.: *Survey on unmanned aerial vehicle networks for civil applications: A communication's viewpoint*, IEEE Communications Surveys and Tutorials, DOI: 10.1109/COMST.2016.2560343, 2016.

[42] Hocraffer, A.; Nam, C. S.: *A meta-analysis of human-system interfaces in unmanned aerial vehicle (UAV) swarm management*, Applied Ergonomics, vol. **58**, pp. 66-80, 2017.

[43] Hou, M.; Zhu, H.; Zhou, M.; Arrabito, G. R.: *Optimizing operator agent interaction in intelligent adaptive interface design: A conceptual framework*, IEEE Transactions on Systems, Man, and Cybernetics, Part C: Applications and Reviews, vol. **41**, pp. 161-178, 2011.

[44] Hristu-Varsakelis, D.; Egerstedt, M.; Krishnaprasad, P.: *On the structural complexity of the motion description language MDLe*, 42nd IEEE Conf. on Decision and Control, pp. 3360–3365, 2003.

[45] Hutchins, A. R.; Cummings, M. L.; Draper, M.; Hughes, T.: *Representing Autonomous Systems Self-Confidence through Competency Boundaries*, In Proceedings of the Human Factors and Ergonomics Society Annual Meeting, SAGE Publications, vol. **59**, pp. 279-283, 2015.

[46] Ivancevic, V.; Yue, Y.: *Hamiltonian dynamics and control of a joint autonomous landair operation*, Nonlinear Dynamics, vol. **84**, pp. 1853-1865, 2016.

[47] Jakubovskis, A.: *Strategic facility location, capacity acquisition, and technology choice decisions under demand uncertainty: Robust vs. non-robust optimization approaches*, European Journal of Operational Research, vol. **260**, pp. 1095-1104, 2017.

[48] Johnson, L. B.; Choi, H. L.; Ponda, S. S.; How, J. P.: *Decentralized task allocation using local information consistency assumptions*, AIAA Journal of Aerospace Information Systems, vol. **14**, pp. 103-122, 2017.

[49] Jones, E. G.; Dias, M. B.; Stentz, A.: *Time-extended multi-robot coordination for domains with intra-path constraints*, Autonomous robots, vol. **30**, pp. 41-56, 2011.

[50] Kaddouh, B. Y.; Crowther, W. J.; Hollingsworth, P.: *Dynamic resource allocation for efficient sharing of services from heterogeneous autonomous vehicles*, Journal of Aerospace Information Systems, vol. **13**, pp. 450-474, 2016.

[51] Kaul, L.; Zlot, R.; Bosse, M.: *Continuous time three dimensional mapping for micro aerial vehicles with a passively actuated rotating laser scanner*, Journal of Field Robotics, vol. **33**, pp. 103-132, 2016.

[52] Kendoul, F.: *Survey of advances in guidance, navigation, and control of unmanned rotorcraft systems*, Journal of Field Robotics, vol. **29**, pp. 315-378, 2012.

[53] Khosiawan, Y.; Park, Y. S.; Moon, I.; Nilakantan, J. M.; Nielsen, I.: *Task scheduling system for UAV operations in indoor environment*, arXiv preprint arXiv:1604.06223, 2016.

[54] Kriegel, M.; Schulte, A.: *Work system analysis of the integration of autonomous functions and intelligent operator assistance in UAV Guidance*, In RTO-Meeting Proceedings HFM-135. Biarritz, France, 2006.

[55] Lemon, O.; Bracy, A.; Gruenstein, A.; Peters, S.: *The WITAS multimodal dialogue system I*, In INTERSPEECH, pp. 1559-1562, 2001.

[56] Liemhetcharat, S.; Veloso, M.: *Weighted synergy graphs for effective team formation with heterogeneous ad hoc agents*, Artificial Intelligence, vol. **208**, pp. 41-65, 2014.

[57] Littman, M.: *A tutorial on partially observable Markov decision process*, Journal of Mathematical Psychology, vol. **53**, pp. 119–125, 2009.

[58] Liu, L., Orzay, N.; Topai, U.; Murray, R.M.: *Synthesis of reactive switching protocols from temporal logic specification*, IEEE Transactions on Automatic Control, vol. **58**, pp. 1771-1785, 2013.

[59] Liu, H.; Lin, M.; Deng, L.: *UAV route planning for aerial photography under interval uncertainties*, Optik-International Journal for Light and Electron Optics, vol. **127**, pp. 9695-9700, 2016.

[60] Luo, J.; Yan, G.: *Fast iterative algorithm for solving MoserVeselov equation*, AIAA Journal of Guidance, Control, and Dynamics, vol. **38**, pp. 949-954, 2015.

[61] Maguire, C.: *Using unmanned aerial vehicles and structure from motion software to monitor coastal erosion in Southeast Florida*, In Open access theses, 81 pages, paper 525, 2014.

[62] Manikas, T. W.; Mitchell, A. T.; Chang, F. C.: *Mission planning analysis using decision diagrams*, In Proceedings of Reed-Muller Workshop, pp. 61-65, 2013.

[63] Martini, S.; Di Baccio, D.; Romero, F. A.; Jimnez, A. V.; Pallottino, L.; Dini, G.; Ollero, A.: *Distributed motion misbehavior detection in teams of heterogeneous aerial robots*, Robotics and Autonomous Systems, vol. **74**, pp. 30-39, 2015.

[64] McCune, R.; Purta, R.; Dobski, M.; Jaworski, A.; Madey, G.; Wei, Y.; Blake, M. B.: *Investigations of DDDAS for command and control of UAV swarms with agent-based modeling*, In Proceedings of the IEEE Winter Simulation Conference: Simulation: Making Decisions in a Complex World, pp. 1467-1478, DOI: 10.1109/WSC.2013.6721531, 2013.

[65] Mosterman, P. J.; Sanabria, D. E.; Bilgin, E.; Zhang, K.; Zander, J.: *Automating humanitarian missions with a heterogeneous fleet of vehicles*, Annual Reviews in Control, vol. **38**, pp. 259-270, 2014.

[66] Murphy, R. R.; Burke, J. L.: *Up from the rubble: Lessons learned about HRI from search and rescue*, In Proceedings of the Human Factors and Ergonomics Society Annual Meeting (SAGE Publications), vol. **49**, pp. 437-441, 2005.

[67] Nolan, P.; Paley, D. A.; Kroeger, K.: *Multi-UAS path planning for non-uniform data collection in precision agriculture*, In IEEE Aerospace Conference, pp. 1-12, 2017.

[68] Nunes, E.; Manner, M.; Mitiche, H.; Gini, M.: *A taxonomy for task allocation problems with temporal and ordering constraints*, Robotics and Autonomous Systems, vol. **90**, pp. 55-70, 2017.

[69] Oh, G.; Kim, Y.; Ahn, J.; Choi, H. L.: *Market-based task assignment for cooperative timing missions in dynamic environments*, Journal of Intelligent and Robotic Systems, pp. 1-27, DOI 10.1007/s10846-017-0493-x, 2017.

[70] Patterson, M. D.; Pate, D. J.; German, B. J.: *Performance flexibility of reconfigurable families of unmanned aerial vehicles*, AIAA Journal of Aircraft, vol. **49**, pp. 1831-1843, 2012.

[71] Purtee, M. D.; Gluck, K. A.; Krusmark, M. A.; Kotte, S. A.; Lefebvre, A. T.: *Verbal protocol analysis for validation of UAV operator model*, In Proceedings of the 25th Interservice/Industry Training, Simulation, and Education Conference, pp. 1741-1750. Orlando, FL: National Defense Industrial Association, 2003.

[72] Raikov, A.: *Convergent networked decision-making using group insights*, Complex and Intelligent Systems, vol. **1**, pp. 57-68, 2015.

[73] Saadou-Yaye, A.; Aruz, J.: *μ-Autonomy: intelligent command of movable objects*, Procedia Computer Science, vol. **61**, pp. 500-506, 2015.

[74] Schumann, B.; Ferraro, M.; Surendra, A.; Scanlan, J. P.; Fangohr, H.: *Better design decisions through operational modeling during the early design phases*, AIAA Journal of Aerospace Information Systems, vol. **11**, pp. 195-210, 2014.

[75] Schwartz, B., Ngele, L., Angerer, A., MacDonald, B. A.: *Towards a graphical language for quadrotor missions*, arXiv preprint arXiv:1412.1961, 2014.

[76] Sen, S. D.; Adams, J. A.: *An influence diagram based multi-criteria decision making framework for multirobot coalition formation*, Autonomous Agents and Multi-Agent Systems, vol. **29**, pp. 1061-1090, 2015.

[77] Seyboth, G. S.; Allgower, F. : *Synchronized model matching: a novel approach to cooperative control of nonlinear multi-agent systems*, Preprints of the 19th World Congress of The International Federation of Automatic Control, Cape Town, South Africa, pp. 1985-1990, 2014.

[78] Silva, D. C.; Abreu, P. H.; Reis, L. P.; Oliveira, E.: *Development of a flexible language for mission description for multi-robot missions*, Information Sciences, vol. **288**, pp. 27-44, 2014.

[79] Soler, M.; Olivares, A.; Staffetti, E.: *Multiphase optimal control framework for commercial aircraft 4D flight planning problems*, AIAA Journal of Aircraft, vol. **52**, pp. 274-286, 2014.

[80] Song, B. D.; Kim, J.; Morrison, J. R.: *Rolling horizon path planning of an autonomous system of UAVs for persistent cooperative service: MILP*

formulation and efficient heuristics, Journal of Intelligent and Robotic Systems, vol. **84**, pp. 241-258, 2016.

[81] Summers, T. H.; Akella, M. R.; Mears, M. J.: *Coordinated standoff tracking of moving targets: control laws and information architectures*, AIAA Journal of Guidance, Control and Dynamics, vol. **32**, pp. 56–69, 2009.

[82] Tracol, M.; Desharnais, J.; Zhioua, A.: *Computing distances between probabilistic automata*, 9th Workshop on quantitative aspects of programming languages, pp. 148-162DOI 10.4204/EPTCS.57.11, 2011.

[83] Truong, T. V. A.; Hattenberger, G.; Ronfle-Nadaud, C.: *The cooperation between unmanned aerial vehicles using a mission planner*, In 11th IEEE International Conference on Industrial Informatics (INDIN), pp. 797-803, 2013.

[84] Turner, D. J.: *Multi-sensor, multi-temporal, and ultra-high resolution environmental remote sensing from UAVs*, PhD thesis, Univ. of Tasmania, pp. 163, 2015.

[85] Ulusoy, A.; Smith, S. L.; Ding, X. C.; Belta, C.; Rus, D. : *Optimal multi-robot path planning with temporal logic constraints*, IROS IEEE/RSJ Int. Conference on Intelligent Robots and Systems, pp. 3087-3092, 2011.

[86] Ure, N. K.; Chowdhary, G.; How, J. P.; Vavrina, M. A.; Vian J.: *Health aware planning under uncertainty for UAV missions in heterogeneous teams*, European Control Conference, Zurich, Switzerland, pp. 3312-3319, 2013.

[87] VanderBerg,J. P.; Patil, S.; Alterovitz, R. : *Motion planning under uncertainty using differential dynamic programming in Belief space*, Int. Symp. of Robotics Research, pp. 473-490, 2011.

[88] Varela, G.; Caamano, P.; Orjales, F.; Deibe, A.; Lopez-Pena, F.; Duro, R. J.: *Autonomous UAV based search operations using constrained sampling Evolutionary Algorithms*, Neurocomputing, vol. **132**, pp. 54-67, 2014.

[89] Vilar, R. G.; Shin, H. S.: *Communication-aware task assignment for UAV cooperation in urban environments*, 2nd IFAC Workshop on Research, Education and Development of Unmanned Aerial Systems, Compiegne, France, IFAC Proceedings, vol. **46**, pp. 352-359, 2013.

[90] Wei, Y.; Madey, G. R.; Blake, M. B.: *Agent-based simulation for UAV swarm mission planning and execution*, In Proceedings of the Agent-Directed Simulation Symposium, Society for Computer Simulation International, 2013.

[91] Xargay, E.; Kaminer, I.; Pascoal, A.; Hovakimyan, N.; Dobrokhodov, V.; Cichella, V.; Aguiar, A. P.; Ghabcheloo, R.: *Time-critical cooperative*

path following of multiple unmanned aerial vehicles over time-varying networks, AIAA Journal of Guidance, Control and Dynamics, vol. **36**, pp. 499-516, 2013.

[92] Xiang, Y.; Xuan, Z.; Tang, M.; Zhang, J.; Sun, M.: *3D space detection and coverage of wireless sensor network based on spatial correlation*, Journal of Network and Computer Applications, vol. **61**, pp. 93-101, 2016.

[93] Yazicioglu, A. Y.; Abbas, W.; Egerstedt, M.: *Graph distances and controllability of networks*, IEEE Transactions on Automatic Control, vol. **61**, pp. 4125-4130, 2016.

[94] Zervoudakis, F.; Rosenblum, D. S.; Elbaum, S.; Finkelstein, A.: *Cascading verification: an integrated method for domain-specific model checking*, In Proceedings of the 9th ACM Joint Meeting on Foundations of Software Engineering, pp. 400-410, 2013.

CHAPTER 3

Orienteering and Coverage

3.1 INTRODUCTION

In this chapter, operations concern generic aerial robotic problems, such as **orienteering**, and **coverage** in applications such as surveillance, search and rescue, geo-location, exploration, monitoring, mapping, etc. The use of UAVs, instead of ground robots, provides several advantages. The capacity to fly allows avoiding obstacles and having a bird's-eye view. A Minkowski sum of an obstacle expands the obstacle according to the UAV's size, while simultaneously, the UAV shrinks to a reference point. In large outdoors, the Minkowski sum represents the inaccessible area for the UAV. If the Minkowski sums of different obstacles intersect, then they will be merged into one to form a closed inaccessible area for the UAV. The space outside the union region is regarded as the free space for the UAV, and consequently, the UAV can follow paths in it [59]. If they have a small size, they can also navigate in narrow outdoor and indoor environments and they represent only a limited invasive impact. Before proceeding further on orienteering and coverage, some basic features of operational research involved in UAV missions are presented. These are the vehicle routing problem, traveling salesperson problem, postperson problem and knapsack problem. The material in this section is not intended to be exhaustive, but rather to provide a sufficient introduction to those who may be unfamiliar with such topics.

3.2 OPERATIONAL RESEARCH PRELIMINARIES

This section presents some fundamental algorithms in UAV mission planning such as the vehicle routing problem, the traveling salesperson problem and its variants, the postperson problem be it Chinese or rural, and finally the knapsack method.

3.2.1 General vehicle routing problem

The **general vehicle routing problem** (GVRP) is the problem of finding distinct feasible tours maximizing the profit determined by the accumulated revenue of all orders served by a vehicle reduced by the cost for operating the tours [39]. In the GVRP, a transportation request is specified by a non-empty set of a pickup, delivery and/or service location which have to be visited in a particular sequence by the same vehicle, the time windows in which these locations have to be visited and the revenue gained when the transportation request is served. The GVRP is a combined load acceptance and routing problem which generalizes the vehicle routing problem and **pick-up and delivery problem** (PDP) [88].

Definition 73 *A **tour** of a vehicle is a journey starting at the vehicle's start location and ending at its final location, passing all other locations the vehicle has to visit in the correct sequence, and passing all locations belonging to each transportation request assigned to the vehicle in the correct respective sequence.*

Definition 74 *A tour is **feasible** if and only if for all orders assigned to the tour compatibility constraints hold at each point in the tour time window and capacity restrictions hold.*

The objective is to find distinct feasible tours maximizing the profit which is determined by the accumulated revenue of all served transportation requests, reduced by the accumulated costs for operating these tours.

The most widely studied VRPs are the **capacitated VRP** and the **vehicle routing problem with time windows** (VRPTW). **Variable neighborhood search** (VNS) is a meta-heuristic based on the idea of systematically changing the neighborhood structure during the search. **VNS** systematically exploits the following observation:

1. A **local optimum** with respect to one neighborhood structure is not necessary as for another.
2. A **global optimum** is a local optimum with respect to all possible neighborhood structures.
3. For many problems **local optima** with respect to one or several neighborhoods are relatively close to each other.

3.2.2 Traveling salesperson problem

The most basic strategic problem is often how to choose the order of UAV way-points in a set of possible locations.

Problem 75 ***Traveling salesperson problem (TSP)**: Given a set of n points in the plane, TSP asks if there exists a tour for the UAV that visits all these points exactly once.*

In general, the TSP includes two different kinds:

1. The **symmetric traveling salesperson problem** (STSP): there is only one way between two adjacent cities, i.e., the distance between cities A and B is equal to the distance between cities B and A.
2. The **asymmetric traveling salesperson problem** (ATSP): there is no such symmetry and it is possible to have two different costs or distances between two cities.

TSP is a representative of a large class of problems known as combinatorial problems. Among them, TSP is one of the most important, since it is easy to describe but difficult to solve. There are many variations to the TSP where a tour of minimum length has to be found that passes through every target location precisely once. Some are listed below:

1. The **Euclidean traveling salesperson problem** (ETSP) is a TSP where the distances between the vertices are precisely the Euclidean distances of the target locations in the plane.
2. The **Dubins traveling salesperson problem** (DTSP) refers to a kinematically constrained vehicle, such as an airplane at a constant altitude.
3. The **traveling salesperson problem with neighborhoods** (TSPN) extends the TSP to the case where each vertex of the tour is allowed to move in a given region. This approach takes into account the communication range or the sensor footprint of the UAV.
4. The **Euclidean traveling salesperson problem with neighborhoods** (ETSPN) seeks for a shortest Euclidean path passing through the regions.
5. The **traveling repairman problem**, also known as delivery man problem or minimum latency problem, aims to find a tour, or a Hamiltonian cycle, on a given graph that minimizes the total traveling time plus the waiting time sum across every customer (relatively to some fixed node).
6. The **dynamic traveling repairman problem** (DTRP) in which the UAV is required to visit a dynamically changing set of targets [16] is an example of distributed task allocation problems.
7. The **k-traveling repairman problem** (k-TRP): The traveling repairman problem with differentiated waiting times is a variation of the traveling repairman problem and traveling salesperson problem with cumulative costs. The waiting time, or latency, of customer k is the total time involved in the path 1 to k in the tour and can also be thought of as delay for service. The traveling time is sometimes viewed as latency of the repairman. Given an undirected graph $G = (V, E)$ and a source vertex $s \in V$, the k-TRP, also known as the minimum latency problem, asks

for k tours, each starting at s and together covering all the vertices such that the sum of the latencies experienced by the customers is minimum. The latency of a customer p is defined to be the distance traveled before visiting p for the first time.

At the higher decision making level, the dynamics of the UAV are usually not taken into account and the mission planner might typically choose to solve the ETSP. The first step determines the order in which the way-points should be visited by the UAV. At the lower level, a path planner takes as an input this way-point ordering and designs feasible trajectory between the way-points respecting the dynamics of the UAV. Even if each problem is solved optimally, however, the separation into two successive steps can be inefficient since the sequence of points chosen by the TSP algorithm is often hard to follow for the physical system. In order to improve the performance of the UAV system, mission planning and path planning steps should be integrated [33].

3.2.2.1 Deterministic traveling salesperson

A salesperson has to visit several cities (or road junctions). Starting at a certain city, the salesperson wants to find a route of minimum length which traverses each of the destination cities exactly once leading back to his starting point. Modeling the problem as a complete graph with n vertices, the salesperson wishes to make a tour or **Hamiltonian cycle**, visiting each cycle exactly once and finishing at the initial city. A TSP instance is given by a complete graph G on a node set $V = \{1, 2, .., m\}$ for some integer m and by a cost function assigning a cost c_{ij} to the arc (i, j) for any i, j in V. The salesperson wishes to make the tour whose total cost is minimum where the total cost is the sum of the individual costs along the edges of the tour.

Remark 76 *There are different approaches for solving the traveling salesperson problem. Classical methods consist of heuristic and exact methods. Heuristic methods like* ***cutting planes*** *and* ***branch and bound*** *can only optimally solve small problems, whereas* ***Markov chains****, and* ***tabu search*** *are good for large problems. Besides, some algorithms based on greedy principles such as* ***nearest neighbor*** *and* ***spanning tree*** *can be introduced as efficient solving methods. New methods such as* ***nature-based optimization*** *algorithms,* ***evolutionary computation****,* ***neural networks****,* ***ant systems****,* ***particle swarm optimization****,* ***simulated annealing****, bee colony optimization, intelligent water drops algorithms, and artificial immune systems are among solving techniques inspired by observing nature [24].*

G has a traveling salesperson tour with cost at most k. Given a TSP instance with m nodes, any tour passing once through any city is a feasible solution and its cost leads to an upper bound to the least possible cost. Algorithms that construct in polynomial time with respect to m feasible solutions, and thus upper bounds for the optimum value, are called heuristics. In general, these algorithms produce solutions but without any quality guarantee as to how far is their cost from the least possible one. If it can be shown that the cost of the returned solution is always less than k times the least possible cost, for some real number $k > 1$, the heuristic is called a $k-$approximation algorithm. The data consist of weights assigned to the edges of a finite complete graph, and the objective is to find a Hamiltonian cycle, a cycle passing through all the vertices, of the graph while having the minimum total weight. $c(A)$ denotes the total cost of the edges in the subset $A \subseteq E$:

$$c(A) = \sum_{(u,v) \in A} c(u, v) \tag{3.1}$$

In many practical situations, the least costly way to go from a place u to a place w is to go directly, with no intermediate steps. The cost function c satisfies the triangle inequality if for all the vertices, $u, v, w \in V$

$$c(u, w) \leq c(u, v) + c(v, w) \tag{3.2}$$

This triangle inequality is satisfied in many applications, not all, it depends on the chosen cost. In this case, the minimum spanning tree can be used to create a tour whose cost is no more than twice that of the minimum tree weight, as long as the cost function satifies the triangle inequality. The pseudocode of the TSP approach can be presented in Algorithm 2.

Lines 1 to 5 in Procedure MST-PRIM of Algorithm 2 set the key of each vertex to ∞ (except for the root r, whose key is set to 0 so that it will be the first vertex processed), set the parent of each vertex to NULL and initialize the min-priority queue Q to contain all the vertices. The algorithm maintains the following three-part loop invariant. Prior to each iteration of the while loop of lines 6 to 11 [17].

1. $A = \{(\nu, v, \pi) : \nu \in V - \{r\} - Q\}$.

2. The vertices already placed into the minimum spanning tree are those in $V - Q$.

3. For all vertices $v \in Q$, if $v.\pi \neq NULL$, then $v.key < \infty$ and $v.key$ is the weight of a light edge (ν, v, π) connecting v to some vertex, already placed into the minimum spanning tree.

Line 7 in the procedure identifies a vertex $u \in Q$ incident on a light edge that crosses the cut $(V - Q, Q)$. Removing u from the set Q adds it to the set $V - Q$ of vertices in the tree, thus adding $(u, u.\pi)$ to A. The for loop of lines 8 to 11 updates the *key* and π attributes of every vertex v adjacent to u but not in the tree, thereby maintaining the third part of the loop invariant.

A generalization of the TSP is the **dynamic traveling repairman problem** (DTRP) for dynamic systems [45]. In the DTRP approach, customers are arising dynamically and randomly in a bounded region R and when customers arrive, they wait for the repairman to visit their location and offer a service

Algorithm 2 TSP with Triangle Inequality

1. Select a vertex $r \in G, V$ to be a root vertex
2. Compute a minimum spanning tree T for G from root r using MST-PRIM(G, c, r)
3. Let H be a list of vertices, ordered according to when they are first visited in a pre-order tree walk of T
4. A state transition rule is applied to incrementally build a solution
5. Return the Hamiltonian cycle H

Minimum Spanning Trees: Procedure MST-PRIM(G, c, r)

1. For each $u \in G, V$
2. u.key $= \infty$
3. u.π = NULL
4. r.key = 0
5. Q = G. V
6. While Q $\neq$ 0
7. u= EXTRACT-MIN(Q)
8. for each $v \in G.Adj\,[u]$
9. if $v \in Q$ and $w(u, v) < v.key$
10. v.π = u
11. v.key = w(u,v)

that will take a certain random amount of time. The repairman is modeled as a dynamic system whose output space contains R and the objective is the average time a customer has to wait to be serviced. Such problems appear in search and rescue or surveillance missions [13, 15]. In [30], a scenario is considered in which the target points are generated dynamically, with only prior statistics on their location and a policy is designed to minimize the expected time a target waits to be visited. This formulation is a variation of the DTRP, with the addition of differential constraints on the UAV motion. The analysis can be broken into two limiting cases:

1. **Light load**: targets are generated sporadically. The challenge is to design loitering policies with the property that when a target appears, the expected wait for the closest UAV to arrive is minimized. It reduces to a choice of waiting locations, and solutions are known from the locational optimization literature. These results are applicable to coverage problems in which the UAVs spread out uniformly to comb the environment efficiently.

2. **Heavy load**: targets are generated rapidly. The optimal policy for non-holonomic vehicles relies on Euclidean traveling salesperson tours through large sets of targets. It is strongly related to works concerned with the generation of efficient cooperative strategies of several UAVs moving through given target points and possibly avoiding obstacles or threats.

3.2.2.2 *Stochastic traveling salesperson*

In UAV applications, the traveling salesperson problems are often not deterministic, some of the parameters are not known with certainty at the decision making moment. The stochastic model has been used to represent this uncertainty, including the consideration of probability in the presence of customers, the demand level and the service time at customer's site usually assuming a known distribution governs some of the problem's parameters. In the case of a reconnaissance mission planning, before assigning the UAV to fly over dangerous targets, the flight time, fuel usage, forbidden areas on the flight path are often uncertain, only the belief degree of these quantities can be obtained [99]. For solving vehicle routing problems exactly, algorithms based on **column generation** and **Lagrangian relaxation** are the state of the art. Typically, the master program is a set-partitioning and the sub-problem a variant of the shortest path problem with resource constraints [86].

The **traveling salesperson problem** (TSP) and the **shortest path problem** (SPP), with deterministic arc costs but uncertain topologies, assume that any arc exists with a certain probability, and there are only two possible scenarios for a route connected by a set of arcs: success or failure. The non-existence or failure of any single arc within the selected route a-priori would result in the failure of the entire route due to the lack of recourse. To measure the risk of failure and to find the lowest-cost route with an acceptable level of reliability, the **Bernoulli trial** of each arc's existence is assumed independent of each other for convenience. In risk-constrained stochastic network flow problems, these models are useful in situations where both the travel cost and the reliability are important to the decision makers. For example, in the case that a city is struck by an earthquake, a quick emergency response is required to send the rescue workers and humanitarian supplies to the damage zones via the most efficient route. The consequence of not considering the possibility that the roads may be broken due to uncertain factors such as aftershocks, heavy traffic and weather conditions, however, could be crucial as significant resources would be trapped and not be delivered in time. Therefore, a more reliable route needs to be planned in advance to prevent the route failure with a desired level of confidence. In addition, there is always a tradeoff between minimizing the total cost and maximizing the reliability. Solving the proposed models would benefit decision makers to find a good balance in between. By setting different confidence levels, the corresponding optimal costs can be obtained and therefore the tradeoff between the total cost and the risk level can be achieved. The cost could be significantly saved by slightly decreasing the route reliability, and on the other hand, the risk could be largely reduced by selecting another route with only a little higher total cost. While considering independent failures of all arcs, the number of scenarios of a reliable routing problem would exponentially increase as the number of arcs in the input network increases, and the problem of a large-sized or even moderate-sized network would be very difficult to be solved [44].

A **stochastic traveling salesperson problem** is studied where the n targets are randomly and independently sampled from a uniform distribution. The dynamic version of the TSP is stated as follows:

Problem 77 ***Dynamic stochastic traveling salesperson problem:*** *Given a stochastic process that generates target points, is there a policy that guarantees that the number of unvisited points does not diverge over time? If such stable policies exist, what is the minimum expected time that a newly generated target waits before being visited by the UAV?*

The **stochastic traveling salesperson problem** can consider the scenario when n targets points are stochastically generated according to a uniform probability distribution function. The proposed recursive **bead-tiling algorithm** to visit these points consists of a sequence of phases, during each phase, a closed path is constructed [74]. The methodology is presented in Algorithm 3. This process is iterated and at each phase, meta-beads composed of two neighboring meta-beads from the previous phase are considered. After the last recursive phase, the leftover targets are visited using the alternating algorithm.

The **stochastic routing problem** is the generalization of the arc routing methods. Consider a complete graph $G = (V, E)$ on n nodes on which a routing problem is defined. If every possible subset of the node set V may

Algorithm 3 Stochastic TSP Algorithm

1. In the first phase of the algorithm, a tour is constructed with the following properties:
   (a) It visits all non-empty beads once.
   (b) It visits all rows in sequence top to down, alternating between left to right and right to left passes, and visiting all non-empty beads in a row.
   (c) When visiting a non-empty bead, it visits at least one target in it.
2. Instead of considering single beads, meta-beads composed of two beads each and proceed in a way similar to the first phase, are now considered with the following properties:
   (a) The tour visits all non-empty meta-beads once.
   (b) It visits all meta-beads rows in sequence top to down, alternating between left to right and right to left passes, and visiting all non-empty meta-beads in a row.
   (c) When visiting a non-empty meta-bead, it visits at least one target in it.

---

or may not be present on any given instance of the optimization problem, then there are 2^n possible instances of the problem, all the possible subsets of V. Suppose instance S has probability $prob(S)$ of occurring. Given a method $\mathbb{U}$ for updating the a priori solution f to the full-scale optimization problem on the original graph $G = (V, E)$, $\mathbb{U}$ will then produce for problem instance S a feasible solution $t_f(S)$ with value $L_f(S)$. In the case of the TSP, $t_f(S)$ would be a tour through a subset S of nodes and $L_f(S)$ the length of that tour. Then, given that the updating method $\mathbb{U}$ has already been selected, the natural choice for the a priori solution f is to minimize the expected cost:

$$E[L_f] = \sum_{S \subseteq V} prob(S) L_f(S) \tag{3.3}$$

with the summation being over all the subsets of V. The weighted average over all problem instances of the values $L_f(S)$ obtained is minimized by applying the updating method $\mathbb{U}$ to the a priori solution f.

The **probabilistic TSP** is essentially a traveling salesperson problem in which the number of points to be visited in each problem instance is a random variable. Considering a problem of routing through a set of n known points, on any given instance of the problem, only a subset S consisting of $|S| = k$ out of n points $(0 \leq k \leq n)$ must be visited. Ideally, the tour should be re-optimized for every instance by re-optimization in many cases. Unfortunately, re-optimization might turn out to be too time consuming. Instead, an a priori tour should be found through all n points. On any given instance of the problem the k points present will then be visited in the same order. The updating method for the probabilistic TSP is therefore to visit the point on every problem instance in the same order as in the a priori tour, skipping those points that are not present in that problem instance. The **probabilistic traveling salesperson problem** is to design an a priori route for each UAV, in which the route is followed exactly, simply skipping locations not requiring a visit. The goal is to find an a priori tour with the least expected cost. The probabilistic TSP represents a strategic planning model in which stochastic factors are considered explicitly. If n nodes are assumed spread over a bounded area, each node has a given probability of requiring a visit. The probability of requiring a visit is the **coverage problem**. Service requirements are assumed independent across nodes. The a priori tour is one in which the nodes are visited with the least expected length. The a priori tour is one in which the nodes are visited in the order given by the tour and nodes requiring a visit are simply skipped [60].

3.2.3 Postperson problem

The postperson problem is divided into two topics: the Chinese postperson problem and the rural postperson problem.

3.2.3.1 Chinese postperson problem

This section focuses on the **Chinese postperson problem** and its variations which involves constructing a tour of the road network traveling along each road with the shortest distance. Starting at a given point, the postperson tries to find a route of minimum length to traverse each street at least once and leading back to the post office. The Chinese postperson problem seeks an optimal path that visits a pre-defined subset of graph edges at least once. An optimal path is defined as the lowest cost coverage path given the current graph information. Typically, the road network is mapped to an undirected graph $G = (V, E)$ and edge weights $w : E \rightarrow \mathbb{R}^+$, where the roads are represented by the edge set E and road crossings are represented by the node set V. Each edge is weighted with the length of the road or the amount of time needed to pass it. The Chinese postperson problem algorithm involves first constructing an even graph from the road network graph. This even graph has a set of vertices with an even number of edges attached to them. This is required as any traverse of the junction by approaching on one road and leaving on another, which means that only an even number of edges will produce an entry and exit pair for the tour. As the road network graph roads may have junctions with an odd number of edges, some roads are chosen for duplication in the graph. The technique chooses a set of roads with the shortest combined length to minimize duplication. The tour of the even graph is calculated by determining the Euler tour of the graph, which visits every edge exactly once or twice for duplicated edge. The Chinese postperson problem works well for applications where it is necessary to traverse every part of the space.

Algorithm 4 Chinese Postperson Problem Algorithm

1. **Input** : s (start vertex), G (connected graph where each edge has a cost value)
2. **Output** : P (value if tour found or empty is no tour found)
3. For $i \in 2, .., n-1$ do
4. if sEven(G), then P=FindEulerCycle(G,s)
5. else
6. $$\begin{aligned} O &= FindOddVertices(G) \\ O' &= FindAllPairsShortestPath(O) \\ Mate &= FindMinMatching(O') \\ G' &= (G, Mate) \\ P &= FindEulerCycle(G', s) \end{aligned}$$
7. end
8. Return P

The environment is initially known in the form of a prior map. This prior map is converted into a graph structure with goal locations as nodes in the graph and paths between goals as edges in the graph. The first step in solving the coverage problem is to assume the prior map is accurate and generate a tour that covers all the edges in the graph. This Chinese postperson problem pseudo-code can be represented as Algorithm 4. Its optimal tour consists of traversing all the edges in the graph at least once and starting and ending at the same node.

The first step in Algorithm 4 is to calculate the degree of each vertex. If all the vertices have even degree, then the algorithm finds an Euler tour using the End-Pairing technique pairing. If any vertices have odd degree, a minimum weighted matching among the odd degree vertices is found using an all pairs shortest path graph of the odd vertices. Because the matching algorithm requires a complete graph, all pairs' shortest path algorithm is a way to optimally connect all the odd vertices. The matching finds the minimum cost set of edges that connect the odd nodes. Finally, the algorithm finds a tour on the new Eulerian graph. The end-pairing technique is used to generate the Eulerian cycle from the graph, consisting of two steps:

1. First, it builds cycles that intersect at least one vertex.
2. Next, the cycles are merged together two at a time by adding one cycle onto another at the intersecting vertex.

Algorithm 5 Cycle Building Algorithm

1. **Input** : s (start vertex), G (graph)
2. **Output** : C (cycle found)
3. Begin
4. $C = s$
5. $i = s$
6. e = NextEdgeHeuristic(G, s, i)
7. i=OtherEndPoint(e, i)
8. While $i \neq s$ do
9. e = NextEdgeHeuristic(G, s, i)
10. i=OtherEndPoint(e, i)
11. C = [C; i];
12. RemoveEdge (G, e)
13. End
14. Return C

The cycle building step is shown in Algorithm 5. During each step of the algorithm, edges are added to a path sequence and removed from the graph until the starting node of the path is encountered. In the original end-pairing algorithm, the heuristic for choosing the next edge to add to the sequence consisted of picking a random edge incident to the current node. To maintain a small coverage environment, edges are chosen in such a way that the path travels away from the start and then travels back always visiting the farthest unvisited edges until it reaches the start. Essentially, the coverage path should be always walking along the boundary of the coverage sub-graph. This will allow the edges around the start to be as connected as possible while separating the coverage and travel sub-graphs.

When environmental changes are found online, re-planning is done on the updated graph with different starting and ending vertices. To remedy this disparity, an artificial edge is added from the current UAV location c to the ending vertex s in the graph. This edge (c, s) is assigned a large cost value to prevent it from being doubled in the solution. Using this modified graph, a tour from s to s is found. The edge (c, s) is then deleted from the graph and from the tour. The algorithm adjusts the coverage path to start at the current location and travel to the end location.

Algorithm 6 Online Coverage Algorithm

1. **Input**: s (start vertex), c (current vertex), G=(C,T) (graph where each edge has a label and a cost value), C is the subset of coverage edges, T is a subset of travel edges, OTP is a subset of optimal travel paths.
2. **Output** : tour found P
3. Begin
4. $G' = G$
5. If $c \neq s$, then $G' = [G, (c, s, INF]]$
6. If sConnected(C), then $P = CPP(s, G')$
7. else$P = RPP(s, G', OTP)$
8. If $c \neq s$ and $P \neq []$, then
9. RemoveEdge $[P, (c, s, INF)]$
10. end
11. Return P

Remark 78 *As shown in Algorithm 6, if the unvisited edges in the new prob-*

lem are connected, the Chinese postperson problem algorithm (CPP) is run otherwise, the rural postperson problem (RPP) is run.

The **Reeb graph** can be used as input to the Chinese postperson problem to calculate a Eulerian circuit, which consists of a closed path traversing every cell at least once. The Reeb graph is a construction which originated in Morse theory to study a real-valued function defined on a topological space. The structure of a Morse function can be made explicit by plotting the evolution of the component of the level set. The Reeb graph is a fundamental data structure that encodes the topology of a shape. It is obtained by contracting to a point the connected components of the level-sets (also called contours) of a function defined on a mesh. Reeb graphs can determine whether a surface has been reconstructed correctly, indicate problem areas, and can be used to encode and animate a model. The Reeb graph has been used in various applications to study noisy data which creates a desire to define a measure of similarity between these structures. A Eulerian circuit can be achieved by doubling selected edges of the Reeb graph, although no edge needs to be duplicated more than once. The Eulerian circuit is the solution of the linear programming problem:

$$\text{Minimize } z = \sum_{e \in E} c_e.x_e \tag{3.4}$$

subject to :

$$\sum_{e \in E} a_{ne}.x_e - 2w_n = k_n; \forall n \in V, x_e \in \mathbb{N}, \forall e \in E; w_n \in \mathbb{N} \tag{3.5}$$

where $\sum_{e \in E} a_{ne}.x_e$ is the number of added edges to node $n \in V$. To be Eulerian, an odd number of edges must be added to nodes with odd degree and an even number of edges must be added to nodes with even degree; a_{ne} is equal to 1 if node n meets edge e and 0 otherwise; x_e is the number of added copies of edge e in the solution; w_n is an integer variable that will force $\sum_{e \in E} a_{ne}.x_e$ to be odd for odd nodes and even for even nodes; k_n is 1 for nodes with odd degree and 0 otherwise; c_e is a real number representing the cost of edge e. To prevent repeat coverage, cells corresponding to doubled Reeb graph edges are split into non-overlapping top and bottom sub-cells. In [101], a cell-splitting scheme guarantee that sub-cells share the same critical points as their parent, by interpolating between the bounding critical points and between the top and bottom boundaries of the original cell. At the end of the analysis phase, the resulting Eulerian circuit outlines a cyclic path through all connected cells in the environment.

For the road network search using multiple UAVs, a variation of the typical Chinese postperson problem algorithm is required, so that it can consider the operational and physical characteristics of the UAV in the search problem. Since the fixed-wing UAV cannot change its heading angle instantaneously due to the physical constraint, the trajectory has to meet the speed and turn limits of the UAV.

Different metrics can be used, such as the Euclidean distance between the vertices, the number of connected coverage components in the graph, the number of optimal travel paths in the partition set, the number of branches in the search tree, the percentage of re-planning calls that are Chinese postperson problem rather than rural postperson problem calls, and the computation time in seconds [7]. Computational requirements are often a limiting factor in the capability of real time trajectory optimization algorithm. As UAVs perform more complex missions with higher levels of autonomy, trajectory optimization algorithms need to be efficient and have the flexibility to adapt to different missions.

The Chinese postperson problem has a lot of variations such as the **capacitated Chinese postperson problem** (CCPP) which capacitates the total edge cost or the **rural Chinese postperson problem** which visits certain roads but not necessarily all of them, the **windy Chinese postperson problem** which has different values for the same edge according to the direction. The **k-CPP** deals with the deployment of several postmen.

Problem 79 *The k-CPP can be formulated as follows: given a connected edge-weighted graph G and integers p and k, decide whether there are at least k closed walks such that every edge of G is contained in at least one of them and the total weight of the edges in the walks is at most p?*

Min-Max k-CPP (MM-k-CPP) algorithms are described for multi-agent road network search. **MM k-CPP** is a variation of k-CPP which considers the route of the similar length. This objective can be required if the UAV should finish road search mission with the minimum mission completion time.

3.2.3.2 Rural postperson problem

In many practical applications, it is not necessary to traverse every part of the space. The **rural postperson problem** (RPP) seeks a tour that traverses a required subset of the graph edges using the extra graph edges as travel links between the required edges. Optimal solutions exist that formulate the rural postperson problem as an integer linear program and solve it using **branch and bound**. Another approach introduces new dominance relations, such as computing the minimum spanning tree on the connected components in a graph to solve large problem instances. Additionally, many TSP heuristics have been extended to the rural postperson problem [98]. There are two sets of graph edges: required and optional. The required edges are defined as coverage edges and the optional as travel edges. Any solution would include all coverage edges and some combinations of travel edges.

Definition 80 *A* ***coverage or travel vertex*** *is a vertex in the graph incident to only coverage or travel edges, respectively. A border vertex is a vertex* *in the graph incident to at least one coverage edge and at least one travel edge. A travel path is a sequence of travel segments and travel vertices connecting a pair of border vertices.*

Branch and bound is a method of iterating through a set of solutions until an optimal solution is found:

1. In the **branching step**, the algorithm forms n branches of sub-problems where each sub-problem is a node in the branch-and-bound tree. The solution to any sub-problem could lead to a solution to the original problem.

2. In the **bounding step**, the algorithm computes a lower bound for the sub-problem.

These lower bounds enable branch and bound to guarantee a solution is optimal without having to search through all possible sub-problems. Branch and bound is a general method for handling hard problems that are slight deviations for low complexity problems. Finally, an **optimal travel path** (OTP) is a travel path connecting a pair of border vertices such that it is the lowest cost path between the vertices, and the vertices are not in the same cluster. Optimal travel paths are shortest paths between clusters of coverage segments that do not cut through any part of a coverage cluster. All the optimal travel paths are computed by finding the lowest cost path p_{ij}between each pair of border vertices v_i and v_j in different clusters. If p_{ij} is a travel path, it is saved as an optimal travel path. If it is not a travel path, then v_i and v_j do not have an optimal travel path between them (i.e., $p_{ij} = NULL$). The optimal travel path becomes the partition set. The iterations are set within the branch and bound framework, at each branch step, the algorithm generates a new sub-problem by either including or excluding an optimal travel path.

Pseudo-code of the rural postperson problem is presented in Algorithm 7. At the beginning, cost 0 is assigned to the unlabeled **optimal travel paths** and solves the problem with all the optimal travel paths as required edges using the Chinese postperson problem are shown in Algorithm 4. This problem and the Chinese postperson problem costs are pushed onto a priority queue. The Chinese postperson problem cost is the lower bound on the problem since all the optimal travel paths have zero cost. While the queue is not empty, the lowest cost sub-problem is selected from the queue. For the sub-problem, the algorithm selects an unlabeled optimal travel paths P_{ij} with the best path cost. By employing this strategy of closing the optimal travel path with the highest path cost, the aim is to increase the lower bound with the highest amount, which may help focus the search to the correct branch and prevent extraneous explorations. Once an optimal travel path P_{ij} is selected, two branches are generated:

1. The first branch includes P_{ij} in the solution, this time with real path cost assigned.

Algorithm 7 Rural Postperson Problem

1. **Input**: s (start vertex), G=(C,T) (graph where each edge has a label and a cost value), C is the subset of coverage edges, T is a subset of travel edges, OTP is a subset of optimal travel paths.
2. **Output** : P (tour found)
3. Begin
4. $PQ = []$
5. $G' = [G, OTP]$ where $\forall OTP, cost(p_{ij}) = 0$
6. $P = CPP(s, G')$, add to $PQ\,(PQ, [G', P])$
7. While ! is Empty (PQ) do
8. $[G', P]$= PopLowestCost(PQ);
9. P_{ij}= FindMaxOTP(G')
10. If $P_{ij} = []$, then return P;
11. G" = IncludeEdge (G', P_{ij})
12. P1 = CPP(s, G')
13. AddToPQ(pq, [G", P]
14. RemoveEdge e(G', P_{ij})
15. P2 = CPP(s, G")
16. AddToPQ(pq, [G", P2]
17. end
18. Return P

2. The second branch omits P_{ij} from the solution.

A solution to each branch is found using the CPP algorithm. Because each solution is generated with a cost of 0 assigned to the unlabeled OTP in the sub-problem, the costs of the inclusion and exclusion CPP solutions lower bounds on the cost of the RPP problem with and without using P_{ij} for travel respectively. These new sub-problems are added to the priority queue and the algorithm iterates until the lowest cost problem in the queue contains no optimal travel path. The solution to this problem is the optimal solution to the rural postperson problem since it has either included or excluded every single optimal travel path in the solution and has a path cost that is equal to or lower than the lower bounds of the other branches. The branch and bound algorithms for the RPP is an exponential algorithm with a complexity $O\left(|V|^3 2^t\right)$ where t is the number of optimal travel path and $|V|$ is the number of vertices in the graph.

While most research on arc routing problems focuses on the static environment, there has been work that address dynamic graphs such as the **dynamic rural postperson problem** (DRPP). Dynamic changes occur when the environment differs from the original map. There are two categories of planners that handle these differences [67]:

1. **Contingency planners** model the uncertainty in the environment and plan for all possible scenarios.

2. **Assumptive planners** presume the perceived world state is correct and plan based on this assumption.

If disparities arise, the perceived state is corrected and replanning occurs. The lower complexity assumptive planning is used in order to generate solutions quickly. In the presented planner, an initial plan is found based on the graph of the environment. As the UAV uncovers differences between the map and environment during traversal the algorithm propagates them into the graph structure. This may require a simple graph modification such as adding, removing or changing the cost of an edge. But it can also result in more significant graph restructuring. These changes may convert the initial planning problem into an entirely different problem. For the coverage problems, most changes in the environment are discovered when the UAV is actively traversing the space. These online changes are typically detected when the UAV is not at the starting location, but at a middle location along the coverage path. At this point, some of the edges have already been visited. Because it is not necessary to revisit the edges that are already traversed, the visited edges in the previous plan are converted to travel edges.

Cluster Algorithm The algorithm is based on the cluster first, route second. In the first step, the edge set E is divided into k clusters and then a tour for each cluster is computed. This algorithm pseudo-code can be represented as a constructive heuristic method and described by Algorithm 8.

Algorithm 8 Cluster Algorithm

1. Determine the set of k representative edges $f_1, .., f_k$ of cluster F_i for each vehicle. Let f_1 be the edge having the maximum distance from the depot and f_2 be the edge having maximum distance from f_1. The rest of successive edges are successively determined by maximizing the minimum distance to the already existing representatives. Then, the remaining edges are assigned to the cluster according to the weighted distance between e and f_i. Consider the distance between representative edges and depot, number of assigned edge to the cluster F_i and cost of the cluster.

2. Include edges for connectivity. Add edges between every vertex and depot and determine minimum spanning tree which includes original edges in each cluster for connection between the edges.

3. The rural Chinese postperson problem: Compute Chinese postperson problem route of required subset of edges out of total edges by using the conventional CPP.

Rural Chinese Postperson Problem Unlike the cluster algorithm, the first route algorithm follows a route first, cluster second. In a first step, postperson tour which cover all edges is computed, and then this tour is divided by k tour segments which have the similar length. This method is described in Algorithm 9.

3.2.4 Knapsack problem

Differently from ground vehicles, the UAV has to fly along the road only to cover a certain edge which is not connected. This modified search problem can be formulated as a **multi-choice multi-dimensional knapsack problem** which is to find an optimal solution minimizing flight time. Classical multidimensional knapsack problem is to pick up items for knapsacks for maximum total values so that the total resource required does not exceed the resource constraint of knapsacks. For applying multi-dimensional knapsack problem to the road network search, UAVs are assumed as the knapsacks, the roads to be searched are resources and limited flight time or energy of each UAV is capacity of knapsacks. Multi-dimensional knapsack problem formulation allows to consider the limitations of each UAV flight time and different types of roads, vehicles and minimum turning radius and get the sub-optimal solution of the coordinated road search assignment. Moreover, for fixed-wing UAVs, the Dubins path planning produces the shortest and flyable paths taking into consideration their dynamical constraints, thus the Dubins path is used to calculate the cost function of the modified search problem [6].

The classical knapsack problem is defined as follows:

Algorithm 9 Rural Chinese Postperson Algorithm

1. Compute an optimal postperson route C^* using the conventional CPP.
2. Compute splitting nodes: (k-1) splitting nodes $v_{p_1}, ..., v_{p_{k-1}}$ on C^* are determined in such a way that they mark tour segments of C^* approximately having the same length. Approximated tour segment length, L_j is computed by using the shortest path tour lower bounded s_{max},

$$s_{max} = \frac{1}{2} \max_{e=u,v \in E} w(SP(v_1, u)) + w(e) + w(SP(v_1, v_1)) \quad (3.6)$$

$$L_j = \frac{j}{k} \left(w(C^* - 2s_{max}) \right) + s_{max}, 1 \leq k \leq N - 1 \quad (3.7)$$

   where N denotes the number of UAVs, $w(\alpha)$ represents the distance of the sub-tour α and SP represents the shortest path between nodes considering road network. Then, the splitting node v_{p_j} is determined as being the last node such that $w(C^*_{v_{p_j}}) \leq L_j, C^*_{v_n}$ is the sub-tour of C^* starting at the depot node and ending at v_n.
3. k-postmen tours: construct k tours $C = (C_1, ..., C_k)$ by connecting tour segments with shortest paths to the depot node.

Problem 81 ***Knapsack problem:*** *Given an knapsack capacity $C > 0$ and a set $I = \{1, \dots, n\}$ of items, with profits $P_i \geq 0$ and weights $w_i \geq 0$, the knapsack problem asks for a maximum profit subset of items whose total weight does not exceed the capacity. The problem can be formulated using the following* ***mixed-integer linear program*** *(MILP):*

$$\max \left\{ \sum_{i \in I} p_i x_i : \sum_{i \in I} w_i x_i \leq C, x_i \in \{0, 1\}, i \in I \right\} \quad (3.8)$$

where each variable x_i takes the value 1 *if and only if item i is inserted in the knapsack.*

The two classic approaches for solving this problem exactly are branch and bound and dynamic programming [34]. These algorithms work in two phases: in the forward phase, the optimal value of the profit function is calculated and in the backtracking phase, an actual solution is determined using this optimal profit. Similar to the TSP, the knapsack problem has online and offline versions. Assuming the weights are identical among items, the offline version of the knapsack problem becomes trivial: a greedy algorithm can be used that selects the items with the maximal value one by one until no more items can be added. However, when the values of the items are not fully known ahead of time, the knapsack problem with identical weights is an interesting problem with many applications. The profitability of an object type is the ratio of its profit to weight. In the UAV domain, many objective functions may be incorporated, such as time, energy, flight time. Different and more complex models can be discussed including the binary multiple criteria knapsack problem, problems with many constraints and multi-period as well as time-dependent models.

3.3 ORIENTEERING

The problem of routing one UAV over **points of interest** (POI) can be formulated as an **orienteering problem**. The **orienteering problem** is a generalization of the traveling salesperson problem. It is defined on a graph in which the vertices represent geographical locations at which a reward can be collected. The orienteering problem includes the traveling salesperson problem and the knapsack problem [95].

Definition 82 *The **orienteering problem** is defined on a network where each node represents a point of interest and each arc represents travel between two nodes. Each node can be associated with a weight and each arc with a travel time. The goal is to find a maximum prize path with a travel time. The orienteering problem can be generalized to cases with time window constraints.*

Definition 83 *A **time window** is a time interval, which starts with the earliest time a task can start, and ends with the latest time the task can end. If the earliest time is not given, the latest time is referred to as the deadline. A time window is said to be closed if both start and end times are given.*

3.3.1 Orienteering problem formulation

The orienteering problems represent a family of vehicle routing problems which take into account practical situations where giving service to customers is optional and produces a profit if it is done within a time limit. Representing customers by vertices of a graph, the orienteering problem aims at selecting a subset of vertices and designing a route not longer than a pre-specified time distance limitation, to maximize the total collected profit. A team of uncapacitated vehicles located at a depot must give service to a set of regular customers, while another of optional customers is available to be potentially serviced. Each optional customer generates a profit that is collected if it is serviced. The problem aims at designing vehicle routes with a time duration not longer than a pre-specified limit, serving all regular customers and some of the optional customers to maximize the total profit collected [72].

3.3.1.1 Nominal orienteering problem

First, the nominal orienteering problem is considered in which all input parameters are assumed to be deterministic. If N represents the set of targets and $|N|$ its cardinality, the depot location, representing the UAV recovery point is denoted by the vertex $0 \notin N$ and $N^+ = N \cup \{0\}$, each target $i \in N$ is associated to a value p_i and the orienteering problem is formulated on a complete graph $G = (N^+, A)$ with $|N|+1$ vertices. A weight f_{ij}, representing the expected fuel consumption between targets i, j plus the expected fuel required to record target j, is associated to each arc $(i,j) \in A$. The fuel capacity of the UAV is denoted by F. A binary decision variable x_{ij} is introduced for every arc $(i,j) \in A$, $x_{ij} = 1$ if $arc(i,j)$ is used in the tour. An auxiliary variable u_i is introduced to denote the position of vertex i in the tour. The goal is to find a tour of maximum profit, feasible with respect to the fuel constraint which starts and ends at the recovery point. Based on these definitions, the formulation of the **nominal orienteering problem** is the following:

Problem 84 ***Orienteering problem*** *(OP)*

$$\max \sum_{i \in N} p_i \sum_{j \in N^+\{i\}} x_{ij} \tag{3.9}$$

subject to

$$\sum_{(i,j) \in A} f_{ij} x_{ij} \leq F \quad \textit{Capacity constraints} \tag{3.10}$$

$$\sum_{i \in N} x_{0i} = \sum_{i \in N} x_{i0} = 1 \quad \textit{The tour starts and ends at the depot} \tag{3.11}$$

$$\sum_{i \in N^+\{j\}} x_{ji} = \sum_{i \in N^+\{j\}} x_{ij} \leq 1, \forall j \in N \tag{3.12}$$

Flow conservation ensures that a vertex is visited at most once.

$$u_i - u_j + 1 \leq (1 - x_{ij})|N|, \forall i,j \in N \quad \textit{Prevents the construction of subtours} \tag{3.13}$$

$$1 \leq u_i \leq |N|, \forall i \in N \quad \textit{Boundary constraints} \tag{3.14}$$

$$x_{ij} \in \{0,1\}, \forall i,j \in N \quad \textit{Integrality constraints} \tag{3.15}$$

Remark 85 *The orienteering problem can be considered as a variant of the TSP. In contrast to the TSP, in which the goal is to minimize the tour length to visit all the targets, the orienteering problem objective is to maximize the total sum of the collected rewards while the reward collecting tour does not exceed the specified travel budget. Thus, the orienteering problem is a more suitable formulation for cases where visiting all the targets is unfeasible with the given travel budget.*

In operations research, the environmental coverage solution uses the representation of the environment as a graph, and using algorithms such as the traveling salesperson or postperson problems to generate optimal solutions. In the graph representation, nodes in the graph are locations in the environment and edges in the graph are the paths between the locations. Each edge has a cost assigned to it where the cost can represent measurements such as Euclidean distance between locations, terrain traversability, travel time or a combination of several metrics. Additionally, each edge is undirected. One possible approach of resolution is to seek a path that visits all the edges or a designated edge subset of the graph [61].

In order to improve the effectiveness of a reconnaissance mission, it is important to visit the largest number of interesting target locations possible, taking into account operational constraints related to fuel usage, weather conditions and endurance of the UAV. Given the uncertainty in the operational environment, robust planning solutions are required [31]. Some locations can be more relevant than others in terms of **information gathering** and therefore priorities are usually assigned to the locations. In order to optimize data collection, the UAV should fly a tour including target locations of higher priority, starting and ending at the recovery point. The tour planning can be modeled as the orienteering problem, wherein the target locations correspond to the nodes, profits are associated to the nodes to model the target location priorities, the arcs represent the flight path from one target location to the other and fuel consumption for such a flight path is modeled by the weight on the associated arc. The depot represents the recovery point of the UAV. Often there are obstacles present in the area being covered. A common technique is to consider obstacles in the environment as static. This allows the coverage area to be decomposed into unobstructed cells. Each cell is then treated individually and covered independently. The major disadvantage of this approach is the requirement of an a priori map of the environment and the obstacles within. The decomposed cells may well be too small for under-actuated vehicles to move around in. Other limitations include excessive overlap in the coverage and wasted time traveling from one cell to the other.

Approaches which deal with dynamic obstacles detect that an obstacle is present which should not have been there. In the cases of large obstacles, it might be better to split up the coverage area into cells. Another example is a **Voronoi** diagram where the paths are edges in the graph and the path intersections are nodes. This is one way to generate optimal paths for some of the problems in continuous space coverage [40]. In general, the routing approach consists in reducing a general routing problem to the shortest path problem. The specification of routes is a problem in its own right. If the route network is modeled as a directed graph, then the routing problem is the discrete problem of finding paths in a graph, which must be solved before the speed profile is sought [73].

3.3.1.2 Robust orienteering problem

In reality, the input parameters of a UAV planning problem may be uncertain. Since a UAV operates in a dynamic but uncertain environment, effective mission planning should be able to deal with environment changes and with changing expectations. In fact, weather circumstances like wind have a great impact on the fuel consumption of the UAV. Since replanning costs can be significant, it is important to generate UAV tour plans that are robust. So the **robust orienteering problem** (ROP) is introduced. In UAV mission planning, **sustainability** (robustness) of an initial plan is highly valued. More specifically, the flight plan that is constructed before the actual start of the UAV flight should be designed in such a way that the probability of visiting all planned targets is sufficiently high. The ROP is suitable for designing the initial plan of the UAV, since it provides a tool to balance the probability of feasibility of the initial plan against the profit value of the planned tour. The actual fuel realizations are not known yet at the planning stage. It will depend on these fuel realizations whether or not all planned targets can be visited during the actual flight of the UAV. In this paragraph, a formal description of the **ROP** is given. The ROP explicitly considers uncertainty in the weight of the arcs f_{ij}, whose realizations are assumed to lie in the interval $\left[\bar{f_{ij}} - \sigma_{i,j}, \bar{f_{ij}} + \sigma_{i,j}\right]$ where $\bar{f_{ij}}$ is the expected fuel consumption on the arc from target i to j. When modeling real life instances of the UAV problem, the expected fuel consumption $\bar{f_{ij}}$ is based on the current weather circumstances: the speed and the direction of the wind. Consequently, the possible correlations between the weights of the arcs are already captured in these expected values and thus the deviations from the expected values are assumed to be uncorrelated noise. The formulation of the ROP which incorporates robustness against fuel uncertainty is the following:

Problem 86 ***Robust orienteering problem*** *(ROP)*

$$\max \sum_{i \in N} p_i \sum_{j \in N^+\{i\}} x_{ij} \tag{3.16}$$

subject to

$$\sum_{(i,j) \in A} \bar{f_{ij}} x_{ij} + \sum_{s \in S} \rho_s \left\| y^s \right\|_s^* \leq F \quad \textit{Capacity constraints} \tag{3.17}$$

$$\sum_{s \in S} y_{ij}^s = \sigma_{ij} x_{ij}, \forall (i,j) \in A \tag{3.18}$$

$$\sum_{i \in N} x_{0i} = \sum_{i \in N} x_{i0} = 1 \quad \textit{The tour starts and ends at the depot} \tag{3.19}$$

$$\textstyle\sum_{i \in N^+\{j\}} x_{ji} = \sum_{i \in N^+\{j\}} x_{ij} \leq 1, \forall j \in N \tag{3.20}$$

Flow conservation ensures that a vertex is visited at most once.

$$u_i - u_j + 1 \leq (1 - x_{ij})|N|, \forall i,j \in N \quad \textit{Prevents the construction of sub-tours} \tag{3.21}$$

$$1 \leq u_i \leq |N|, \forall i \in N \quad \textit{Boundary constraints} \tag{3.22}$$

$$x_{ij} \in \{0,1\}, \forall i,j \in N \quad \textit{Integrality constraints} \tag{3.23}$$

$$y_{ij}^s \in \mathbb{R}, \forall s \in S, (i,j) \in A \tag{3.24}$$

Constraint ((3.17)) makes the problem in general non-linear. However, for the uncertainty sets such as the L^∞, L^2, L^1 ball and the intersections defined by these balls, the problem remain tractable. A planning approach that complements the robust UAV tour with agility principles can be introduced. Three different policies can be used:

1. The first approach selects the **most profitable target** out of the set of available targets.
2. The second one selects the **target with the highest ratio** between the profit value and the total expected fuel required, to fly to and record the target and fly back to the recovery point.
3. The third policy is a **re-optimization policy** where the expected fuel consumption is used to find the remaining part of the tour that would be optimal for the deterministic case with nominal fuel consumption. At each target, only the first target of the optimal nominal tour is selected as the target to visit next.

Additional details on this implementation can be found in [31]. The ROP can be extended with **agility principles** used to make decisions during the flight of the UAV in which all fuel realizations that have been revealed so far are taken into account. First, given the fuel realizations, the initial tour will be followed as long as possible. This might imply that the UAV has to return to its recovery point before all planned targets are reached. In case of beneficial fuel realization, the resulting extra fuel capacity can be exploited to increase the total profit value obtained during the entire mission. Since visiting the planned targets is of primary interest, additional targets are considered only after reaching the final target of the planned tour. The uncertain sets are designed to find a solution by taking a certain part of the uncertain parameters into consideration. Consequently robust optimization allows to tune the level of conservatism applied by the choice of the uncertainty set.

3.3.1.3 UAV team orienteering problem

In the team orienteering problem, a team of m UAVs are scheduled to serve a set of nodes, each node is associated with a reward. The objective is to maximize the total received reward while the travel time of each route must be not more than a time limit. The team orienteering problem is defined on a complete graph $G = (V, E)$ where $V = \{0, 1, \dots, n+1\}$ is the set of nodes and $E = \{(iV, j)|i, j \in\}$ is the set of edges, c_{ij} is the travel time of edge $(i, j) \in E$ and r_i is the reward of node i, a feasible path must start at node 0 and finish at node $n+1$, and its travel time cannot exceed a time limit T_{max}.

Problem 87 ***Team orienteering problem:*** *Let $R(x_k)$ be the total received reward of a path $x_k \in \Omega$. The problem is to maximize the total received reward:*

$$\max \sum_{x_k \in \Omega} R(x_k) y_k \tag{3.25}$$

subject to:

$$\sum_{x_k \in \Omega} a_{ik} y_k \leq 1; \quad i = 1, \ldots, n \tag{3.26}$$

$$\sum_{x_k \in \Omega} y_k \leq m; y_k \in \{0, 1\} \tag{3.27}$$

where Ω *is the set of all feasible paths,* $a_{ik} = 1$ *if path* x_k *visits node* i*, otherwise* $a_{ik} = 0$*,* $y_k = 1$ *if path* $x_k \in \Omega$ *is traveled otherwise* $y_k = 0$*.*

The first constraint means that each node only can be visited at most once, and the second constraint ensures that there are at most m paths in a feasible solution. Several exact algorithms have been developed, however, there is no algorithm currently able to find an optimal solution. An alternative approach is meta-heuristic, which aims to yield a satisfactory solution within reasonable time [49].

An extension of this problem is the multi-constraint team orienteering problem with multiple time windows. Often, soft time windows are used for which a penalty is incurred in the objective function in case of time window violations. However, in the case of UAV mission planning, there are hard time windows. In a dynamic situation, a UAV tour should be designed by taking into account the possibility that during the flight emergency recordings, new targets become known and should take place: time-sensitive targets, which have priority over the foreseen targets. They are only worthwhile visiting within a pre-defined time limit, the emergency reaction time defined by the mission type. Therefore, the UAV will be sent to such a target as soon as it appears if remotely possible, considering the uncertainty in the time required to travel to the target. In this dynamic situation, at each moment of the flight, the UAV is either on its way to a foreseen target, on its way to a new target that has just appeared, recording or waiting to start recording at a target, or on its way back to the depot. Profits are obtained by recording foreseen targets that are reached in time and for which the recording time could be completed, without interruption. The available mission time will be partly devoted to new targets, and partly to foreseen targets. In case a planned tour to foreseen targets is purposely located in the proximity of the locations where new targets are expected to appear, the UAV will likely be in time to record a new target if it appears. On the other hand, when the planned tour does not take the possible locations of new targets into account beforehand, the expected profit to be obtained from foreseen targets might be higher. It is therefore worthwhile to consider in advance how much emphasis should be put on the possible appearance of these new, time-sensitive targets, when designing the planned tour: the online stochastic UAV mission planning problem with time windows and time-sensitive targets.

A different class of fast algorithms based on a decomposition of the orienteering problem into a knapsack problem (assignment of tasks with values to a UAV with limited task budget) and a subsequent traveling salesperson problem (selection of most efficient route for assigned tasks) is developed in [27]. Fast knapsack algorithms are based on selecting tasks in terms of their marginal value per additional resource cost. For orienteering problems, the resource cost is hard to evaluate because it requires the solution of the subsequent traveling salesperson problem. An approach based on spanning trees which allow for estimation of increased resource costs, leading to a fast algorithm for selecting tasks to be performed by each UAV is thus developed in Algorithm 10.

Algorithm 10 Orienteering Knapsack Algorithm

1. Initialization: The attention should be restricted to all tasks that are within round-trip distance of the source node. The domain is partitioned circularly and equally and the total reward of tasks present in each sector is calculated. The goal is to start the trees with some geographic diversity, with anticipation of directions where the higher value tasks exist.

2. Tree growing: Starting from source vertex, a tree using a greedy knapsack approach is grown. A reward to connection cost ratio is used to select the appropriate next vertex. After no tree can add additional vertices without exceeding its budget estimate, the cost of a tour is refined for each tree, estimating the cost of a tour using the topological order imposed from a depth-first traversal of the tree.

3. Tour construction: The resulting tree from the first stage, tree-growing, identifies a set of vertices S with tasks that should be performed.

4. Tour improvement: The reward to incremental cost ratio of each unselected vertex with respect to the tours, and insert vertex in order of biggest cost ratio as long as the cost of the tour meets the budget constraint.

5. Tour refinement: After more vertices are inserted into the tour, the obtained tour may not be an optimal TSP tour on this enlarged vertex subset. In this case, a new tour must be found for the current vertex subset and repeat the tour improvement step until no further improvements can be made. This step is not needed in most cases, but it is a simple step.

The **Dubins orienteering problem** is a generalization of the orienteering problem for fixed-wing UAVs [68]. Its objective is to maximize the total collected rewards by visiting a subset of the given target locations by a fixed-wing UAV, while the length of the collecting tour does not exceed a given travel budget. Therefore, a solution of the Dubins orienteering problem requires determining particular heading angles at the target locations to minimize the length of Dubins maneuvers between the targets. Regarding computational complexity, the Dubins orienteering problem can be considered as more challenging than the Euclidean orienteering problem, as changing only one heading angle or target location in the reward collecting path usually enforces the change of all heading angles of nearby connected target locations.

The proposed solution in [32] starts by constructing a tour using the available information, and hence determining the first target to be visited. After completing recording at this target and if no time-sensitive target has appeared, the next foreseen target to visit is determined by re-planning the tour based on past travel and recording time realizations. This re-planning is based on the **maximum coverage stochastic orienteering problem with time windows** (MCS-OPTW). The MCS-OPTW planning approach provides a path from the current location to the depot, containing only foreseen targets. The next target planned to be visited is the first target in this path. In executing this re-planning procedure, the MCS-OPTW balances two objectives: maximizing the expected profit obtained by recording foreseen targets and maximizing the expected **weighted location coverage** (WLC) of the path. The weighted location coverage relates to the distance of the arcs to locations where new targets are expected to appear. Hence, by this second objective, the maximum coverage stochastic orienteering problem with time windows selects foreseen targets such that the UAV will be sent in the direction of areas where new targets are expected to appear. In both objectives, the expected values are determined based on pre-defined probability distributions of the travel and recording times.

3.3.2 UAV sensor selection

The **UAV sensor selection and routing problem** is a generalization of the **orienteering problem**.

Problem 88 ***UAV sensor selection and routing problem:*** *A single UAV begins at a starting location and must reach a designated destination prior to time T. Along with the starting and ending points, a set of locations exists with an associated benefit that may be collected by the UAV. Find the optimal path for a UAV orienteering with a given sensor.*

Mission planning can be viewed as a complex version of path planning where the objective is to visit a sequence of targets to achieve the objectives of the mission. The integrated sensor selection and routing model can be defined as a **mixed integer linear programming formulation** [61]. A successful path planning algorithm should produce a path that is not constrained by end points or heading that utilizes the full capability of the aircraft's sensors and that satisfies the dynamic constraints on the UAV.

In [43], a path planning method for sensing a group of closely spaced targets is developed that utilizes the planning flexibility provided by the sensor footprint, while operating within dynamic constraints of the UAV. The path planning objective is to minimize the path length required to view all of the targets. In addressing problems of this nature, three technical challenges must be addressed:

1. Coupling between path segments,
2. Utilization of the sensor footprint, and
3. Determination of the viewing order of the targets.

For Dubins vehicles, capabilities can be provided by discrete time paths which are built by assembling primitive turn and straight segments to form a flyable path. For this work, each primitive segment in a discrete step path is of specified length and is either a turn or a straight line. Assembling the left turn, right turn and straight primitives creates a tree of flyable paths. Thus the objective for the path planner is to search the path tree for the branch that accomplishes the desired objectives in the shortest distance. Other parametric curves can be used such as Cartesian polynomials, different kinds of splines, Pythagorean hodograph, etc. [17]. The learning real-time A^* algorithm can be used to learn which branch of a defined path tree best accomplishes the desired path planning objectives.

Another example follows.

Problem 89 *Given a set of stationary ground targets in a terrain (natural, urban or mixed), the objective is to compute a path for the reconnaissance UAV so that it can photograph all targets in minimum time, because terrain features can occlude visibility.*

As a result, in order for a target to be photographed, the UAV must be located where both the target is in close enough range to satisfy the camera's resolution and the target is not blocked by terrain. For a given target, the set of all such UAV positions is called the **target's visibility region**. The UAV path planning can be complicated by wind, airspace constraints, dynamic constraints and the UAV body itself occluding visibility. However, under simplifying assumptions, if the UAV is modeled as a Dubins vehicle, the targets visibility regions can be approximated by polygons and the path is a closed tour [63]. Then the 2D reconnaissance path planning for a fixed-wing UAV, also called Dubins vehicle because of its limited curvature, can be reduced to the following:

Problem 90 *For a Dubins vehicle, find a shortest planar closed tour that visits at least one point in each of a set of polygons. This is referenced to as the* ***polygon visiting Dubins traveling salesperson problem*** *(PVDTSP).*

Sampling-based road-map methods operate by sampling finite discrete sets of poses (positions and configurations) in the target visibility regions in order to approximate a polygon visiting Dubins traveling salesperson problem instance by a **finite-one in set traveling salesperson problem** (FOTSP).

The FOTSP is the problem of finding a minimum cost closed path that passes through at least one vertex in each of a finite collection of clusters, the clusters being mutually exclusive finite vertex sets. Once a road-map has been constructed, the algorithm converts the FOTSP instance into an **asymmetric traveling salesperson problem** (ATSP) instance to solve a standard solver can be applied.

The desired behavior for a fixed-wing UAV is to maximize sensor coverage of the target. The objective function that drives this behavior is a weighted sum of separate objectives:

$$\begin{aligned} J = \int_{t_0}^{t_f} \left(W_1 u_1^2 + W_2 u_2^2 + W_3 u_3^2 + W_4 u_4^2\right) + \\ + \int_{t_0}^{t_f} W_5 \left[(r_x - r_x^d)^2 + (r_y - r_y^d)^2 + (r_z - r_z^d)^2\right] \end{aligned} \tag{3.28}$$

where W_{1-5} are given weights, u_{1-4} are the control inputs, (r_x, r_y, r_z) the actual 3-D position and (r_x^d, r_y^d, r_z^d). The first four terms penalize control effort and the fifth term weighs the square of the distance to the target. Constraints on the problem include the equations of motion and dynamic limitations. The target position and velocity vectors are continuously provided to the algorithm. Between path planning computations, the winds can be assumed as constant. A standard **vehicle routing problem** can be considered but with demands which are probabilistic in nature rather than deterministic. The problem is then to determine a fixed set of routes of minimal expected total length which corresponds to the expected total length of the fixed set of routes plus the expected value of extra travel distances that might be required. The extra distance will be due to the possibility of demand on one or more routes may occasionally exceed the capacity of a vehicle and force it to go back to the depot before continuing on its route.

The following two solutions updating methods can be defined:

1. Under method $\mathbb{U}_a$, the UAV visits all the points in the same fixed order as under the a priori tour but serves only location points requiring service during that particular problem instance. The total expected distance traveled corresponds to the fixed length of the a priori tour plus the expected value of the additional distance that must be covered whenever the demand on the route exceeds vehicle capacity.

2. Method $\mathbb{U}_b$ is defined similarly to $\mathbb{U}_a$ with the difference that location points with no demand on a particular instance of the UAV tour are simply skipped.

If each point x_i requiring a visit with a probability p_i independently of the others has a unit demand and the UAV has a capacity q, then an a priori tour through n points has to be found. On any given instance, the subsets of points present will then be visited in the same order as they appear in the a priori tour. Moreover, if the demand on the route exceeds the capacity of the vehicle, the UAV has to go back to the recovery point before continuing on its route. The problem of finding such a tour of minimum expected total length is defined as a **capacitated probabilistic traveling salesperson problem**.

3.4 COVERAGE

A coverage algorithm in robotics is a strategy on how to cover all points in terms of a given area using a set of sensors and a robot. It describes how the UAV should move to cover an area completely, ideally in a safe, energy consuming and time effective manner taking into account the wind effect. Initially, the focus was coverage of structured and semi-structured indoor areas [71, 89]. With the introduction of GPS, the focus has turned to outdoor coverage; however, due to nearby buildings and trees, GPS signals are often corrupted. The unstructured nature of an outdoor environment makes covering an outdoor area with all its obstacles and simultaneously performing reliable localization a difficult task [62].

Remark 91 *The* ***coverage*** *of an unknown environment is also known as the* ***sweeping problem****, or mapping of an unknown environment. Basically, the problem can either be solved by providing ability for localization and map building first, or by directly deriving an algorithm that performs sweeping without explicit mapping of the area. Instead of a* ***measure of coverage****, an average event detection time can be used for evaluating the algorithm.*

It is a challenge to develop a system that can handle any coverage tasks in indoor and outdoor environments with various UAV platforms with different kinematic and dynamic constraints. Assessing each task and finding the most suitable algorithm for that task is necessary. Yet, some of these tasks can be grouped together in the same category allowing the usage of the same coverage approach. There are many challenges when working in the outdoor environment, such as weather conditions, lighting conditions and unstructured surroundings. In many real-world applications, there exists the added complication of obstacles within the coverage area. For safety reasons, it is important that the obstacles are avoided. Furthermore, large obstacles in the vicinity of operations tend to corrupt GPS signals, which is most commonly used for outdoor localization. Hence, it is desirable to employ alternative localization techniques.

Remark 92 *In the literature, the term* ***coverage*** *has also been applied to the problem of distributing a group of mobile sensor units within an environment such that their positioning achieves maximum coverage of an area of interest. Most of these approaches assume that the mobile units do not move after reaching their desired positions, unless the configuration of the environment is altered during run-time.*

The traveling salesperson problem aims to identify a least cost Hamiltonian tour on a given network. All nodes of the network must be visited exactly once in this traveling salesperson problem. Identifying a tour over a subset of nodes so that the others are within a reasonable distance of some tour stop can be more desirable. The aim of the **time constrained maximal coverage salesperson problem** is to find a tour visiting a subset of the demand points, so as to maximize the demand covered subject to a time constraint. The demands of the vertices that are on the tour are assumed to be fully covered, while only a percentage of the demand of a vertex is covered if it is not visited but is within a specified distance of some tour stop. Routing of UAVs for information gathering against intruders is a perfect application for this problem [65]. The first problem incorporating the coverage concept into a routing scheme is the covering salesperson problem. It is the problem of identifying a minimum length Hamiltonian tour over a subset of vertices in a way that every vertex not on the tour lies within a certain distance of some visited vertex. A generalization is to have an additional cost incurred for every node visited by the tour, while each node is associated with a weighted demand representing the minimum number of times it has to be covered. It can be classified into three categories: maximizing profit under a distance constraint, minimizing distance under a profit constraint, or a combination of distance minimization and profit maximization.

The coverage planning problem is related to the covering salesperson problem where an agent must visit a neighborhood of each city. Coverage algorithms can be classified as heuristic or complete depending on whether or not they provably guarantee complete coverage of the free space. The following classifications can be proposed [35]:

1. **Classical exact cellular decomposition methods** break the free space down into simple, non-overlapping regions called cells. The union of all the cells exactly fills the free space. These regions, which contain no obstacles, are easy to cover and can be swept by the UAV using simple motions.

2. **Morse-based cellular decomposition approach** is based on critical points of Morse functions. A Morse function is one whose critical points are non-degenerate. A critical point is a value where either the function is not differentiable or all its partial derivatives are zero. By choosing different Morse functions, different cell shapes are obtained.

3. **Landmark-based topological coverage approach** uses simpler landmarks to determine an exact cellular decomposition termed slice decomposition. Due to the use of simpler landmarks, slice decomposition can handle a larger variety of environments.

4. **Grid-based methods** use a representation of the environment decomposed into a collection of uniform grid cells. Most grid-based methods are resolution-complete, that is, their completeness depends on the resolution of the grid map.

5. **Graph-based coverage** is interesting for environments that can be represented as a graph. In particular, it can take into account that the prior map information provided as a graph might be incomplete, accounting for environmental constraints in the environment, such as restrictions in

certain directions in the graph and providing strategies for on-line re-planning when changes in the graph are detected by the UAV's sensors when performing coverage scenarios.

6. **Coverage under uncertainty scenarios** is useful when the lack of a global localization system such as GPS makes the UAV accumulate drift, and hence a growing uncertainty about its pose. Although the topological representations, such as the adjacency graph, are tolerant to localization error, the performance of coverage algorithms, even if using such representations, is still affected.

7. **Maximum weighted coverage problem** is useful when a limited number of bases should be open such that the combined coverage is maximized [50]. It is also known as the maximal covering location problem.

The **swap local search method** iteratively improves an initial feasible solution by closing a sub-set of opened bases and opening a different sub-set of bases with strictly increased coverage. The two sub-sets are chosen such that they have the same cardinality. The simplification steps, and the resulting instance transformations, can be characterized as follows:

1. Remove unnecessary bases and demand points,

2. Create a bijection between bases and demand points,

3. Use the inherent symmetry of the coverage problem, and

4. Simplify the structure of the instance.

Another classification for coverage can be proposed into three classes: **barrier coverage**, **perimeter coverage** and **area coverage**. Using UAVs needs a coverage path planning algorithm and a coordinated patrolling plan [3]. **Perimeter surveillance** missions can be approached as a patrolling mission along a defined path: the perimeter. On the other hand, **area surveillance** missions could be divided in two different problems: an area coverage path planning problem and a patrolling mission along that path. A frequency-based patrolling approach involves the frequency of visits as the parameter to optimize, if the area has to be covered again and again. Coverage path planning tries to build an efficient path for the UAVs which ensures that every point in the area can be monitored from at least one position along the path. The challenge is to maximize the area to cover with a given amount of sensors, assuming a defined coverage range for each sensor.

3.4.1 Barrier coverage

Among intruder surveillance applications, barrier coverage is a widely known coverage model to detect intruders. A barrier is a line of sensors across the entire field of interest. The sensing ranges of two neighbor sensors in the barrier are overlapped and thus the intruders are guaranteed to be detected. Barrier coverage minimizes the probability of undetected intruder passing through the arrangement in a static arrangement of UAVs forming a barrier [85].

Definition 93 *A path P is said to be 1-covered if it intercepts at least one distinct sensor.*

Definition 94 *A sensor network is said to be strongly 1-barrier covered if:*

$$P(\textit{any crossing path is 1-covered}) = 1 \tag{3.29}$$

Barrier coverage of sensor networks provides sensor barriers that guard boundaries of critical infrastructures [13]. Many considerations can be taken into account. Some approaches examined **weak/strong barrier** and **k−barrier** in terms of barrier detection capability, others focused on the network lifetime of barriers and attempted to extend a barrier's network lifetime by reducing the number of barrier members. Others were concerned about the cost of barrier construction and took advantage of the mobility of mobile camera sensors to reduce the number of sensors required for barrier construction. The k-barrier detects intruders by constructing k weak/strong barriers. The relationship between the number of sensors required and the success rate of the barrier construction can be analyzed to find the optimal number of sensors to scatter in a random deployment context. Collaboration and information fusion between neighboring sensors reduce the number of barrier members and prolong the network lifetime of barriers using a **sleep-wake-up schedule**.

3.4.1.1 Barrier coverage approach

The problem of barrier coverage is classified on the basis of the type of region [100]:

1. Barrier coverage along a landmark can be formulated along a line or a point on it. The problem of barrier coverage along line W with direction $\bar{\theta}$ is formulated as follows:

$$W = \left\{p \in \mathbb{R}^2 : u^T p = d_1\right\}, \bar{\theta} = \beta + \pi/2 \tag{3.30}$$

   where $u = [\cos\beta, \sin\beta]$ is a unit vector with a given $\beta \in [-\pi/2, \pi/2]$ measured with respect to x-axis and d_1 is a given scalar associated with W. The UAVs are supposed to make a barrier of length L from W and are evenly deployed maximizing the length L.

2. Barrier coverage between two landmarks: In this problem, a barrier of sensors is supposed to ensure coverage between two landmarks. The problem of barrier coverage between two landmarks L_i and L_j is formulated as follows.

Problem 95 *Let u be a unit vector associated with L_i and L_j with $u = \frac{L_i - L_j}{\|L_i - L_j\|}$. The unit vector $u = [\cos\beta, \sin\beta]^T$ for some $\beta \in [-\pi/2, \pi/2]$ characterizes the bearing of L_j relative to L_i. An associated scalar is also defined as $\bar{\theta} = \beta + \pi/2$. A line L using L_i and L_j is defined as: $L = \{p \in \mathbb{R}^2 : (L_j - L_i)^T u^{\perp} = 0\}$. n points h_i are defined on L as: Let there be n UAVs and two distinct landmarks $L1$ and $L2$. Then, a decentralized control law for barrier coverage between the landmarks is formulated if for almost all initial sensor positions, there exists a permutation $\{z_1, z_2, \ldots, z_n\}$ of the set $\{1, 2, \ldots, n\}$ such that the following condition holds: $lim_{k\to\infty} \|p_{z_i}(kT) - h_i\|$.*

The problem of barrier coverage can be categorized on the basis of approaches used:

1. **Nearest neighbor rule**: In the basic method, one simply stores each training instance in memory. The power of the method comes from the retrieval process. Given a new test instance, one finds the stored training case that is nearest according to some distance measure, notes the class of the retrieved case, and predicts the new instance will have the same class. Many variants exist on this basic algorithm. Another version retrieves the k-closest instances and bases predictions on a weighted vote, incorporating the distance of each stored instance from the test case; such techniques are often referred to as k-nearest neighbor algorithms. However, the simplicity of this method does not mean it lacks power.

2. **Artificial potential field**: The main feature of this approach is its scalar potential field which represents both a repulsive force from obstacles and an attractive force to the goal. Therefore, a UAV's path from its starting position to the goal is found by threading the valleys of the potential field. The principle is quite simple and gives good results in many cases. However, since local minima of the potential are sometimes produced, the UAV is trapped before it can reach the goal in such cases. In addition, the generation of a potential field may require a large computation time in a complicated environment that includes concave objects. However, many possibilities of hybridation exist allowing avoiding this local minima.

3. **Virtual force field**: This technique is predicated on the gravitational force field. It lies in the integration of two known concepts: certainty grids for obstacle representation and potential fields for navigation. It is popular because of its simplicity, online adaptive nature and real time prompting.

4. **Generalized non-uniform coverage**: This approach may concern the problem of border patrol, or adaptive sampling in 2-D or 3-D environments where the non-uniformity in the sampled field is dominant in

one dimension. This is also closely related to information gathering and sensor array optimization problems.

The set of n UAVs initially situated at arbitrary positions $x_1(0), \ldots, x_n(0)$ is located in the interval $[0,1]$. The density of information at each point is measured by a function $\rho : [0,1] \rightarrow (0,\infty)$, this function being bounded: $\rho_{min} \leq \rho \leq \rho_{max}$. The metric is defined by $d_\rho(a,b) = \int_a^b \rho(z)dz$. This metric expands regions where ρ is large and shrinks regions where ρ is small.

Definition 96 ***Set of points coverage:*** *The coverage of a set of points $x_1, \ldots, x_n$ relative to the density field ρ is defined as:*

$$\Phi(x_1, \ldots, x_n, \rho) = \max_{y \in [0,1]} \min_{i=1,;n} d_\rho(y, x_i) \tag{3.31}$$

The best (smallest) possible coverage is given by:

$$\Phi^* = \inf_{(x_1,\ldots,x_n) \in [0,1]^n} \Phi(x_1, \ldots, x_n, \rho) \tag{3.32}$$

One possible control law can be found in [58], using the fact that the non-uniform coverage problem can be made uniform by a transformation.

3.4.1.2 Sensor deployment and coverage

In coverage problem, sensors must be placed in a certain area to retrieve information about the environment. A model of the sensors needs to be elaborated since how the surroundings are perceived by the UAVs strongly affects the evolution of the algorithm. Most of the existing results on coverage consider sensors with symmetric, omnidirectional field of view and only recently agents with anisotropic and vision-based sensors have been considered. The major challenge in the real-world implementation of coverage algorithms lies in the communication and information exchange among the UAVs. Some definitions useful for the sequel are introduced.

Definition 97 ***Field of view:*** *The field of view of the optical sensor is defined as the extent of observable world that it can collect at any given time.*

In a planar application, the set $\mathbb{R}^2$ is partitioned as follows:

$$\begin{aligned} R_1 &= \left\{s \in \mathbb{R}^2 : s_x \leq 0\right\} \\ R_2 &= \left\{s \in \mathbb{R}^2 : s_x > 0 \text{ and } \|s\| \leq 1\right\} \\ R_3 &= \left\{s \in \mathbb{R}^2 : s_x > 0 \text{ and } \|s\| > 1\right\} \end{aligned} \tag{3.33}$$

Definition 98 ***Visibility:*** *The visibility is defined as:*

$$vis_{I_3}(s) = \left\{ \begin{array}{cc} 0 & s \in R_1 \\ s_x & s \in R_2 \\ \frac{s_x}{\|s\|} & s \in R_3 \end{array} \right\} \tag{3.34}$$

The visibility of all the points in R_1 is zero because the region R_1 lays on the back of the agent. The points in R_2 are on the front of the agent and close to it, so the visibility increases with the distance from the agent and with the centrality in the agent's field of view.

The problem of barrier coverage by stationary wireless sensors that are assisted by a UAV with the capacity to move sensors is considered. Assume that n sensors are initially arbitrarily distributed on a line segment barrier. Each sensor is said to cover the portion of the barrier that intersects with its sensing area. Owing to incorrect initial position, or the death of some of the sensors, the barrier is not completely covered by the sensors. Assume that n sensors $s_1, s_2, \ldots, s_n$ are distributed on the line segment $[0, L]$ of length L with endpoints 0 and L in locations $x_1 \leq x_2 \leq \cdots \leq x_n$. The range of all sensors is assumed to be identical, and is equal to a positive real number $r > 0$. Thus sensor s_i in position x_i defines a closed interval $[x_i - r, x_i + r]$ of length $2r$ centered at the current position x_i of the sensor, in which it can detect an intruding object or an event of interest. The total range of the sensors is sufficient to cover the entire line segment $[0, L]$, i.e., $2rn \geq L$. A gap is a closed sub-interval g of $[0, L]$ such that no point in g is within the range of a sensor. Clearly, an initial placement of the sensors may have gaps. The sensors provide complete coverage of $[0; L]$ if they leave no gaps [25].

3.4.2 Perimeter coverage

3.4.2.1 Coverage of a circle

In the coverage of a circle using a network of UAVs considered as mobile sensors with non-identical maximum velocities, the goal is to deploy the UAVs on the circle such that the largest arrival time from the mobile sensor network to any point on the circle is minimized. This problem is motivated by the facts that in practice the assumption of UAVs with identical moving speed often cannot be satisfied, and events taking place in the mission domain only last for a finite time period. When the sensing range of mobile sensors is negligible with respect to the length of a circle, reduction of the largest arrival time from a sensor network to the points on the circle will increase the possibility of capturing the events taking place on the circle before they fade away. To drive the sensors to the optimal locations such that the overall sensing performance of the sensor network is optimized, gradient descent coverage control laws based on **Voronoi** partition are developed for mobile sensors with limited sensing and communication capabilities to minimize a locational optimization function. A Voronoi diagram is a subdivision of a Euclidean space according to a given finite set of generating points such that each generating point is assigned a Voronoi cell containing the space which is closer to this generating point than to any other.

Voronoi diagrams

Definition 99 ***Planar ordinary Voronoi diagram:*** *Given a set of finite number of distinct points in the Euclidean plane, all locations in that space are associated with the closest member with respect to the Euclidean distance. The result is a* ***tessellation*** *of that plane into a set of regions associated with members of the point set. This tessellation is called a planar ordinary Voronoi diagram generated by the point set and the regions constituting the Voronoi diagram are Voronoi polygons. Voronoi edges are the boundaries of the polygons. An end-point of a Voronoi edge is a Voronoi vertex.*

Remark 100 *Other metrics than the Euclidean distance can be used in that definition. These Voronoi diagrams can be generalized in a variety of ways: weighting of the points, considering regions associated with sub-sets of points rather than individual points, including obstacles in the space, considering regions associated with sets of geometric features other than points, examining Voronoi diagrams on networks and on moving points [64].*

Centroidal Voronoi diagrams

Definition 101 ***Centroid:*** *Given a region $U \subset \mathbb{R}^N$ and a density function ρ defined in U, the centroid z^* of U is defined by:*

$$z^* = \frac{\int_U y\rho(y)dy}{\int_U \rho(y)dy} \tag{3.35}$$

The ***Voronoi regions*** $\left\{\hat{U}_i\right\}$, *given k points $z_i, i = 1, \ldots, k$ in U, are defined as:*

$$\bar{U}_i = \{s \in U; |x - z_i| < |x - z_j|; j = 1, \ldots, k, j \neq i\} \tag{3.36}$$

The ***centroidal Voronoi tessellation*** *corresponds to the situation where:*

$$z_i = z_i^*, i = 1, \ldots, k \tag{3.37}$$

Identification of bubbles

Definition 102 ***Bubbles*** *are the set of points of a mesh such that the value of a field ρ (for example the density) is below a pre-defined threshold for all points in the set and each point of the set can be reached starting from any point of the same set by moving along the directions of the model.*

Two bubbles never share points, otherwise they are fused in a single bubble. The identification of such type of bubbles is like the problem of **cluster-labeling.** A cluster-labeling procedure assigns unique labels to each distinct cluster of lattice points. Once a bubble has been identified with a unique label, its center of mass can be identified. Delaunay triangulation, involving only topological properties, can be computed starting from the circumcenters of the triangles. To differentiate nodes that lie at the boundary between two

Voronoi cells from those that lie at the vertices, the number of gaps along the frame are counted. It is necessary to distinguish between nearest neighbors and second neighbors. Finally, the comparison of two different triangulations leverages a suitable isomorphism linking two triangulation graphs. Given the adjacency list, the differences can be checked and the location of topological changes can be extracted [11].

A benchmark problem of 1-D coverage is the uniform coverage problem in which the distance between neighboring agents is required to reach a consensus. The sensing performance of a homogeneous sensor network is maximized when the sensors are uniformly deployed on a line or circle provided that the information density of all points on the line or circle is identical. In contrast, when the **density of information** is not uniform over the mission space, uniform coverage is generally undesirable and more sensors should be deployed in areas with high information density. Problems that are closely related to the coverage control problem on a circle include circular formation and multi-agent consensus. In the circular formation problem, a team of mobile agents is required to form a formation on a circle and the desired distance between neighboring agents is generally prescribed a priori. In contrast, in the coverage control problem the desired distance between sensors is unknown beforehand and depends on the coverage cost function to be optimized. In [79], a distributed coverage control scheme is developed for heterogeneous mobile sensor networks on a circle to minimize the coverage cost function while preserving the mobile sensors order on the circle. The difficulties caused by the heterogeneity of mobile sensors maximum velocities are:

1. For a network of mobile sensors with identical maximum speed, the optimal configuration is uniform deployment of the sensors on the circle. However, it is still unclear under what conditions the coverage cost function is minimized when a network of heterogeneous mobile sensors is deployed.

2. Different constraints are imposed on the mobile sensors' control inputs due to the existence of non-identical maximum velocity for each sensor. This complicates the proof of mobile sensors' order preservation and convergence analysis of the distributed coverage control scheme, by taking into consideration input constraints and order preservation of the mobile sensors.

3. Environmental conditions, such as wind and lightning conditions, should also be taken into account.

3.4.2.2 *Dynamic boundary coverage*

Bayesian search focuses on how to estimate the target's motion and position based on probability theory. The assumptions of Bayesian search are that the search area can be divided into finite cells/graphs and that each cell represents individual **probability of detection** (PD). The goal is to determine the optimal path to find the lost or moving target according to the **probability distribution function** (PDF). The three steps of **Bayesian** search for a target are as follows:

1. Compute prior PDF according to motion information (e.g., flight dynamics and drift data).

2. Compute posterior PDF according to sensor information.

3. Move to the highest probability cell, scan this area, and update the posterior PDF as the prior PDF.

The three steps are repeated until the target is found. Two optimization objectives are as follows: to maximize the probability of detection or minimize the **expected time to detection** (ETTD). The advantages of Bayesian search are that the imperfect sensing-detection and target's motion can be modeled as a probability distribution and the probability of each cell is updated according to real-time sensing data [90].

In **dynamic boundary coverage task**, a team of UAVs must allocate themselves around the boundary of a region or object according to a desired configuration or density. It is dynamic boundary coverage, in which UAVs asynchronously join a boundary and later leave it to recharge or perform other tasks. In **stochastic coverage schemes** (SCS), UAVs probabilistically choose positions on the boundary.

1. Stochastic coverage schemes enable a probabilistic analysis of the graph for different classes of inputs identified by the joint PDF of UAV positions.

2. Stochastic coverage schemes allow to model natural phenomena, such as **random sequential adsorption** (RSA), the clustering of ants around a food item, and Renyi Parking, the process by which a team of cars parks without collisions in a parking lot.

Each UAV can locally sense its environment and communicate with other UAVs nearby. UAVs can distinguish between other robots and a boundary of interest, but they lack global localization: highly limited on-board power may preclude the use of GPS, or they may operate in GPS-denied environments. The UAVs also lack prior information about their environment. Each UAV exhibits random motion that may be programmed, for instance to perform probabilistic search and tracking tasks, or that arises from inherent sensor and actuator noise. This random motion produces uncertainty in the locations of UAV encounters with a boundary. When a UAV attaches to the boundary, it selects an interval of the boundary of length R lying completely within the boundary.

Definition 103 *A* ***stochastic coverage scheme*** *is the choice of multiple* *random points on the boundary. Formally, a stochastic coverage scheme is a 1-D* ***point process*** *(PP) realized on the boundary.*

A special case of a point process involves UAVs attaching to a boundary at predefined locations. In the **Poisson point process** (PPP), UAVs attach independently to the boundary, one of its generalizations is the Markov Process. The independent attachments in PPP make them easy to analyze. On the other hand, interactions between UAVs are harder to handle and require generalizations of PPP [53]. The Poisson Voronoi diagram is an infinite Voronoi diagram refering to the situation in which points are located in space at random, according to the homogeneous PPP.

3.4.3 Area coverage

3.4.3.1 Preliminaries

The **area coverage** problem with UAVs can be solved from two different approaches:

1. In the online coverage algorithms, the area to cover is unknown a priori, and step-by-step, has to discover obstacles, compute their paths and avoid collisions. Both Voronoi spatial partitioning and coverage can be handled in a distributed manner, with minimal communication overhead.
2. In the offline algorithms, the UAVs have a map of the area and the obstacles, and can plan the path to cover the whole area.

The usefulness of a map not only depends on its quality but also on the application. In some domains certain errors are negligible or not so important. That is why there is not one measurement for map quality. Different attributes of a map should be measured separately and weighed according to the needs of the application [76]. Those attributes can include:

1. **Coverage**: how much area was traversed/visited.
2. **Resolution quality**: to what level/detail are features visible.
3. **Global accuracy**: correctness of positions of features in the global reference frame.
4. **Relative accuracy**: correctness of feature positions after correcting (the initial error of) the map reference frame.
5. **Local consistencies**: correctness of positions of different local groups of features relative to each other.
6. **Brokenness**: how often is the map broken, i.e., how many partitions of it are misaligned with respect to each other by rotational offsets.

The objective of area coverage algorithm studies is:

Problem 104 *Given a region in the plane, and given the shape of a cutter, find the shortest path for the cutter such that every point within the region is covered by the cutter at some position along the path.*

In a multi-UAV system, each UAV is equipped with communication, navigation and sensing capabilities allowing it to become a node of the mobile sensor network which can be steered to the region of interest. The coordination problem involves two challenges:

1. What information should the UAVs interchange?

2. How should each UAV read that information?

Coverage path planning is the determination of a path that a UAV must take in order to pass over each point in an environment.

Definition 105 ***Coverage and k-coverage:*** *Given a set of n UAVs* $U = \{u_1, u_2, \ldots, u_n\}$ *in a 2-D area* X. *Each UAV* $u_i (i = 1, \ldots, n)$ *is located at coordinate* (x_i, y_i) *inside* X *and has a sensing range of* r_i *called sensing radius. Any point* X *is said to be covered by* u_i *if it is within the sensing range of* u_i *and any point in* X *is said to be* $k-$*covered if it is within at least* k *UAVs' sensing ranges.*

The goal of the directional $k-$coverage problem in camera sensor network is to have an object captured by k cameras.

Definition 106 ***Connectivity:*** *When two UAVs* u_i *and* u_j *are located inside* X, u_i *and* u_j *are connected if they can communicate with each other.*

Definition 107 ***General sensibility:*** *The general sensibility* $S(u_i, P)$ *of* u_i *at an arbitrary point* P *is defined as:*

$$S(u_i, P) = \frac{\lambda}{d(u_i, P)^K} \tag{3.38}$$

where $d(u_i, P)$ *is the Euclidean distance between the UAV* u_i *and the point* P, *and positive constants* λ *and* K *are sensor technology-dependent parameters.*

When a UAV has only to reach some point in an orbit around a location to cover a surrounding region, by placing and sequencing orbit centers minimizing their number and inter-orbit motion costs, efficient coverage may be obtained. At a high level, the strategy may be described as [26]:

1. Tessellate regions to be covered with sensing-footprint-sized circles;

2. Provide visitation points to graph tour generator, producing an orbit sequence;

3. Choose the nearest point on the tour for each vehicle and tour direction;
4. Follow the orbit sequence with each vehicle in parallel;
5. Return UAVs to their start locations.

A class of complex goals impose **temporal constraints** on the trajectories for a given system, referred to also as temporal goals. They can be described using a formal framework such as **linear temporal logic** (LTL), **computation tree logic** and $\mu-$**calculus**. The specification language, the discrete abstraction of the UAV model and the planning framework depend on the particular problem being solved and the kind of guarantees required. Unfortunately, only linear approximations of the UAV dynamics can be incorporated. Multi-layered planning is used for safety analysis of hybrid systems with reachability specifications and motion planning involving complex models and environments.

Coverage path planning Coverage path planning determines the path that ensures a complete coverage in free workspace. Since the UAV has to fly over all points in the free workspace, the coverage problem is related to the covering salesperson problem:

1. **Static**: it is a measure of how a static configuration of UAVs covers a domain or samples a probability distribution.
2. **Dynamic**: it is a measure of how well the points on the trajectories of the sensor cover a domain. Coverage gets better and better as every point in the domain is visited or is close to being visited by an agent.
3. **Uniform**: uses metric inspired by the ergodic theory of dynamical system. The behavior of an algorithm that attempts to achieve uniform coverage is multi-scale.
4. In **persistent monitoring**, the goal can be to patrol the whole mission domain while driving the uncertainty of all targets in the mission domain to zero [88]. The uncertainty at each target point is assumed to evolve non-linearly in time. Given a closed path, multi-agent persistent monitoring with the minimum patrol period can be achieved by optimizing the agent's moving speed and initial locations on the path [98].

Features of large size are guaranteed to be detected first, followed by features of smaller and smaller size [81, 83, 88].

Definition 108 *A system is said to exhibit* **ergodic dynamics** *if it visits every sub set of the phase space with a probability equal to the measure of that subset. For a good coverage of a stationary target, this translates to requiring that the amount of time spent by the mobile sensors in an arbitrary set be proportional to the probability of finding the target in that set. For good coverage* *of a moving target, this translates to requiring that the amount of time spent in certain tube sets be proportional to the probability of finding the target in the tube sets.*

A model is assured for the motion of the targets to construct these tube sets and define appropriate metrics for coverage. The model for the target motion can be approximate and the dynamics of targets for which precise knowledge is not available can be captured using stochastic models. Using these metrics for uniform coverage, centralized feedback control laws are derived for the motion of the mobile sensors. For applications in environmental monitoring with a mobile sensor network, it is often important to generate accurate spatio-temporal maps of scalar fields, such as temperature or pollutant concentration. Sometimes, it is important to map the boundary of a region.

A different strategy towards coverage can be presented as the environment is partitioned with the objective of minimizing the amount of rotation the UAV has to perform, rather than minimizing the total traveled distance. An improved strategy is to reduce the total length of the coverage path. In another approach, a set of pre-calculated motion strategies are chosen in order to minimize repeating coverage. A grid-based approach can be used for planning a complete coverage path using a spanning tree formulation. The primary requirement for a solution to exist is that the environment must be decomposable into a grid with a pre-determined resolution. Coverage can also be achieved by partitioning the environment into a grid. The spanning tree techniques can be extended to achieve efficient grid-based coverage using a team of multiple UAVs [102]. In general, most of the techniques used for the distributed coverage of some region are based on cellular decomposition. The area to be covered is divided between the agents based on their relative locations. Two methods for cooperative coverage, one probabilistic and the other based on an exact cellular decomposition, can be discussed. Domains with non-uniform traversability can also be examined [70].

To facilitate the means for autonomous inspection, a UAV must be equipped with the necessary accurate control units, the appropriate sensor systems and the relevant global path planning intelligence. Such path planning algorithms should be able to lead to the quick computation of efficient paths that result in full coverage of the structure to be inspected, while respecting the on-board sensor limitations as well as the vehicle motion constraints that may apply. Due to the nature of the coverage problem, its inherent difficulties still pose hard limitations on the performance, efficiency and practical applicability of the proposed solutions especially when 3-D structures are considered. The path planning algorithm employs a two-step optimization paradigm to compute good viewpoints that together provide full coverage while leading to a connecting path that has a low cost. In order to enable path planning for real 3-D structures, advanced algorithms employ a two-step optimization scheme proved to be more versatile with respect to the inspection scenario.

1. Compute the minimal set of viewpoints that covers the whole structure: **art gallery problem** (AGP).
2. Compute the shortest connecting tour over all these viewpoints: traveling salesperson problem.

A recent application of these concepts allows some redundancy in the AGP such that it is able to improve the path in a post-processing step. This algorithm can deal with 3-D scenarios. Close-to-optimal solutions are derived at the inherently large cost of computational efficiency. Within every iteration, the set of updated viewpoint configurations is selected such that combining all viewpoints full coverage is achieved, and the cost-to-travel between the corresponding vehicle configuration and the neighboring viewpoint configurations gets reduced. Subsequently, the optimally connecting and collision-free tour is re-computed. The proposed approach selects one admissible viewpoint for every triangle in the mesh of the structure to be inspected. In order to compute viewpoints that allow low-cost connections, an iterative re-sampling scheme is employed. Between each re-sampling, the best path connecting the current viewpoints is re-computed. The quality of the viewpoints is assessed by the cost to connect to their respective neighbors on the latest tour. This cost is minimized in the subsequent re-sampling, resulting in locally optimized paths. Initialization of the viewpoints is arbitrarily done such that full coverage is provided with, at this stage, non-optimized viewpoints [18].

Remark 109 *A large class of surveillance problems for passive targets in mapped regions can be converted to traveling salesperson problems, which are solved offline to provide several alternative solutions stored offline for online use [46]. An end-to-end framework is constructed, integrating existing algorithms where available, and developing others where necessary, to solve this class of problems. In an instance of the traveling salesperson problem, the distances between any pair of n points are given. The problem is to find the shortest closed path (tour) visiting every point exactly once. This problem is often encountered in robotics and has been traditionally solved in two steps with the common layered controller architecture for UAVs [57, 74].*

A strategic level regarding the choice of the way-points is made before the operational decisions regarding routing of UAVs between the different way-points. These two decisions are strongly related by factors such as UAV range and endurance, topography and communications, and the mission requirements. Different approaches have been developed such as a method which creates a spanning tree and generate the coverage paths as the boundary around it. Many different types of basic path patterns have been proposed for coverage algorithms. The most common patterns are parallel swath, also known as parallel milling, or zig-zag patterns. The seed spreader algorithm describes an efficient, deterministic, and complete coverage strategy for simple regions, by having the robot move in **back and forth**, or **lawnmower motion**, or **sweeping motions**. The following standard search patterns are:

1. **Lawnmower Search** which consists in flying along straight lines with 180 degree turns at the end. Based on the sweep direction, there are two types of lawnmowers:
   (a) **Parallel Track Search** if the search area is large and level, only the approximate location of the target is known and uniform coverage is desired;
   (b) **Creeping Line Search** if the search area is narrow and long and the probable location of the target is thought to be on either side of the search track;
2. **Spiral Search** and **Expanding Square Search** if the search area is small and the position of the target is known within close limits;
3. **Sector Search** used similarly to the expanding square search, it offers several advantages: concentrated coverage near the center of the search area, easier to fly than the expanding square search, and view of the search area from many angles;
4. **Contour Search** used to patrol obstacles, often assumed to be polygonal.

The **Boustrophedon cellular decomposition** (BCD) algorithm is an extension to the seed spreader algorithm which guaranteed complete coverage of bounded environments, with a variety of control Morse functions. Other typical patterns include inward and outward spirals, random walks and wall following or contour following. Unfortunately, these planners need absolute localization or not taking UAV kinematics into account and not dealing efficiently with obstacles. In terms of tasks, the UAV motion is not the primary one but it is a necessity to perform the main coverage task. When multiple UAVs are used, a previous decomposition of the field to cover is required. Two approaches are commonly used: exact cell decompositions and approximate cell decomposition. After this task, the path for every vehicle to cover the area assigned is computed.

3.4.3.2 Boustrophedon cellular decomposition

The Boustrophedon cellular decomposition allows to obtain a coverage path. This technique implies the area division in smaller subareas that can be covered with a simple back and forth method. This method can be extended to a multi-UAV system with limited communications [23]. The Boustrophedon decomposition is a generalization of the trapezoidal decomposition valid for non-polygonal obstacles. The exact cellular decomposition is the union of non-intersecting cells composing the target environment. Each cell is covered by simple back and forth motions. Once each cell is covered, then the entire environment is covered. Therefore, coverage is reduced to finding an exhaustive path through a graph which represents the adjacency relationships of the cells in the Boustrophedon decomposition. The approach is resolved by Algorithm 11 [97].

Algorithm 11 Boustrophedon Cellular Decomposition

1. Decompose the accessible area of the workspace into non-overlapping cells.

2. Construct an adjacent graph where each vertex is a cell and each edge connects two vertices corresponding to two adjacent cells.

3. Determine an exhaustive walk through the adjacent graph based on a depth-first-like graph search algorithm such that each vertex is visited at least once. Let V be the list that represents a consecutive sequence of vertices of a walk through the adjacent graph, as follows:

   (a) Start with any cell resulting from the decomposition. Add it into V and mark it as visited.

   (b) Move to the first counterclockwise unvisited cell in the neighboring cells of the current cell. Add this cell into V and mark it as visited.

   (c) Repeat the preceding step until reaching a cell whose neighboring cells have all been visited.

   (d) Backtrack and add each visited cell into V until a cell with an unvisited neighboring cell is reached. Go to Step 3b.

   (e) If no cell with an unvisited neighboring cell is found during the backtracking process, then the walk is the consecutive sequence of vertices in V.

4. As V is determined, drive the UAV to start at the first cell corresponding to the first vertex of V and move it to the next cell based on V. Perform the coverage task by using a Boustrophedon motion only when the UAV enters an unvisited cell. Repeat moving and covering until the cell corresponding to the final vertex in V is reached.

Remark 110 *One good strategy is to have long motions in parallel with one of the edges of the workspace assuming that there is a bounding rectangle with an edge collinear with one of the edges from the polygon that addresses the workspace. To have a pattern with straight lines that are as long as possible reduces the number of changes in direction so as to reduce exploration time. Another one is to take the direction of the dominant wind into consideration in order to avoid the lateral wind effect. An alternative method is to align the direction of coverage directly with the distribution of the free space, under the assumption that the length of sweep lines will be maximized along the dominant axis of the free space. Given the eigenspace decomposition for free space pixel* *coordinates, the direction of coverage is set orthogonally to the heading of the eigenvector with the largest eigenvalue.*

A crucial parameter required by this algorithm is the **coverage footprint** which measures the spacing between consecutive parallel sweep lines in the back and forth motion path. Different factors contribute to the definition of the coverage footprint, depending on the intended application: the footprint width determines whether the UAV will arrive at the top or the bottom boundary of a completed cell. Another extra sweep line can be added to avoid this problem. The Boustrophedon family of algorithms ensures the complete coverage of an unknown environment, although none of these algorithms provide any guarantees on the optimality of the coverage path.

Remark 111 *A complete coverage strategy for unknown rectilinear environments using a square robot with contact sensing can perform an online decomposition of free space, where each resulting rectangular cell could be covered using back-and-forth seed spreader motions that are parallel to the walls of the environment.*

The stochastic trajectory optimization motion planning is used in [36] to re-shape the nominal coverage path so that it adapts to the actual target structure perceived on-site via on-board sensors. This algorithm explores the space around an initial trajectory by generating noisy trajectories, which are then combined to produce an updated trajectory with lower cost in each iteration. It optimizes a cost function based on a combination of smoothness and application-specific costs, such as obstacles, constraints, etc. General costs for which derivatives are not available can be included in the cost function as this algorithm does not use gradient information.

3.4.3.3 Spiral path

An alternative to the Boustrophedon motion is the spiral motion pattern. These types of motion are applicable to multiple scenarios: regular or irregular areas, low or high density of obstacles. The spiral is defined by three parameters: initial point, initial direction, and pattern direction. Different exploration areas can be studied in order to determine these parameters. The spiral algorithm has been designed to start inside the area and perform counter-clockwise rotations to the perimeter of the area. Beginning within the area is useful for performing a search in a completely unknown terrain, but it can be inefficient if the UAV is initially deployed on the border of the field.

1. The initial point (C_x, C_y) is the node where the UAV starts the spiral path. The location of this node depends on the features of the area (shape or parity) and it has an influence on other features of the spiral (initial direction, pattern direction, or final point).

2. The initial direction defines the first movement of the UAV. The algorithm uses only two values for this parameter: clockwise when the area

is a square or a rectangle with more columns than rows, and counter-clockwise when the area is a rectangle with more rows than columns. These initial selections are defined in this way in order to reduce the number of turns.

3. The pattern direction is defined as the sense of movement where the path gets closer to the area limits. This ensures that each loop of the spiral is closer to the boundary than the previous one. If the motion is in other directions, the distance to the area limits will not change. The distance between the path and the boundary only decreases when the robot is moving in the pattern direction.

For example, in [5] the problem of locating a UGV using a UAV is considered, maximizing the probability of finding the ground vehicle. The length of the spiral between the UAV and ground vehicle at time $T = 0$ is an important parameter. Knowing the description of the spiral $r = m\theta$, its length L is given by:

$$L = \frac{m}{2}\theta_{max}\sqrt{1+\theta_{max}^2} + \frac{1}{2}ln\left|\theta_{max} + \sqrt{1+\theta_{max}^2}\right| \tag{3.39}$$

With this equation and the UAV's average speed V_a it is possible to determine the time for the UAV to navigate to the start point of the ground vehicle:

$$t = \frac{L}{V_a} \tag{3.40}$$

The location of the ground vehicle after the elapsed time needs to be predicted. One method for mapping the possible location includes describing circles with radius r:

$$r = V_g t \tag{3.41}$$

The focal length f can be determined as follows:

$$f = \frac{ccd_y/2}{\tan(\pi/7)} \tag{3.42}$$

The angle of view of the camera can be defined as:

$$\begin{aligned} aov_x &= 2\arctan ccd_x/(2f) \\ aov_y &= 2\arctan ccd_y/(2f) \end{aligned} \tag{3.43}$$

Now, the field of view can be described as:

$$\begin{aligned} fov_x &= h\tan aov_x \\ fov_y &= h\tan aov_y \end{aligned} \tag{3.44}$$

The maximizing factor to determine is the height of the UAV:

$$h = \max\left(\frac{fov_x}{\tan(aov_x)}, \frac{fov_y}{\tan(aov_y)}\right) \tag{3.45}$$

Maximizing the probability of capturing the ground vehicle implies that the field of view must be greater or equal to the circular area describing the possible location of the ground vehicle.

3-D terrain reconstruction 3-D terrain reconstruction is a two-level problem, the first one considers the acquisition of the aerial images and the second one considers the 3-D reconstruction. The camera is assumed to be mounted on the UAV to obtain images fully covering the area of interest. As these images will be later used for terrain reconstruction, several considerations arise:

1. **Overlapping**: Consecutive pictures should have a given percentage of overlapping. The greater the overlap is, the higher the accuracy of the 3-D model will be.
2. **Time contiguity**: The quality of the 3-D texture will be higher when the pictures of contiguous areas of the terrain are taken at similar time. Otherwise, uncorrelated shadows or visual differences may appear, leading to a more difficult reconstruction and a less quality texture.
3. **Orientation**: It is desired to have the pictures taken in the same orientation because it leads to a simplification in the 3-D reconstruction phase.

The UAV will be always heading to the same direction and will move sideways and backwards when it is required. The pictures' orientation will be the same for all pictures taken, facilitating the reconstruction problem. The algorithm returns a path that will be traversed as a zigzag or back and forth motion composed by longitudinal (the rows), transverse and possibly slightly diagonal moves [87]. In order to generate back and forth motion, the distance between two rows is needed and its calculation depends on the defined vertical overlap and the camera footprint. Let v be the vertical overlap and let w be the width of the camera footprint. The distance d among rows is the vertical distance between both footprints. Taking into account the vertical overlapping d is calculated as follows:

$$d = w.(1 - v) \tag{3.46}$$

The number of turns, n, to photograph a given polygon depends on the values of d, w, l_s where l_s is the length of the optimal line sweep direction given by Algorithm 12 for a single convex polygon coverage.

For each segment, two turning points are needed, thus leading to a total number of turns given by:

$$n = \left\{ \begin{array}{ll} 2.[z/d] & \text{if } z \bmod d \leq w/2 \\ 2.([z/d] + 1) & \text{if } z \bmod d \leq w/2 \end{array} \right\} \tag{3.47}$$

where $z = l_s - w/2$. The number of turns depends on z because d is fixed at the problem's formulation according to the resolution required for the images. A coverage alternative is defined according to two criteria: if the current direction of the line sweep or the opposite one is considered; and the way the coverage path is constructed: clockwise (the first turn is made to the right) or counterclockwise (the first turn is made to the left).

Algorithm 12 Optimal Line Sweep Direction

1. distance(e,v): Euclidean distance between edge e and vertex v
2. **For all** edges in the polygon **do**
   (a) max-dist-edge =0.
   (b) **for all** vertex in the polygon **do**
      i. **If** distance(edge, vertex) $\geq$ max-dist-edge **then**
      ii. max-dist-edge = distance (edge, vertex)
      iii. opposed-vertex = vertex
      iv. **end if**
   (c) **end for**.
3. **If** (max-dist-edge $\leq$ optimal-dist) or (is first edge) **then**
   (a) optimal-dist=max-dist-edge
   (b) line-sweep=direction FROM edge TO opposed vertex
   (c) **end if**
4. **End for**

Signal-searching approach The **signal-searching approach** is layered in two procedures: coverage area optimization and coverage path planning. The overall procedure for the task proposed is as follows:

1. Find a solution to the small area enclosed rectangle for the workspace shape.
2. Compute the obtained bounding rectangle orientation.
3. Perform a point sampling of the field according to sensor coverage range.
4. Compute the motion pattern with a determined orientation.

The pseudo-code of this procedure is given in Algorithm 13 [38].

If an unshaped field is considered, the main direction used in the discretization process becomes highly relevant for reducing the number of cells to be visited by the UAV during the mission. Indeed, the area to cover might be optimized if this angle is considered as exploring orientation. Moreover, this angle can have an influence on the mission definition because it can be used together with the navigation information for guidance purposes. That is, it allows for smoother paths, and it ensures that the UAV goes from way-point to way-point in a straight line. The **small area enclosed rectangle** problem refers to finding the smallest rectangle (i.e., the minimum area rectangle) that

Algorithm 13 Signal Searching Algorithm

1. $r \leftarrow$ Effective area to cover (polyline)
2. $\Phi \leftarrow$ Extract orientation(r)
3. $x \leftarrow$ Waypoint sampling (r, sensor range)
4. $p \leftarrow$ Motion pattern (x, Φ)
5. Return p

encloses a polygon regardless of its shape. This approach can be solved in $O(n)$ with methods such as rotating calipers or minimum bounding rectangle. The UAV's heading during the searching is obtained from the small area enclosed rectangle main direction [93].

3.4.3.4 Distributed coverage

Distributed coverage control is widely studied on continuous domains. A multi-UAV spanning tree coverage formulation can be applied over a general cell-based representation of free space. Others have applied genetic algorithms and visual landmarks to improve the speed of coverage. An information-theoretic path planner on a hexagonal grid representation of free space was also investigated, and a variant of the Chinese postperson problem solved the problem of boundary coverage, in which the objective was to cover the immediate area around the boundary of obstacles within the environment. Another possible approach is to employ potential fields to drive each UAV away from the nearby UAVs and obstacles. Alternatively, a prevailing approach is to model the underlying locational optimization problem as a continuous p-median problem and to employ Lloyd's algorithm as presented in Algorithm 14. As such, the UAVs are driven onto a local optimum, i.e., a centroidal Voronoi configuration, where each point in the space is assigned to the nearest agent, and each UAV is located at the center of mass of its own region. Later on, this method was extended for UAVs with distance-limited sensing and communications and limited power, as well as for heterogeneous UAVs covering non-convex regions. Also, the requirement of sensing density functions was relaxed by incorporating methods from adaptive control and learning. Distributed coverage control was studied on discrete spaces represented as graphs. One possible approach is to achieve a centroidal Voronoi partition of the graph via pairwise gossip algorithms or via asynchronous greedy updates.

Definition 112 *The* ***Voronoi region*** V_i^* *of a sensor is defined by all points which are* closer *in the sense of the considered distance measure to that sensor than any other. For the Euclidean distance measure, the Voronoi region* V_i

associated with its generator p_i is:

$$V_i = \{q \in \mathbf{Q}, \|q - p_i\| \leq \|q - p_i\|, \forall j \neq i\} \tag{3.48}$$

Definition 113 *The* ***Voronoi partition*** V_i^* *of agent* i*, for the anisotropic case is*

$$V_i^* = \left\{q \in \mathbf{Q}, \|q - p_i\|_{L_i} \leq \|q - p_i\|_{L_i}, \forall j \neq i\right\} \tag{3.49}$$

The anisotropic Voronoi partition is not only determined by the sensor's position but also the sensor's orientation [41]

Definition 114 ***Centroidal Voronoi configuration:*** *Given the set of points* $\mathbf{P} \in \mathbf{Q}$*,* $C_{V_i^*}$ *is the center of mass (centroid) of an anisotropic Voronoi partition. A Voronoi tessellation is called an anisotropic centroidal Voronoi configuration if*

$$p_i = C_{V_i}; \forall i \tag{3.50}$$

i.e., the points P *serve as generators and also centroids for the anisotropic Voronoi tessellations.*

One problem with a simple centroidal Voronoi transformation algorithm is that it requires a convex and obstacle-free environment to ensure that the motion is always possible to the centroids, otherwise it may not converge to a steady state.

Alternatively, distributed coverage control on discrete spaces can be studied in a game theoretic framework. Sensors with variable footprints can achieve power-aware optimal coverage on a discretized space, and a team of heterogeneous mobile agents can be driven on a graph to maximize the number of covered nodes [103]. A multi-UAV coverage solution for a rectilinear polygonal environment focused on assigning partitions of the environment proportionally based on each UAV's capabilities. The algorithm divided the free space into simple regions, and focused on selecting a per-region coverage pattern that minimized the number of turns.

Optimal coverage Optimal coverage algorithms for general time-varying density functions can be used to influence a team of UAVs, making possible human operators interaction with large teams of UAVs. The density function could represent the probability of a lost person being at a certain point in an area, for a search and rescue mission. In order to talk about optimal coverage, a cost ϕ must be associated to a UAV configuration that describes how well a given area is being covered.

Problem 115 *The coverage problem involves placing* n *UAVs in* $D \subset \mathbb{R}^2$ *where* $p_i \in D, i = 1, \ldots, n$ *is the position of the* i^{th} *UAV. Moreover, the domain itself is divided into regions of dominance* $P_1, \ldots, P_n$*, forming a proper partition of* D*, where the* i^{th} *UAV has to cover* P_i*. The associated density function* $\phi(q, t)$ *assumed to be bounded and continuously differentiable in both arguments, captures the relative importance of a point* $q \in D$ *at time* t*.*

Algorithm 14 Lloyd's Method

1. Given a polygon P and a set of k generating points at positions z_i.
2. Loop
   (a) Construct the Voronoi diagram $\{V_i\}$ for the points $\{z_i\}$.
   (b) Compute the centroid $\{c_i\}$ for each Voronoi cell $\{V_i\}$.
   (c) Set each point $\{z_i\}$ to the position of its associated centroid $\{c_i\}$.
   (d) If this new set of points meets some convergence criterion, terminate.
3. end Loop

For non-slowly varying density function, timing information must be included in the motion of the UAVs, for example as in [55]:

$$\frac{d}{dt}\left(\sum_{i=1}^{n}\int_{v_i}\|q-c_i\|^2\,\phi(q,t)dq\right)=0 \tag{3.51}$$

The mass m_i and center of mass c_i of the ith Voronoi cell V_i is

$$\begin{aligned} m_{i,t}(p,t) &= \int_{V_i}\dot{\phi}(q,t)dq \\ c_{i,t} &= \frac{1}{m_{i,t}}\left(\int_{V_i}q\dot{\phi}(q,t)dq\right) \end{aligned} \tag{3.52}$$

The solution can be given by:

$$\dot{p}_i = c_{i,t} - \left(k+\frac{m_{i,t}}{m_{i,0}}\right)(p_i - c_{i,0}) \tag{3.53}$$

Zermelo-Voronoi diagram In many applications of UAV, significant insight can be gleaned from data structures associated with **Voronoi-like partitioning** [10]. A typical application can be the following: given a number of landing sites, divide the area into distinct non-overlapping cells (one for each landing site) such that the corresponding site in the cell is the closest one (in terms of time) to land for any UAV flying over this cell in the presence of winds. A similar application that fits in the same framework is the task of sub-dividing the plane into **guard/safety** zones such that a **guard/rescue aircraft** residing within each particular zone can reach all points in its assigned zone faster than any other guard/rescuer outside its zone. This is the **generalized minimum distance problems** where the relevant metric is the minimum intercept or arrival time. Area surveillance missions can also be addressed using a frequency based approach where the objective implies to optimize the elapsed time between two consecutive visits to any position known as the refresh time [4].

The construction of **generalized Voronoi diagrams** with time as the distance metric is in general a difficult task for two reasons:

1. The distance metric is not symmetric and it may not be expressible in closed form.

2. Such problems fall under the general case of partition problems for which the UAV's dynamics must be taken into account.

It is assumed that the UAV's motion is affected by the presence of temporally varying winds. Since the generalized distance of this Voronoi-like partition problem is the minimum time to go of the Zermelo problem, this partition of the configuration space is known as **Zermelo-Voronoi diagram** (ZVD). This problem deals with a special partition of the Euclidean plane with respect to a generalized distance function. The characterization of this Voronoi-like partition takes into account the proximity relations between a UAV that travels in the presence of winds and the set of Voronoi generators. The question of determining the generator from a given set which is the closest in terms of arrival time, to the agent at a particular instant of time, reduces the problem of determining the set of the Zermelo-Voronoi partition that the UAV resides in at the given instant of time: this is the **point location problem**. The **dynamic Voronoi diagram problem** associates the standard Voronoi diagram with a time-varying transformation as in the case of a time-varying winds. The **dual Zermelo-Voronoi diagram** problem leads to a partition problem similar to the Zermelo-Voronoi diagram with the difference that the generalized distance of the dual Zermelo-Voronoi diagram is the minimum time of the Zermelo problem from a Voronoi generator to a point in the plane. The minimum time of the Zermelo navigation problem is not a symmetric function with respect to the initial and final configurations.

Environment with polygonal obstacles Solutions of the coverage control problem based on partitioning the mission space overlook the fact that the overall sensing performance may be improved by sharing the observations made by multiple sensors. In addition, many approaches assume uniform sensing quality and an unlimited sensing range. A number of solution techniques are also based on a centralized controller, which is inconsistent with the distributed communication and computation structure of sensor networks. Moreover, the combinatorial complexity of the problem constrains the application of such schemes to limited-size networks. Finally, another issue that appears to be neglected is the movement of sensors, which not only impacts sensing performance but also influences wireless communication: because of the limited on-board power and computational capacity, a sensor network is not only required to sense but also to collect and transmit data as well. For this reason, both sensing quality and communication performance need to be jointly considered when controlling the deployment of sensors.

Problem 116 ***Coverage with polygonal obstacles:*** *The mission space* $\Omega \subset \mathbb{R}^2$ *is a non-self-intersecting polygon. The mission space may contain obstacles which can interfere with the movement of the sensor nodes and the propagation of event signals. The boundaries of these obstacles are modeled as* m *non-self-intersecting polygons properly contained in* Ω.

$$\max_s \int_\Omega R(x)P(x,s)dx$$

subject to $s_i \in F, i = 1, \dots, N$ *where*

$$P(x,s) = 1 - \prod_{i=1}^{N} \hat{p}_i(x,s_i)$$

and

$$\hat{p}_i(x,s_i) = \left\{ \begin{array}{ll} p_i(x,s_i) & \textit{if } x \in V(s_i) \\ \tilde{p}_i(x,s_i) & \textit{if } x \in \bar{V}(s_i) \end{array} \right\} \tag{3.54}$$

with

$$p_i(x,s_i) = p_0 \exp^{-\lambda_i \|x-s_i\|}$$

and $\tilde{p}_i(x,s_i) \leq p_i(x,s_i)$

A gradient-based motion control scheme is developed in [105] to maximize the joint detection probability of random events in such mission spaces, taking into account the discontinuities that are introduced by obstacles in the sensing probability models. The optimization scheme requires only local information at each node. A modified objective function is also proposed which allows a more balanced coverage when necessary.

Area coverage in non-convex environment Non-convex domains pose non-convex optimization problems with non-convex constraints. An approach based on the Lloyd algorithm and tangent bug algorithm is presented in [19]. The tangent bug algorithm is a local path planner with obstacle avoidance behavior. The control strategy is composed of two layers of abstraction:

1. The Lloyd algorithm provides goal updates based on successive computation of Voronoi regions and their centroids on the upper level. The geodesic distance helps keeping the UAV inside the environment region on its way to the target. The geodesic distance measure calculates the paths along the boundaries and avoids the obstacles.

2. Tangent bug algorithm plans the UAV path to the next centroid target position on the lower level. It is a simple but efficient sensor-based planner capable of handling unknown environments by using a range sensor. This algorithm shows both characteristic behaviors:

(a) Motion-toward-target, a form of gradient descent.

(b) Boundary-following, a form of exploration of obstacle boundary.

The description of the implemented navigation algorithm is detailed in Algorithm 15 while the tangent bug subroutine is given in Algorithm 16. The proposed control strategy computes the Lloyd algorithm using only the virtual generators.

Algorithm 15 Coverage for Non-Convex Environment

1. Set of UAVs $i = 1, \ldots, n$ with initial positions p_i^s in environment Q and each UAV is provided with:
   (a) Localization and knowledge of Q
   (b) Voronoi region computation
   (c) subroutine tangent bug algorithm
2. initialize at time $T_i^s = 0, g_i^{real} \leftarrow p_i^s, g_i^{virt} \leftarrow p_i^s$
3. Loop
   (a) Acquire positions p_i and $\left\{g_j^{virt}\right\}_{j=1}^{k}, j \neq i$ of k neighbors
   (b) Construct the local Voronoi region V_i associated with g_i^{virt}
   (c) Compute the mass centroid C_{V_i} of the Voronoi region and update virtual target position $t_i^{virt} \leftarrow C_{V_i}$
   (d) Run tangent bug algorithm
4. end Loop
5. compute the final Voronoi region associated with g_i^{real}

In distributed mobile sensing applications, networks of agents that are heterogeneous respecting both actuation as well as body and sensory footprint are often modeled by recourse to power diagrams: generalized Voronoi diagrams with additive weights. In [9], power diagrams are used for identifying collision-free multi-robot configurations, and a constrained optimization framework is proposed combining coverage control and collision avoidance for fully actuated disk-shaped robots.

Most weighted Voronoi cells assume the correct weightings are known a priori. In [69], an algorithm is proposed to adapt trust weightings online using only comparisons between a UAV's sensor measurements, and those of its neighbors. A measure of sensor discrepancy is integrated into a cost function for the team, and used to derive an adaptation law for each UAV to change its trust weightings on line, while simultaneously performing a Voronoi-based

Algorithm 16 Tangent Bug

1. Set of n UAVs in environment Q and each UAV $i = 1, \ldots, n$ is provided with:
   (a) Obstacle avoidance: sensing and computation
   (b) Virtual target t_i^{virt} and $var \leftarrow t_i^{virt}$
2. Loop 2
   (a) If V_i is an obstacle boundary Voronoi region, then
   (b) project t_i^{virt} to point p_i^* onto ∂Q and set $var \leftarrow p_i^*$
   (c) end if
   (d) Update real target position $t_i^{real} \leftarrow var$
   (e) Execute next motion step toward real target t_i^{real}, apply obstacle avoidance to drive to next position p_i
   (f) Update real generator position $g_i^{real} \leftarrow p_i$
   (g) simulate next motion step toward virtual target t_i^{virt} and update virtual generator position g_i^{virt}
3. end Loop
4. return virtual generator g_i^{virt}

coverage control algorithm. The weightings serve as an adaptive way to assess trust between UAVs and improve the overall sensing quality of the team.

Given, a convex or non-convex shaped area $A \subset \mathbb{R}^2$ decomposed approximately by a finite set of regular cells $C = \{c_1, \dots, c_n\}$ such that, $A \approx \bigcup_{c \in C} c$; a coverage trajectory P with a finite set of continuous way-points p, which can be written as $P = \bigcup_{p \in P} p$, where the way-points correspond to the centroid of a corresponding cell, and consequently a cell corresponds to an image sample, thus $dim(P) = dim(C)$; a team of UAVs with attitude and position control, and capable of way-point navigation. Each UAV is characterized by a position in $[X, Y, Z]$ and orientation $[\phi, \theta, \psi]$. The variable of interest to minimize is the number of turns performed in P, which correspond to the number of rotations made by a UAV around the z-axis (yaw movements). The area coverage problem oriented to mosaicking missions can be abstractly described as follows [92]:

Problem 117 *For each quad-rotor of the team, an optimal trajectory can be computed by:*

$$min_\psi \left(J(\psi) = K_1 \times \sum_{i=1}^{m} \psi_k^i + K_2 \qquad k \in \{0, \pi/4, \pi/2, 3\pi/4\} \right) \tag{3.55}$$

where $\psi_{\pm 3\pi/4} > \psi_{\pm \pi/2} > \psi_{\pm \pi/4} > \psi_0$ *and* $K_i \in \mathbb{R}$ *are weights such that* $K_2 > K_1$.

The geometric nature of the coverage problem necessitates the proper partitioning of the space among the nodes, using a proper metric so that each node's control action should be dependent on the part of the space instead of requiring global knowledge of the state of the network. Assuming homogeneous sensor networks, the region A can be partitioned and assigned among the n nodes into polygonal cells, $V_i, i = 1, \dots, n$. The tessellation is based on the standard Euclidean metric defined as:

$$V_i = \{q \in A; \|q - p_i\| \leq \|q - p_j\|, j = 1, \dots, n\} \tag{3.56}$$

indicating assignment of points in A to the nearest node in the Euclidean sense.

Definition 118 *The* ***geodesic Voronoi partitioning*** *is defined as:*

$$V_i^g = \{q \in A; d(q - p_i) \leq d(q - p_j), j = 1, \dots, n\} \tag{3.57}$$

It assigns parts of the space under surveillance among the nodes based on the geodesic rather than the Euclidean distance. In general, the boundaries of the geodesic Voronoi cells are comprised by line segments and hyperbolas. The geodesic Voronoi diagram consists of a tessellation of the environment A, while each geodesic Voronoi cell V_i^g is always a compact set, contrary to the cells produce by the Euclidean ones, when dealing with non-convex environments [84].

Expanding grid coverage In this paragraph, an approach is considered to coordinate a team of UAVs without a central supervision, by using only local interactions between the UAVs [7]. When this decentralized approach is used, much of the communication overhead is saved, the hardware of the UAVs can be simple and better modularity is achieved. A properly designed system should be readily scalable, achieving **reliability** and **robustness** through **redundancy**. A team must cover an unknown region in the grid that possibly expands over time. This problem is strongly related to the problem of distributed search after mobile or evading targets. In general, most of the techniques used for the task of a distributed coverage use some sort of cellular decomposition [51].

In the **cooperative cleaners** case study, a team of robots cooperate in order to reach the common goal of cleaning a dirty floor. A cleaning robot has a bounded amount of memory; it can only observe the state of the floor in its immediate surroundings and decide on its movement based on these observations. Consequently, the overall topology of the floor contamination is unknown to the robots. The robots use an indirect means of communication based on signals and sensing, and the desired goal of cleaning the whole floor is thus an emerging property of multi-robot cooperation.

1. In the **static cooperative cleaners problem**, the dirty shape of the floor does not grow due to spreading of contamination. The shape of the floor is a region in $\mathbb{R}^2$ represented as an undirected graph G. Let V be the set of vertices in G. Each element in V is a tile of the floor and is represented as a pair $v = (x, y)$. Let E be the set of edges in G, each edge is a pair of vertices (v, w) such that v and w are connected through a 4-neighbor relation. The dirty floor F_t is a sub-graph of G, where t represent the time. In the initial state, G is assumed to be a single connected component without holes or obstacles, and $F_0 = G$. All the cleaning robots are identical and there is no explicit communication between the robots (only broadcast and sensing actions in the local environment are allowed). All the cleaning robots start and finish their task in the same vertex. In addition, the whole system should support fault-tolerance: Even if almost all the robots cease to work before completion of the mission, the remaining ones will eventually complete the mission [82].

2. In the **dynamic cooperative cleaners problem**, the time is discrete. Let the undirected graph $G(V, E)$ denote a 2-D integer grid $\mathbb{Z}^2$ whose vertices have a binary property called contamination. Let $Cont_t(v)$ be the contamination state of the vertex v at time t, taking either the value 1 or 0. let F_t be the contamination state of the vertex at time t,

$$F_t = v \in G | Cont_t(v) = 1 \tag{3.58}$$

F_0 is assumed to be a single connected component. This algorithm will preserve this property along its evolution. Let a team of k UAVs that can move on the grid G be placed at time t_0 on F_0 at point $P_0 \in F_t$. Each UAV is equipped with a sensor capable of telling the contamination status of all vertices in the digital sphere of diameter 7. In all the vertices their **Manhattan distance** from the UAV is equal or smaller than 3. A UAV is also aware of other UAVs which are located in these vertices and all of them agree on a common direction. Each vertex may contain any number of UAVs simultaneously. When a UAV moves to a vertex v, it has the possibility of cleaning this tile (i.e., causing $Cont(v)$ to become 0). The UAVs do not have any prior knowledge of the shape or size of the subgraph F_0 except that it is a single and simply connected component. The contaminated region F_t is assumed to be surrounded at its boundary by a rubber-like elastic barrier, dynamically reshaping itself to fit the evolution of the contaminated region over time. This barrier is intended to guarantee the preservation of the simple connectivity of F_t crucial for the operation of the UAVs, due to their limited memory. When a UAV cleans a contaminated vertex, the barrier retreats in order to fit the void previously occupied by the cleaned vertex. In every step, the contamination spreads, that is if $t = nd$ for some positive integer n, then

$$\forall v \in F_t, \forall u \in \{4 - Neighbors(v)\}, Cont_{t+1}(u) = 1 \tag{3.59}$$

Here the term $4 - Neighbors(v)$ simply means that the 4 vertices are adjacent to vertex v. While the contamination spreads, the elastic barrier stretches while preserving the simple connectivity of the region. For the UAVs traveling along the vertices of F, the barrier signals the boundary of the contaminated region. The UAV's goal is to clean G by eliminating the contamination entirely. No central control is allowed and the system is fully decentralized: all the UAVs are identical and no explicit communication between the UAVs is allowed. An important advantage of this approach in addition to the simplicity of the UAVs is **fault tolerance**; even if some UAVs are no longer in good flying condition before completion, the remaining ones will eventually complete the mission, if possible. A cleaning algorithm is proposed in [7], for exploring and cleaning an unknown contaminated sub-grid F, expanding every d time steps. This algorithm is based on a constant traversal of the contaminated region, preserving the connectivity until the region is cleaned entirely. Until the conditions of completion of mission are met, each UAV goes through the following sequence of commands. The pseudo-code is presented in Algorithm 17.

Remark 119 *Any waiting UAVs may become active again if the conditions change. More details on this implementation can be found in [7].*

Algorithm 17 Cleaning Algorithm

1. First, each UAV calculates its desired destination at the current time.
2. Then each UAV calculates whether it should give a priority to another UAV located at the same vertex and wishes to move to the same destination.
3. When two or more UAVs are located at the same vertex and wish to move towards the same direction, the UAV that had entered the vertex first gets to leave the vertex, while the other UAVs wait.
4. Before actually moving, each UAV that had obtained a permission to move must now locally synchronize its movement with its neighbors, in order to avoid simultaneous movements which may damage the connectivity of the region.
5. When a UAV is not delayed by any other agent, it executes its desired movement.

Coverage control Three related problems of UAV movement in arbitrary dimensions are: coverage, search, and navigation. A UAV is asked to accomplish a motion-related task in an unknown environment whose geometry is learned by the UAV during navigation [20]. In the standard coverage control problem, the goal of the UAV team is to reach asymptotically a configuration where the agent positions $lim_{t \to \inf} p_i(t); i \in [n]$ minimize the following performance measure capturing the quality of coverage of certain events:

$$E_n(p) = E\left[min_{i \in [n]} f(\|p_i - Z\|)\right] \tag{3.60}$$

where $f : \mathbb{R}_{\geq 0} \to \mathbb{R}_{\geq 0}$ is an increasing continuously differentiable function. The random variable Z represents the location of an event of interest occurring in the workspace. To interpret relation (3.60), the cost of servicing an event at location z with a UAV at location π is measured by $f(\|p_i - Z\|)$, and an event must be serviced by the UAV closest to the location of this event. For example, in monitoring applications, $f(\|p_i - Z\|)$ can measure the degradation of the sensing performance with the distance to the event. In vehicle routing problems, this cost might be the time it takes a UAV to travel to the event location, i.e., $f(\|p_i - Z\|) = \|p_i - Z\| / v_i$, assuming enough time between successive events [56]. Typically, partitioning is done so as to optimize a cost function which measures the quality of service provided over all of the regions. Coverage control additionally optimizes the positioning of UAVs inside a region. Many algorithms for partitioning and coverage control in robotic networks build on Lloyd's algorithm on optimal quantizer selection through **centering and partitioning**. The basic pseudo-code is presented in Algorithm 14. There are also multi-agent partitioning algorithms built on

market principles or auctions. Distributed Lloyd methods are built around separate partitioning and centering steps [29].

Way-point coverage of time-varying graphs Dynamic graphs represent the UAV's highly dynamic networks. The **dynamic map visitation problem** (DMVP) is considered, in which a team of UAVs must visit a collection of critical locations as quickly as possible, in an environment that may change rapidly and unpredictably during the UAVs' navigation. The DMVP applies formulations of highly dynamic graphs or **time-varying graphs** (TVGs), to a graph navigation problem. When incorporating dynamics into dynamic map visitation problem, there are many options for how to constrain or model the dynamics of the graph. Dynamics can be deterministic or stochastic. The deterministic approach is also relevant for situations in which some prediction of changes is feasible, the graph must be connected at all times. Indeed, for complete map visitation to be possible, every critical location must be eventually reachable. However, in UAVs' application environments, at any given time the way-point graph may be disconnected. There are three classes of time-varying graphs, each of which places constraints on edge dynamics: edges must reappear eventually; edges must appear within some time bound; and edge appearances are periodic. These classes have proven to be critical to the time-varying graphs taxonomy.

Definition 120 *A **dynamic graph** is a five-tuple $G = (V, E, \tau, \rho, \Xi)$, where $\tau \subset \mathbb{T}$ is the lifetime of the system, the presence function $\rho(e,t) = 1$ means that edge $e \in E$ is available at time $t \in \tau$, and the latency function $\Xi(e,t)$ gives the time it takes to cross e if starting at time t. The graph $G = (V, E)$ is called the underlying graph of G, with $|V| = n$.*

The discrete case in which $\mathbb{T} = \mathbb{N}$, edges are undirected is considered, and all edges have uniform travel cost $\Xi(e,t) = 1$ at all times. If agent a is at u, and edge (u, v) is available at time τ , then agent a can take (u, v) during this time step, visiting v at time $\tau + 1$. As a traverses G, a both visits and covers the vertices in its traversal. A temporal sub-graph of a TVG G results from restricting the lifetime τ of G to some $\tau' \subset \tau$. A static snapshot is a temporal sub-graph throughout which the availability of each edge does not change, i.e., edges are static.

Definition 121 *$J = \{(e_1, t_1), \ldots, (e_k, t_k)\}$ is a **journey** $\Leftrightarrow \{e_1, \ldots, e_k\}$ is a walk in G (called the **underlying walk** of J), $\rho(e_i, t_i) = 1$ and $t_{i+1} \geq t_i + \Xi(e_i, t_i)$ for all $i < k$. The **topological length** of J is k, the number of edges traversed. The temporal length is the duration of the journey: (arrival date) - (departure date).*

Problem 122 *Given a dynamic graph G and a set of starting locations S for k agents in G, the time-varying graph foremost coverage or **dynamic map visitation problem** (DMVP) is the task of finding journeys starting at time*

0 for each of these k agents such that every node in V is in some journey, and the maximum temporal length among all k journeys is minimized. The decision variant asks whether these journeys can be found such that no journey ends after time t. Let $T = \sum_{i=1}^{m} t_i$.

For the DMVP minimization problem (G, S) and the corresponding decision problem $(G; S; t)$, input is viewed as a sequence of graphs G_i each represented as an adjacency matrix, with an associated integer duration t_i, i.e., $G = (G_1, t_1), (G_2, t_2), \ldots, (G_m, t_m)$, where G_1 appears initially at time zero [1].

Multi-UAV persistent coverage Persistent coverage differs from static and dynamic coverage in that the coverage of the environment persistently decays and the UAVs have to continually move to maintain the desired level, i.e., it requires repetition and redundant actions. Therefore, in this case, the task can generally never be completed. Dynamic coverage is defined by the use of a mobile sensor network and results from the consistent mobility of sensors. While sensors move around, locations that were uncovered at the beginning will be covered at a later time; therefore, a wider area is covered over time, and intruders that might never be discovered in a fixed sensor network can now be detected by mobile sensors.

Remark 123 *The main difference between multi-agent persistent monitoring and dynamic coverage lies in that dynamic coverage task is completed when all points attain satisfactory coverage level while the persistent monitoring would last forever.*

The **coverage level** can be seen as the quality of a measurement and, in this sense, persistent coverage is often addressed as persistent surveillance or environmental monitoring, especially with UAVs. This approach is more dynamic, flexible, and suitable for multiple source localization, but requires resolving many challenging technical problems such as endurance, planning, coordination, communication, cooperation, and navigation of all the vehicles [11]. The solutions to the persistent coverage problem intend to derive results which are applicable for infinite time. In [66], this problem is formulated in discrete time due to the fact that a distributed system requires discrete communications. The UAVs form a network defined by a communication graph $G^{com}(k) = (V(k), E(k))$. The vertices $V(k)$ of the graph are the positions $p_i(k) \in Q$ of UAV $i, i = 1, \ldots, N$ at time $k \geq 1$. An edge $(i, j) \in E(k)$ if $\|p_i(k) - p_j(k)\| \leq r^{com}$, where r^{com} is the communication radius, i.e., the maximum distance between two UAVs at which they can communicate. The neighbors of the UAV i at instant k are $N_i(k) = \{j = 1, \ldots, N; (i, j) \in E(k)\}$. The coverage of the environment, or coverage function or global map is modeled with a time-varying field $Z(q, k)$.

Problem 124 ***Persistent coverage:*** *Let* $Q \subset \mathbb{R}^2$ *be a bounded environment to be persistently covered by a team of N UAVs, assumed to be holonomic:*

$$p_i(k) = p_i(k-1) + u_i(k-1) \quad \textit{where } \|u_i(k)\| \leq u^{max}$$

The aim of the UAV team is to maintain a desired coverage level $Z^*(q) > 0, \forall q \in Q$. The coverage function at each time instant k is given by Equation (3.61):

$$Z(q,k) = d(q)Z(q,k-1) + \alpha(k) \tag{3.61}$$

where $0 < d(q) < 1$ is the decay gain. The UAV increases the value of the coverage by $\alpha(k)$.

In the first step, at each communication time k, each UAV generates its map-to-communicate $Z_i^{com}(k)$ as:

$$Z_i^{com}(k) = dZ_i(k) + \alpha_i(k) \tag{3.62}$$

Each UAV sends its map-to-communicate to its neighbors and receives their maps. With this information, the first step of the update is performed dividing the map into two parts: the coverage area $\Omega_i(k)$ and the rest of the map. The UAV updates each region according to:

$$\begin{array}{c} Z_i^-(k) = Z_i^{com}(k) + \sum_{j \in N_i(k)} \left(\max(Z_j^{com}(k) - dZ_i(k-1), 0\right) \quad \forall q \in \Omega_i(k) \\ Z_i^-(k) = \max\left(Z_j^{com}(k), dZ_i(k-1)\right) \quad \forall q \notin \Omega_i(k) \end{array} \tag{3.63}$$

To counteract the error of estimation, a second updating step is executed. At first, each UAV extracts the region of its coverage area that is overlapped with another UAV and then sends its coverage function in this region to its neighbors:

$$\beta_i(k) = \beta_i(q, p_i(k)) = \alpha_i(q, p_i(k)) \quad \forall q \in \Omega_i^0(k) \tag{3.64}$$

where $\Omega_i^0(k) = \{q \in \Omega_i(k) \bigcup \Omega_j(k) | j \in N_i(k)\}$ is the overlapped area of UAV i with its neighbors. The UAVs exchange the overlapped productions with their neighbors and with the received ones, they perform the final update:

$$Z_i(k) = \left\{ \begin{array}{cc} Z_i^-(k) & \forall q \in \Omega_i(k) \\ Z_i^-(k) - \max_{j \in N_i(k)} b_j(k) + \sum_{j \in N_i(k)} \beta_j(k) & \forall q \notin \Omega_i(k) \end{array} \right\} \tag{3.65}$$

The final step adds the contributions that are not first considered and ends the estimation, as shown in Algorithm 18.

The problem of dispatching UAVs with cameras to monitor road traffic in a large city can be considered. Often, UAVs have limited range and can stay in air only for a limited amount of time. Traffic events such as congestion tend to have strong local correlations, i.e., if the vehicle density at an intersection is high, the same is likely true at intersections that are close-by. Therefore, sequentially visiting intersections following the road network's topological structure may offer little incremental information. As UAVs are not restricted to travel along roads, routes with carefully selected, not necessarily adjacent, intersections can potentially offer much better overall traffic information per unit of traveled distance. Under such settings, the following question then naturally arises: how to plan the best tours for the UAVs so

Algorithm 18 Local Map Update

1. Calculate map-to-communicate $Z_i^{com}(k)$ with relation (3.62)
2. Communicate map to neighbors
3. Update local map $Z_i^-(k)$ with relation (3.63)
4. Extract overlapped production $\beta(q, p_i(k)$ with relation (3.64)
5. Communicate region to neighbors
6. Update local map $Z_i(k)$ with relation (3.65)

that they can collect the maximum amount of traffic information per flight? Due to spatial and temporal variations, such fields can be highly complex and dynamic. However, in applications involving large spatial domains, the underlying spatial domain often does not change. The observation allows to work with the premise that nearby nodes have mostly time-invariant spatial correlations, even though the overall field may change significantly over time. Exploiting these correlations, at any given time, it becomes possible to infer the field's value at a certain node from the values of adjacent nodes [104].

In [80], a multi-vehicle sampling algorithm generates trajectories for non-uniform coverage of a non-stationary spatio-temporal field characterized by varying spatial and temporal de-correlation scales. The sampling algorithm uses a non-linear coordinate transformation that renders the field locally stationary so that existing multi-vehicle control algorithm can be used to provide uniform coverage. When transformed back to original coordinates, the sampling trajectories are concentrated in regions of short spatial and temporal de-correlation scales. Employing a 2-D planner, a 3-D structure is approximated using multiple 2-D layers, treated individually.

Continuous target coverage Design of coordination algorithms for non-convex environments requires tackling several issues such as signal attenuation or visibility loss because of the obstacles. Different kinds of sensing devices have been classified in Euclidean/ Geodesic footprint (non-visibility) and visibility sensors. A transformation of non-convex domains into convex ones was one of the first techniques. Then the real trajectories of the UAVs are obtained through the inverse transformation. For a frontier-based exploration of a non-convex environment, the latter is transformed into a star-shaped domain. Another solution based on Lloyd's algorithm and a path planning method for deploying a group of nodes in a concave environment can also be presented. However, this method is not effective for all types of environments, since it maximizes the coverage of the convex hull of the allowable environment rather than the coverage of the environment itself. In another solution, each UAV moves at a direction determined by the repulsive forces received by other UAVs and/or obstacles. However, this control scheme may lead to sub-optimal topologies. This fact motivated the incorporation of an attractive force to the centroid of the respective Voronoi cell. The geodesic Voronoi diagram allows deployment in non-convex environments. The algorithm assumes that the sensing performance degrades according to the square of the geodesic, rather than the Euclidean distance. In another control strategy, each node is assumed to move to the projection of the geometric centroid of its geodesic Voronoi cell onto the boundary of the environment, the domain of interest considered unknown and therefore an **entropy metric** is used as a density function that allows the nodes to explore and cover the area at the same time. However, in the field of area coverage in non-convex environments, the visibility-based Voronoi diagram was introduced and an algorithm based on Lloyd's algorithm is implemented for a team of UAVs with unlimited range, omnidirectional visibility-based sensors. The 3-D case of that problem is also addressed for a team of UAVs equipped with omnidirectional visibility sensors of infinite sensing range. Non-smooth optimization techniques for the coordination of a homogeneous UAV swarm by minimizing the sensing uncertainty, given that the performance of the visibility-based sensors is reduced according to the square of the distance [48].

Partition of environment coverage The distributed territory partitioning problem for UAV networks consists of designing individual control and communication laws such that the team will divide a space into territories. Typically, partitioning optimizes a cost function which measures the quality of service provided by the team. Coverage control additionally optimizes the positioning of UAVs inside a territory. A distributed coverage control algorithm can be described for a team of UAVs to optimize the response time of the team to service requests in an environment represented by a graph. Optimality is defined with reference to a cost function which depends on the locations of the UAVs and geodesic distances in the graph. As with all multi-UAVs coordination applications, the challenge comes from reducing the communication requirements: the proposed algorithm requires only gossip communication, that is, asynchronous and unreliable pairwise communication [28].

Problem 125 *Given a team of N UAVs with limited sensing and communication capabilities, and a discretized environment, partition the environment into smaller regions and assign one region to each agent. The goal is to optimize the quality of coverage, as measured by a cost functional which depends on the current partition and the positions of the agents.*

Coverage algorithms for UAV deployment and environment partitioning can be described as dynamical systems on a space of partitions. **centering and partitioning** Lloyd's algorithm is an approach to facility location and environment partitioning problems. The Lloyd's algorithm computes centroidal

Voronoi partitions as optimal configurations of an important class of objective functions, called multi-center functions.

Definition 126 *An **N-partition** of Q, denoted by $v = (v_i)_{i=1}^N$ is an ordered collection of N sub-sets of Q with the following properties:*

1. $\bigcup_{i\in\{1,\ldots,N\}} v_i = Q$
2. *$int(v_i) \cap int(v_j)$ is empty for all $i, j \in \{1, \ldots, N\}$ with $i \neq j$ and*
3. *each set $v_i, i \in \{i = 1, \ldots, N\}$, is closed and has non-empty interior*

The set of $N-$partitions of Q is denoted by V_N.

Definition 127 *Let Q be a totally ordered set, and $p_i \in \mathbb{C}$ the **generalized centroid** of p_i is defined as:*

$$Cd(p_i) = min\left\{argmin_{i=1..N} H_1(h, p_i)\right\} \tag{3.66}$$

where $H_1(h, p_i)$ the one-center function is defined as:

$$H_1(h, p_i) = \sum_{k \in p_i} d_{p_i}(h, k)\phi_k \tag{3.67}$$

where ϕ_k is a bounded positive weight function.

Let $p = (p_1, ..., p_N) \in Q^N$ denote the position of N UAVs in the environment Q. Given a team of N UAVs and an $N-$partition, each UAV is naturally in one-to-one correspondence with a component of the partition; v_i is the dominance region of UAV $i \in \{1, \ldots, N\}$. On Q, a density function is defined to be a bounded measurable positive function: $Q \to \mathbb{R}_{\geq 0}$ and a performance function to be a locally Lipschitz, monotone increasing and convex function $f : \mathbb{R}_{\geq 0} \to \mathbb{R}_{\geq 0}$. With these notions, the multi-center function is defined as:

$$H_{multicenter}(v, p) = \sum_{i=1}^{N} \int_{v_i} f(||p_i - q||)\phi(q)dq \tag{3.68}$$

This function is well-defined because closed sets are measurable. The objective function $H_{multicenter}$ must be minimized with respect to both the partition v and the locations p.

Remark 128 *The distributed coverage law, based upon Lloyd's algorithm, has some important limitations: it is applicable only to UAV networks with synchronized and reliable communication along all edges of the Delaunay graph.*

In order to generate the Delaunay graph, some representative points are chosen on the boundaries of the computational domain. For this given set of boundary points, there exists a unique triangulation for 2-D cases or tetrahedralization for 3-D cases. The grid generated in such a way is defined as a Delaunay graph for the given moving grid problem. The graph covers the whole computational domain for the given configuration, including the interior elements. Such triangulation or tetrahedralization is unique, maximizing the minimum angle of a triangle or tetrahedral. The Delaunay graph provides a unique mapping from the given boundary points to a coarse unstructured grid. There exists a pre-determined common communication schedule for all UAVs and, at each communication round, each UAV must simultaneously and reliably communicate its position. Some questions arise as:

1. Is it possible to optimize UAVs positions and environment partition with asynchronous, unreliable, and delayed communication?

2. What if the communication model is that of gossiping agents, that is, a model in which only a pair of UAVs can communicate at any time?

A **partition-based gossip** approach is proposed, in which the robots' positions essentially play no role and where instead dominance regions are iteratively updated as in Algorithm 19. When two agents with distinct centroids communicate, their dominance regions evolve as follows: the union of the two dominance regions is divided into two new dominance regions by the hyperplane bisecting the segment between the two centroids [21].

Algorithm 19 Gossip Coverage Algorithm

For all $t \in \mathbb{Z}_{\geq 0}$, each agent $i \in 1, ..., N$ maintains in memory a dominance region $v_i(t)$. The collection $(v_1(0), ..., v_N(0))$ is an arbitrary polygonal N-partition of Q. At each $t \in \mathbb{Z}_{\geq 0}$, a pair of communicating regions, $v_i(t)$ and $v_j(t)$, is selected by a deterministic or stochastic process to be determined. Every agent $k \notin i, j$ sets $v_k(t+1) := v_k(t)$. Agents i and j perform the following tasks:

1. Agent i transmits to agent j its dominance region $v_i(t)$ and vice versa
2. Both agents compute the centroids $C_d(v_i(t))$ and $C_d(v_j(t))$
3. If $C_d(v_i(t)) = C_d(v_j(t))$, then
4. $v_i(t+1) := v_i(t)$ and $v_j(t+1) := v_j(t)$
5. Else
6. $v_i(t+1) := (v_i(t) \cup v_j(t)) \cup H_{bisector}(C_d(v_i(t)); C_d(v_j(t)))$
7. $v_j(t+1) := (v_i(t) \cup v_j(t)) \cup H_{bisector}(C_d(v_j(t)); C_d(v_i(t)))$

Detection of coverage holes **Delaunay** triangulation can be constructed to discover the topological properties of a network. Delaunay triangulation is an important data structure in computational geometry, which satisfies the empty circle property: For each side in Delaunay triangulation, a circle passing through the endpoints of this side can be determined without enclosing other points [62]. The proposed method of detect and localize coverage holes consists of four phases:

1. **Detection of coverage holes**: every node detects whether coverage holes exist around it based on a hole detection algorithm.
2. **Merging of coverage holes**: a merging method of holes can be provided to present the global view of a coverage hole by indicating the location and shape of an isolated coverage hole.
3. **Size estimation** of local coverage holes: the inscribed empty circles are used for the estimation of the size of every local coverage hole.
4. **Tree description**: for each isolated coverage hole, line segments are used to connect the centers of each pair of inscribed empty circles. If a separated tree can be recognized, the corresponding coverage hole that contains the tree can be exclusively determined.

Probabilistic approach Let S be the set of all the nodes representing the potential locations the UAV needs to visit. In each problem instance, only the nodes that belong to a subset s need to be visited. The probability that the set s requiring a visit is given by $p(s)$. If an a priori tour τ is available, $L_\tau(s)$ represents the length of the tour in which all the nodes in s are visited according to the order of the a priori tour, skipping nodes not requiring a visit, $E[L_\tau]$ represents the expected length of the a priori tour τ

$$E\left[L_\tau(s)\right] = \sum_{s\in S} p(s)L_\tau(s) \tag{3.69}$$

and $E[L_{PTSP}] = min_\tau\left[E[L_\tau]\right]$ represents the expected length of optimal a priori tour. $L_{TSP}(s)$ represents the length of the optimal TSP over nodes in s, $E[\Sigma]$ represents the expected length of the tour produced using a re-optimization technique in which the optimal TSP tour is produced after the problem instance is known:

$$E[\Sigma] = \sum_{s} p(s)L_{TSP}(s) \tag{3.70}$$

The optimal probabilistic traveling salesperson problem solution over n nodes is supposed to be known; this is an a-priori tour. A parameter $\beta > 1$ is selected. The nodes are clustered according to their order in the a priori tour and their coverage probabilities. The pseudo-code of the clustering is given in Algorithm 20 where $G_i, i = 1, .., m$ represents the ith obtained team. After that, the groups must be routed: the median locations $Y_1, Y_2, ..., Y_m$ as the representatives of the groups $G_1, G_2, ..., G_m$ and the **Christofides heuristics** is used to construct a tour over all the representatives $Y_1, Y_2, ..., Y_m$. The

Christofides heuristic relies on the development of a minimum spanning tree and then solves a matching problem. Then each node is connected to its representative and forms a loop or uses sweep Algorithms 12 and 20 to construct an a priori tour within each group.

Algorithm 20 Cluster Algorithm

1. Let $m = \max\left\{\left\lfloor \frac{\sum_i p_i}{\beta} \right\rfloor, 1\right\}$ where the floor function $\lfloor x \rfloor$ represents the largest integer not exceeding x
2. If $\sum_i p_i \leq \beta$, all the nodes are clustered in one group, otherwise follow step 3
3. Select k nodes until $\sum_i p_i < \frac{1}{m}\sum p_i$ and $\sum_{i=1}^{j} p_i \geq \frac{1}{m}\sum_{i=1}^{j} p_i$. If $\sum_i p_i > \frac{1}{m}\sum p_i$, X_j is split into two nodes $X_j', X"_j$. The coverage probability for X_k' is $\frac{1}{m}\sum_i p_i - \sum_{i=1}^{j-1} p_i$ while the coverage probability for $X"_k'$ is $\frac{1}{m}\sum_{i=1}^{j} p_i - \sum_i p_i$. Repeat this procedure until m groups of nodes are obtained.

3.5 CONCLUSION

The first part of this chapter presents operational research basics such as the traveling salesperson problem, postperson problem (Chinese and rural) and knapsack problem. These approaches are at the root of orienteering and coverage missions.

The second part of this chapter presents the UAV orienteering problem as a generalization of the traveling salesperson problem and the knapsack problem. This is the problem of routing a UAV over points of interest.

The third part of this chapter presents the fundamental ideas in the coverage mission. It is how to cover all points in a given barrier/perimeter/area using one or many UAVs with different sets of sensors.

Bibliography

[1] Aaron, E.; Krizanc, D.; Meyerson, E.: *DMVP: foremost waypoint coverage of time-varying graphs*, In International Workshop on Graph-Theoretic Concepts in Computer Science, Springer International Publishing, pp. 29-41, 2014.

[2] Abdollahzadeh, S.; Navimipour, N. J.: *Deployment strategies in the wireless sensor network: A comprehensive review*, Computer Communications, vol. **91**, pp. 1-16, 2016.

[3] Acevedo, J. J.; Arrue, B. C.; Maza, I.; Ollero, A.: *Distributed approach for coverage and patrolling missions with a team of heterogeneous aerial robots under communication constraints*, International Journal of Advanced Robotic Systems, vol. **10**, pp. 1-13, 2013.

[4] Acevedo J.J., Arrue B.C., Diaz-Banez J.M., Ventura I., Maza I., Ollero A.: *One-to one coordination algorithm for decentralized area partition in surveillance missions with a team of aerial robots*, Journal of Intelligent and Robotic Systems, vol. **74**, pp. 269-285, 2014.

[5] Al-Helal, H.; Sprinkle, J.: *UAV search: Maximizing target acquisition*, In 17th IEEE International Conference and Workshops on Engineering of Computer Based Systems, pp. 9-18, 2010.

[6] Alejo, D.; Cobano, J. A.; Heredia, G.; Ollero, A.: *Collision-free Trajectory Planning Based on Maneuver Selection-Particle Swarm Optimization*, In IEEE International Conference on Unmanned Aircraft Systems, pp. 72-81, 2015.

[7] Altshuler, Y.; Bruckstein, A. M: *Static and expanding grid coverage with ant robots: complexity results*, Theoretical Computer Science, vol. **41**, pp. 4661–4674, 2011.

[8] Arezoumand, R.; Mashohor, S.; Marhaban, M. H.: *Efficient terrain coverage for deploying wireless sensor nodes on multi-robot system*, Intelligent Service Robotics, vol. **9**, pp. 163-175, 2016.

[9] Arslan, O.; Koditschek, D. E.: *Voronoi-based coverage control of heterogeneous disk-shaped robots*, In IEEE International Conference on Robotics and Automation, pp. 4259-4266, 2016.

[10] Bakolas, E.; Tsiotras, P.: *The Zermelo-Voronoi diagram, a dynamic partition problem*, Automatica, vol. **46**, pp. 2059-2067, 2012.

[11] Bayat, B.; Crasta, N.; Crespi, A.; Pascoal, A. M.; Ijspeert, A.: *Environmental monitoring using autonomous vehicles: a survey of recent searching techniques*, Current Opinion in Biotechnology, vol. **45**, pp. 76-84, 2017.

[12] Berger, J.; Lo, N.: *An innovative multi-agent search-and-rescue path planning approach*, Computers and Operations Research, vol. **53**, pp. 24-31, 2015.

[13] Bernard, M.; Kondak, K.; Maza, I.; Ollero, A.: *Autonomous transportation and deployment with aerial robots for search and rescue missions*, Journal of Field Robotics, vol. 28, pp. 914-931, 2011.

[14] Bernardini, S.; Fox, M.; Long, D.: *Combining temporal planning with probabilistic reasoning for autonomous surveillance missions*, Autonomous Robots, pp. 1-23, 2015.

[15] Bertuccelli, L. F.; Pellegrino, N.; Cummings, M. L.: *Choice modeling of relook tasks for UAV search missions*, In American Control Conference (ACC), pp. 2410-2415, 2010.

[16] Bertsimas, D.; VanRyzin, G.: *The Dynamic Traveling Repairman problem*, MIT Sloan paper 3036-89-MS, 2011.

[17] Bestaoui Sebbane, Y.: *Planning and decision making of aerial robots*, Springer, 2014.

[18] Bircher, A.; Kamel, M.; Alexis, K.; Burri, M.; Oettershagen, P.; Omari, S.; Mantel, T.; Siegwart, R.: *Three dimensional coverage path planning via viewpoint resampling and tour optimization for aerial robots*, Autonomous Robots, vol. **40**, pp. 1059-1078, 2016.

[19] Breitenmoser, A.; Schwager, M.; Metzger, J. C.; Siegwart, R.; Rus, D.: *Voronoi coverage of non-convex environments with a group of networked robots*, In IEEE International Conference on Robotics and Automation, pp. 4982-4989, 2010.

[20] Brown Kramer, J.; Sabalka, L.: *Multidimensional Online Robot Motion*, arXiv preprint arXiv:0903.4696, 2009.

[21] Bullo, F.; Carli, R.; Frasca, P.: *Gossip coverage control for robotic networks: Dynamical systems on the space of partitions*, SIAM Journal on Control and Optimization, vol. **50**, pp. 419-447, 2012.

[22] Cheng, C. F.; Tsai, K. T.: *Encircled Belt-Barrier Coverage in Wireless Visual Sensor Networks*, Pervasive and Mobile Computing, http://doi.org/10.1016/j.pmcj.2016.08.005, 2017.

[23] Choset, H.: *Coverage of known spaces: the Boustrophedon cellular decomposition*, Autonomous Robots, vol. **9**, pp. 247-253, 2000.

[24] Cotta, C.; Van Hemert, I.: *Recent advances in evolutionary computation for combinatorial optimization*, Springer, 2008.

[25] Czyzowicz, J.; Kranakis, E.; Krizanc, D.; Narayanan, L.; Opatrny, J.: *Optimal online and offline algorithms for robot-assisted restoration of barrier coverage*, In International Workshop on Approximation and Online Algorithms, Springer International Publishing, pp. 119-131, 2014.

[26] Dille, M.; Singh, S.: *Efficient aerial coverage search in road networks*; In AIAA Guidance, Navigation, and Control (GNC) Conference, p. 5094-5109, 2013.

[27] Ding, H.; Castanon, D.: *Fast algorithms for UAV tasking and routing*, In IEEE Conference on Control Applications (CCA), pp. 368-373, 2016.

[28] Durham, J. W.; Carli, R.; Bullo, F.: *Pairwise optimal discrete coverage control for gossiping robots*, In 49th IEEE Conference on Decision and Control, pp. 7286-7291, 2010.

[29] Durham, J. W.; Carli, R.; Frasca, P.; Bullo, F.: *Discrete partitioning and coverage control for gossiping robots*, IEEE Transactions on Robotics, vol. **28**, pp. 364-378, 2012.

[30] Enright, J.J.; Savla, K.; Frazzoli, E.; Bullo, F.: *Stochastic routing problems for multiple UAV*, AIAA Journal of Guidance, Control and Dynamics, vol. **32**, pp. 1152-116, 2009.

[31] Evers, L.; Dollevoet, T.; Barros, A. I.; Monsuur, H.: *Robust UAV mission planning*, Annals of Operations Research, vol. **222**, pp. 293-315, 2014.

[32] Evers, L.; Barros, A. I.; Monsuur, H.; Wagelmans, A.: *Online stochastic UAV mission planning with time windows and time-sensitive targets* European Journal of Operational Research, vol. **238**, pp. 348-362, 2014.

[33] Frederickson, G.; Wittman, B.: *Speedup in the traveling repairman problem with unit time window*, DOI arXiv:0907.5372 [cs.DS], 2009.

[34] Furini, F.; Ljubic, I.; Sinnl, M.: *An Effective Dynamic Programming Algorithm for the Minimum-Cost Maximal Knapsack Packing Problem*, European Journal of Operational Research, doi: 10.1016/j.ejor.2017.03.061, 2017

[35] Galceran, E.; Carreras, M.: *A survey on coverage path planning for robotics*, Robotics and Autonomous Systems, vol. **61**, pp. 1258-1276, 2013.

[36] Galceran, E.; Campos, R.; Palomeras, N.; Ribas, D.; Carreras, M.; Ridao, P.: *Coverage path planning with real time replanning and surface reconstruction for inspection of three dimensional underwater structures using autonomous underwater vehicles*, Journal of Field Robotics, vol. **32**, pp. 952-983, 2015.

[37] Gao, C.; Zhen, Z.; Gong, H.: *A self-organized search and attack algorithm for multiple unmanned aerial vehicles. Aerospace Science and Technology*, vol. **54**, pp. 229-240, 2016.

[38] Garzon, M.; Valente, J.; Roldan, J. J.; Cancar, L.; Barrientos, A.; Del Cerro, J.: *A multirobot system for distributed area coverage and signal searching in large outdoor scenarios*, Journal of Field Robotics, vol. **33**, pp. 1096-1106, 2016.

[39] Goel, A., Gruhn, V.: *A general vehicle routing problem*, European Journal of Operational Research, vol. **191**, pp. 650-660, 2008.

[40] Guha, S.; Munagala, K.; Shi, P.: *Approximation algorithms for restless bandit problems*, Journal of the ACM, vol. **58**, DOI 10.1145/1870103.1870106, 2010.

[41] Gusrialdi, A.; Hirche, S.; Hatanaka, T.; Fujita, M.: *Voronoi Based Coverage Control with Anisotropic Sensors*, 53rd, In Proceedings of American Control Conference, Seattle, pp. 736-741, 2008.

[42] Hazon, N.; Gonen, M.; Kleb, M.: *Approximation and Heuristic Algorithms for Probabilistic Physical Search on General Graphs*, arXiv preprint arXiv:1509.08088, 2015.

[43] Howlett J.K., McLain T., Goodrich M.A.: *Learning real-time A^* path planner for unmanned air vehicle target sensing*, AIAA Journal of Aerospace Computing, Information and Communication, vol. **23**, pp. 108-122, 2006.

[44] Huang, Z.; Zheng, Q. P.; Pasiliao, E. L.; Simmons, D.: *Exact algorithms on reliable routing problems under uncertain topology using aggregation techniques for exponentially many scenarios*, Annals of Operations Research, vol. 249, pp. 141-162, 2017.

[45] Itani, S.; Frazzoli, E.; Dahleh, M. A.: *Dynamic traveling repair-person problem for dynamic systems*, In IEEE Conference on Decision and Control, pp. 465-470, 2008.

[46] Jung, S.; Ariyur, K. B.: *Enabling operational autonomy for unmanned aerial vehicles with scalability*, AIAA Journal of Aerospace Information Systems, vol. **10**, pp. 517-529, 2013.

[47] Kalra, N.; Martinoli, A.: *Optimal multiplicative Bayesian search for a lost target*, In Distributed Autonomous Robotic Systems, Springer Japan, pp. 91-101, 2006.

[48] Kantaros, Y.; Thanou, M.; Tzes, A.: *Distributed coverage control for concave areas by a heterogeneous robot swarm with visibility sensing constraints*, Automatica, vol. **53**, pp. 195-207, 2015.

[49] Ke, L.; Zhai, L.; Li, J.; Chan, F. T.: *Pareto mimic algorithm: An approach to the team orienteering problem*, Omega, vol. **61**, pp. 155-166, 2016.

[50] Kerkkamp, R. B. O.; Aardal, K.: *A constructive proof of swap local search worst-case instances for the maximum coverage problem*, Operations Research Letters, vol. **44**, pp. 329-335, 2016.

[51] Klein, R.; Kriesel, D.; Langetepe, E.: *A local strategy for cleaning expanding cellular domains by simple robots*, Theoretical Computer Science, vol. **605**, pp. 80-94, 2015.

[52] Kriheli, B.; Levner, E.: *Optimal Search and Detection of Clustered Hidden Targets under Imperfect Inspections*, IFAC Proceedings Volumes, vol. **46**, pp. 1656-1661, 2016.

[53] Kumar, G. P.; Berman, S.: *The probabilistic analysis of the network created by dynamic boundary coverage*, arXiv preprint arXiv:1604.01452, 2016.

[54] Lalish, E., Morgansen, K. A.: *Distributed reactive collision avoidance*, Autonomous Robots, vol. **32**, pp. 207-226, 2012.

[55] Lee, S. G.; Diaz-Mercado, Y.; Egerstedt, M.: *Multirobot control using time-varying density functions*, IEEE Transactions on Robotics, vol. **31**, pp. 489-493, 2015.

[56] Le Ny, J.; Pappas, G. J.: *Adaptive algorithms for coverage control and space partitioning in mobile robotic networks*, arXiv preprint arXiv:1011.0520, 2010.

[57] Le Ny, J.; Feron, E.; Frazzoli, E.: *On the Dubins Traveling salesman Problem*, IEEE Transactions on Automatic Control, vol. **57**, pp. 265–270, 2012.

[58] Leonard, N. E.; Olshevsky, A.: *Nonuniform coverage control on the line*, IEEE Transactions on Automatic Control, vol. **58**, pp. 2743-2755, 2013.

[59] Liu, L.; Zlatanova, S.: *An Approach for Indoor Path Computation among Obstacles that Considers User Dimension*, ISPRS International Journal of Geo-Information, vol. **4**, pp. 2821-2841, 2015.

[60] Lu, X.: *Dynamic and stochastic routing optimization: algorithmic development and analysis*, PhD thesis, Univ. of California, Irvine, 2001.

[61] Mufalli, F., Batta, R.; Nagi, R.: *Simultaneous sensor selection and routing of UAV for complex mission plans*, Computers and Operations Research, vol. **39**, pp. 2787–2799, 2012.

[62] Nourani-Vatani, N.: *Coverage algorithms for under-actuated car-like vehicle in an uncertain environment*, PhD Thesis, Technical University of Denmark, Lyngby, 2006.

[63] Obermeyer, K.; Oberlin, P.; Darbha, S.: *Sampling based path planning for a visual reconnaissance UAV*, AIAA Journal of Guidance, Control and Dynamics, vol. **35**, pp. 619-631, 2012.

[64] Okabe, A.; Boots, B.; Sugihara, K.; Chiu, S.N.: *Spatial tessellations: concepts and applications of Voronoi diagrams*, Wiley, 2000.

[65] Ozbaygin, G.; Yaman, H.; Karasan, O. E.: *time constrained maximal covering salesman problem with weighted demands and partial coverage*, Computers and Operations Research, vol. **76**, pp. 226-237, 2016.

[66] Palacios-Gass, J. M.; Montijano, E.; Sags, C.; Llorente, S.: *Distributed coverage estimation and control for multirobot persistent tasks*, IEEE Transactions on Robotics, vol. **32**, pp. 1444-1460, 2016.

[67] Pastor, E.; Royo, P.; Santamaria, E.; Prats, X.: *In flight contingency management for unmanned aircraft systems*, Journal of Aerospace Computing Information and Communication, vol. **9**, pp. 144-160, 2012.

[68] Penicka, R.; Faigl, J., Vana, P.; Saska, M.: *Dubins Orienteering Problem*, IEEE Robotics and Automation Letters, vol. **2**, pp. 1210-1217, 2017.

[69] Pierson, A.; Schwager, M.: *Adaptive inter-robot trust for robust multi-robot sensor coverage*, In Robotics Research, Springer, pp. 167-183, 2016.

[70] Regev, E.; Altshuler, Y.; Bruckstein, A. M.: *The cooperative cleaners problem in stochastic dynamic environments*, arXiv preprint arXiv:1201.6322, 2012.

[71] Renzaglia, A.; Doitsidis, L.; Martinelli, A.; Kosmatopoulos, E.: *Multi-robot three dimensional coverage of unknown areas*, International Journal of Robotics Research, vol. **31**, pp. 738-752, 2012.

[72] Riera-Ledesma, J.; Salazar-Gonzalez, J.J.: *Solving the team orienteering arc routing problem with a column generation approach*, European journal of operational research, http://dx.doi.org/10.1016/j.ejor.2017.03.027, 2017.

[73] Sadovsky, A. V.; Davis, D.; Isaacson, D. R.: *Efficient computation of separation-compliant speed advisories for air traffic arriving in terminal airspace*, ASME Journal of Dynamic Systems, Measurement, and Control, vol. **136**, pp. 536-547, 2014.

[74] Savla, K., Frazzoli, E., Bullo, F.: *Traveling salesperson problems for the Dubbins vehicle*, IEEE Transactions on Automatic Control, vol. **53**, pp. 1378-1391, 2008.

[75] Schouwenaars, T., Mettler, B., Feron, E.: *Hybrid model for trajectory planning of agile autonomous vehicles.* AIAA Journal on Aerospace Computing, Inforation and Communication, vol. **12**, pp. 629-651, 2004.

[76] Schwertfeger, S.; Birk, A.: *Map evaluation using matched topology graphs*, Autonomous Robots, vol. **40**, pp. 761-787, 2016.

[77] Sharifi, F.; Chamseddine, A.; Mahboubi, H.; Zhang, Y.; Aghdam, A. G.: *A distributed deployment strategy for a network of cooperative autonomous vehicles*, IEEE transactions on control systems technology, vol. **23**, pp. 737-745, 2015.

[78] Smith, S.; Tumova, J.; Belta, C.; Rus, D.: *Optimal path planning for surveillance with temporal logic constraints*, International Journal of Robotics Research, vol. **30**, pp. 1695-1708, 2011.

[79] Song, C.; Liu, L.; Feng, G.; Xu, S.: *Coverage control for heterogeneous mobile sensor networks on a circle*, Automatica, vol. **63**, pp. 349-358, 2016.

[80] Sydney, N.; Paley, D. A.: *Multiple coverage control for a non stationary spatio-temporal field*, Automatica, vol. **50**, pp. 1381-1390, 2014.

[81] Tang, J.; Alam, S.; Lokan, C.; Abbass, H.A.: *A multi-objective approach for dynamic airspace sectorization using agent based and geometric models*, Transportation Research part C, vol. **21**, pp. 89-121, 2012.

[82] Tapia-Tarifa, S. L.: *The Cooperative Cleaners Case Study: Modelling and Analysis in Real-Time ABS*, MS Thesis, Department of Informatics, Univ. of Oslo, 116 pages, 2013.

[83] Temizer, S.: *Planning under uncertainty for dynamic collision avoidance*, PhD Thesis, MIT, MA, USA, 2011.

[84] Thanou, M.; Stergiopoulos, Y.; Tzes, A.: *Distributed coverage using geodesic metric for non-convex environments*, In IEEE International Conference on Robotics and Automation, pp. 933-938, 2013.

[85] Tian, J.; Liang, X.; Wang, G.: *Deployment and reallocation in mobile survivability heterogeneous wireless sensor networks for barrier coverage*, Ad Hoc Networks, vol. **36**, pp. 321-331, 2016.

[86] Tilk, C.; Rothenbacher, A. K.; Gschwind, T.; Irnich, S.: *Asymmetry matters: Dynamic Half-Way Points in Bidirectional Labeling for Solving Shortest Path Problems with Resource Constraints Faster*, European Journal of Operational Research, vol. **261**, pp. 530-539, http://dx.doi.org/10.1016/j.ejor.2017.03.017, 2017.

[87] Torres, M.; Pelta, D. A.; Verdegay, J. L.; Torres, J. C.: *Coverage path planning with unmanned aerial vehicles for 3D terrain reconstruction*, Expert Systems with Applications, vol. **55**, pp. 441-451, 2016.

[88] Toth, P.; Vigo, D.: *The Vehicle Routing Problem*, SIAM, Philadelphia, 2002.

[89] Troiani, C.; Martinelli, A.; Laugier, C.; Scaramuzza, D.: *Low computational-complexity algorithms for vision-aided inertial navigation of micro aerial vehicles*, Robotics and Autonomous Systems, vol. **69**, pp. 80-97, 2015.

[90] Tseng, K. S.; Mettler, B.: *Near-optimal probabilistic search via submodularity and sparse regression*, Autonomous Robots, vol. **41**, pp. 205-229, 2017.

[91] Tuna, G., Gungor, V. C.; Gulez, K.; : *An autonomous wireless sensor network deployment system using mobile robots for human existence detection in case of disasters*, Ad Hoc Networks, vol. **13**, pp. 54-68, 2014.

[92] Valente, J.; Del Cerro, J.; Barrientos, A.; Sanz, D.: *Aerial coverage optimization in precision agriculture management: A musical harmony inspired approach*, Computers and Electronics in Agriculture, vol. **99**, pp. 153-159, 2013.

[93] Valente, J.: *Aerial coverage path planning applied to mapping*, PhD thesis, Universidad politecnica de Madrid, 2014.

[94] VanderBerg, J. P.; Patil, S.; Alterovitz, R.: *Motion planning under uncertainty using differential dynamic programming in Belief space*, International Symposium of Robotics Research, 2011.

[95] Verbeeck, C.; Vansteenwegen, P.; Aghezzaf, E. H.: *Solving the stochastic time-dependent orienteering problem with time windows*, European Journal of Operational Research, vol. **255**, pp. 699-718, 2016.

[96] Vieira, L. F. M.; Almiron, M. G.; Loureiro, A. A.: *Link probability, node degree and coverage in three-dimensional networks*, Ad Hoc Networks, vol. **37**, pp. 153-159, 2016.

[97] Viet, H. H.; Dang, V. H.; Laskar, M. N.; Chung T. C.: *BA*: an online complete coverage algorithm for cleaning robots*, Applied Intelligence, vol. **39**, pp. 217-237, 2013.

[98] Wang, H.F.; Wen, Y.P.: *Time-constrained Chinese postman problems*, Computers and Mathematics with Applications, vol. **44**, pp. 375-387, 2002.

[99] Wang, Z.; Guo, J.; Zheng, M.; Wang, Y.: *Uncertain multiobjective traveling salesman problem*, European Journal of Operational Research, vol. **241**, pp. 478-489, 2015.

[100] Waqqas, A.: *Distributed Navigation of Multi-Robot Systems for Sensing Coverage*, PhD thesis, School of Electrical Engineering and Telecommunications, The University of New South Wales, Australia, 224 pages, 2016.

[101] Xu, A.; Viriyasuthee, C.; Rekleitis, I.: *Optimal complete terrain coverage using an unmanned aerial vehicle*, In IEEE International Conference on Robotics and Automation, pp. 2513-2519, 2011.

[102] Xu, A.; Viriyasuthee, C.; Rekleitis, I.: *Efficient complete coverage of a known arbitrary environment with applications to aerial operations*, Autonomous Robots, vol. **36**, pp. 365-381, 2014.

[103] Yazicioglu, A. Y.; Egerstedt, M.; Shamma, J. S.: *Communication-free distributed coverage for networked systems*, IEEE Transactions on Control of Network Systems, DOI: 10.1109/TCNS.2016.2518083, 2016.

[104] Yu, J.; Schwager, M.; Rus, D.: *Correlated orienteering problem and its application to persistent monitoring tasks*, IEEE Transactions on Robotics, vol. **32**, pp. 1106-1118, 2016.

[105] Zhong, M., Cassandras, C. G.: *Distributed coverage control in sensor network environments with polygonal obstacles*, IFAC Proceedings Volumes, vol. **41**, pp. 4162-4167, 2008.

CHAPTER 4

Deployment, Patrolling, and Foraging

4.1 INTRODUCTION

In this chapter, operations concern generic robotic problems such as **deployment**, **patrolling**, and **foraging**. That is a core area of robotics, artificial intelligence, and operations research.

First, about deployment, finding a distribution of a group of UAVs in a known or unknown environment is one of the challenges in multi-robot systems. One challenge is how to balance the workload among UAVs or partition the environment into regions and assign a UAV to each of the regions. The main issue of target coverage is to complete monitoring and information collection of the target coverage through the scheduling and deployment of distributed sensor nodes within the coverage area. Target coverage research includes: low power consumption, real-time, objective coverage ability, versatility and connectivity of algorithm.

Second, about patrolling, large populations of UAVs are ideal for coverage of a large geographic area. They can maintain coverage of the environment after the dispersion is complete. The size of the environment that can be covered is proportional to the population size. After deployment of the mapping UAVs, controlling the patrolling UAVs amounts to a coverage control problem:

1. How should the UAVs move in order to ensure small worst-case latency in patrolling all areas of the surveyed environment? This may require only simple local strategies that do not involve complicated protocols or computations for coordinating the motion of the mobile components, while still achieving complete coverage, with small latency to surveyed locations. Complete coverage of an unknown region can be achieved by performing a structured exploration by a multi-UAV system with bearing only low-resolution sensors.

2. What local policies should be used for patrolling the cell or region? A natural choice for this task is **least recently visited** approach (LRV), in which each cell keeps track of the time elapsed since its last visit from a patrolling UAV. The patrolling UAV policy directs it to move to the adjacent cell with the smallest latency. This amounts to tracking the visit times of regions [64].

Third, foraging represents the broad class of problems integrating robotic exploration, navigation and object identification, manipulation and transport. In multi-robot systems, foraging is a canonical problem for the study of robot-robot cooperation. The anchoring problem is an important aspect of the connection between symbolic and sensory-based processes in autonomous robotic systems. Anchoring is in fact the problem of how to create, and maintain in time, the connection between the symbol and the signal level representations of the same physical object.

4.2 AERIAL DEPLOYMENT

The deployment problem considers the number of needed UAVs for a specific situation (e.g., objective, scenario, constraints) and their initial locations. The **deployment problem** is deciding the number of UAVs and where they will be initially located before performing the mission using their control strategy. Among different topics, covering an area is an important objective in a deployment problem. In this problem, the environment is partitioned into regions and each UAV of the team should be responsible for covering the events happening inside its assigned region. Placing the UAVs in pre-determined regions on the basis of simple geometric structure as triangular, square, rhombus and hexagon lattice, is simple when the environment is known [81]. The environment may change over time and thus deployment must be repeated periodically in order to re-distribute the UAVs. The **coverage** and **connectivity** is one of the fundamental considerations in deployment strategies. A survey of the coverage and connectivity by considering deployment strategies, sleep scheduling mechanism and adjustable coverage radius is presented in [106]. The problem of multi-agent deployment over a source can also be explored. Based on the heat partial differential equation and extremum seeking, a source seeking control algorithm can be designed to deploy the group of agents around the source.

4.2.1 Deployment problem

To measure the performance of any specific solution, a deployment function represents the quality of the UAV's distribution over the field. Such a function might be defined based on the distance of UAVs to the points in the environment, which must be minimized. Therefore, the deployment problem can be translated to a minimization problem [5]:

1. One category of multi-UAV deployment control schemes is based on **artificial potential field** or **virtual force fields**. Moreover, graph-based non-linear feedback control laws in a group of mobile agents can be used to make the graph stay connected at all times.

2. A second category is based on common coverage control approach through the definition of feedback control laws defined with respect to the centroids of Voronoi regions. A distributed approach for optimally deploying a team of UAVs in a domain can be based on the Lloyd algorithm. Each UAV follows a control law, which is a gradient descent algorithm that minimizes the functional encoding the quality of the deployment. In a non-convex environment, geodesic distance metric to address non-convex region instead of Euclidean distance can be used.

4.2.1.1 Deployment methodology

According to the environment, the state-of-the-art in a deployment problem is divided into continuous and discrete setups. One of the useful pre-processing techniques is modeling the workspace as a form of graph, and then the problem becomes one of graph partitioning. In the graph representation, the deployment problem can be interpreted as a **p-median problem**, in which a limited number of facilities will be assigned to customers. The **p-median problem** is a basic discrete location problem. A problem is classified to belong to the location field if some decision regarding the position of new facilities has to be made. In general, the objective or goal of the location problem is related with the distance between new facilities and other elements of the space where they have to be positioned.

Location models may be divided into three groups: continuous, discrete and network models [70]. Large distances between nodes weaken the communication links, lower the throughput, and increase energy consumption. In optimal node placement for sensor deployment, several heuristics have been proposed to find sub-optimal solutions for a fixed topology, assessing the quality of candidate positions on a structural quality metric such as distance and network connectivity. On the other hand, some schemes have advocated dynamic adjustment of nodes location since the optimality of the initial positions may become void during the operation of the network depending on the network state and various external factors [103]. For a centralized coverage algorithm in a discrete environment, an applied **spanning tree covering** (STC) can be applied on the corresponding grid cell in order to partition the graph and direct the UAVs to cover the environment. The multi-objective problem can be formulated as a mixed linear integer programming with two objectives: finding the best position in the environment to deploy the UAVs on them and minimizing the length of UAVs paths from the initial to the end position. Target overlapping areas and a greedy algorithm are used in [92] to design an optimal deployment algorithm based on goal weights, as a result the greedy algorithm realizes optimal coverage monitoring on discrete targets and ensures the connectivity of node-monitoring network. Unlike round coverage model of a 2-D plane, coverage model of 3-D space chooses node position as center of the sphere and the perceived distance as sphere radius.

Voronoi-based coverage control uniquely combines both deployment and allocation in an intrinsically distributed manner via gradient descent (the **move-to-centroid law**) down a utility function minimizing the expected event sensing cost to adaptively achieve a centroidal Voronoi configuration. The application to homogeneous point UAVs and heterogeneous groups of UAVs with various sensori-motor capabilities differ by recourse to power diagrams **generalized Voronoi diagrams with additive weights**. Voronoi-based coverage control involves collision avoidance for point UAVs since UAVs move in their pairwise disjoint Voronoi cells, but an additional collision avoidance strategy is mandatory for safe navigation of finite size UAVs. Existing work on combining coverage control and collision avoidance generally uses either heuristic approaches based on **repulsive fields** and **reciprocal velocity obstacles** causing UAVs to converge to configurations far from optimal sensing configurations; or the projection of a vector field whenever a UAV reaches the boundary of its partition cell introducing a source of discontinuity [10].

Taxonomy for task allocation Cooperation is the **task assignment problem** (TAP) which assigns a finite number of agents to complete a finite number of tasks as efficiently as possible. This problem can be solved with a centralized or decentralized solution. Task allocation for an individual agent is relatively simple; the difficulty occurs when a decentralized algorithm is used for consensus between all UAVs. The auction-based solution is the **consensus-based auction algorithm** (CBAA), which solves the task assignment problem for single agent tasks that are defined as tasks that require a single agent to complete. The CBAA lets agents make bids for tasks and provides a system for decentralized consensus on assignments, giving a conflict-free solution. The consensus-based bundle algorithm solves an extension of the task assignment problem where agents queue up tasks they will complete: individual agents take available tasks and compute every permutation given their current queue of tasks, where the highest rewarded permutation becomes their bid for that task. In this way, agents continually remove and revise new tasks, as other agents find they can create a more valuable sequence with that task. This algorithm can be extended to allow multi-agent tasks requiring agents to cooperate in completing individual tasks [48].

The following useful definitions are taken from [57]:

Definition 129 *A task t is **decomposable** if it can be represented as a set of sub-tasks σ_t for which satisfying some specified combination (ρ_t) of sub-tasks in σ_t satisfies t. The combination of sub-tasks that satisfy t can be represented by a set of relationships ρ, that may include constraints between sub-tasks or rules about which or how many sub-tasks are required. The pair (σ_t,ρ_t) is also called a **decomposition** of t.*

The term **decomposition** can also be used to refer to the process of decomposing a task.

Definition 130 *A task t is* ***multiply decomposable*** *if there is more than one possible decomposition of t.*

Definition 131 *An* ***elemental (or atomic) task*** *is a task that is not decomposable.*

Definition 132 *A* ***decomposable simple task*** *is a task that can be decomposed into elemental or decomposable simple sub-tasks, provided that there exists no decomposition of the task that is multi(agent)-allocatable.*

Definition 133 *A* ***simple task*** *is either an elemental task or a decomposable simple task.*

Definition 134 *A* ***compound task*** *t is a task that can be decomposed into a set of simple or compound sub-tasks with the requirement that there is exactly one fixed full decomposition for t (i.e., a compound task may not have any multiply decomposable tasks at any decomposition step).*

Definition 135 *A* ***complex task*** *is a multiply decomposable task for which there exists at least one decomposition that is a set of multi[agent]-allocatable sub-tasks. Each sub-task in a complex task's decomposition may be simple, compound, or complex.*

The degree of interdependence is represented with a single categorical variable with four values:

1. **No Dependencies**: These are task allocation problems with simple or compound tasks that have independent agent-task utilities. The effective utility of an agent for a task does not depend on any other tasks or agents in the system.

2. **In-schedule dependencies**: These are task allocation problems with simple or compound tasks for which the agent-task utilities have intra-schedule dependencies. The effective utility of an agent for a task depends on what other tasks that agent is performing. Constraints may exist between tasks on a single agent's schedule, or might affect the overall schedule of the agent.

3. **Cross-schedule dependencies**: These are task allocation problems with simple or compound tasks for which the agent-task utilities have inter-schedule dependencies. The effective utility of an agent for a task depends not only on its own schedule, but also on the schedules of other agents in the system. For this class, allowable dependencies are simple dependencies in that the task decomposition can be optimally predetermined prior to task allocation. Constraints may exist between the schedules of different agents.

4. **Complex dependencies**: The agent-task utilities have inter-schedule dependencies for complex tasks, in addition to any in-schedule and cross-schedule dependencies for simple or compound tasks. The effective utility of an agent for a task depends on the schedules of other agents in the system in a manner that is determined by the particular chosen task decomposition. Thus, the optimal task decomposition cannot be decided prior to task allocation, but must be determined concurrently with task allocation. Furthermore, constraints may exist between the schedules of different agents.

Communication constraints Cooperative UAVs have to move to complete their tasks while maintaining communication among themselves without the aid of a communication infrastructure. Initially deploying and ensuring a mobile ad-hoc network in real and complex environments is difficult since the strength of the connection between two UAVs can change rapidly in time or even disappear. A wrong decision about the number of UAVs and their initial location may greatly jeopardize the mission. A coverage algorithm evaluation using both centralized and random initial deployments concludes that the algorithm convergence was slower using a random initial deployment but tended to lead to better overall coverage for sparse topologies. In real situations, it is necessary to ensure several constraints of the system. If the network supports multi-hop connectivity, this kind of constraints may significantly increase the complexity of the random distribution since it would depend not only on the communication constraints, but also on the number of UAVs and their own position. Moreover, **random deployment** may cause unbalanced deployment, thus increasing the number of needed UAVs and energy depletion. A solution that can cover the deployment area within the maximum coverage time allowed is iteratively determined by varying the number and size of groups based on heuristics. An initial deployment in which UAVs start from a compact configuration works well when the main purpose is to spread the UAVs within area coverage scenarios. **Fault tolerance** in multi-UAVs system can be summarized by **bi-connectivity**, meaning that each pair of nodes in the network has at least two disjoint routes between them. Therefore, the failure at any single node does not partition the network. Despite the positive results provided by bi-connected networks, complementary strategies can be introduced such as **attraction forces**, **redundancy** or **transmission power adaptation** to maintain **mobile ad-hoc networks** (MANET) connectivity. Furthermore, in **wireless sensor networks** (WSN), the bi-connectivity feature can be generalized to multi-connectivity, or **k-connectivity**, $k \in \mathbf{N}$ [29].

UAVs deployment as a decision problem The team comprises a set of N UAVs operating in either single or multiple layers such that each UAV has a radio range R. Each UAV is capable of handling a set of K users making continuous request from a particular demand area. The number of requests S_r comes with an arrival rate λ and mean packet size of each service request is $1/\mu$. The deployment model of the UAVs in the heterogeneous networks may consider two aspects:

1. Single layer with multiple UAVs: the number of UAVs is decided on the basis of provisioning of the connectivity with the macro-base station and the number of user requests from a particular demand area. For the single-layer model, the number of UAVs for connection between the base station and the demand area is calculated as $|N| = Z/R$ where Z is the distance of excessive demand area from the base station and R is the radius range. For the full capacity link between the UAVs and the base station, $|N| = S_r/S_u$ where S_u is the number of service requests a single UAV can handle.

2. Multiple layers model with multiple UAVs in each layer: The altitude is taken into account to distinguish between the UAVs of different layers. Each base station has a limited number of UAVs. In this model, the upper layer UAVs act as the main pivot between the lower layer UAVs and the base station. These pivot UAVs can support a number of UAVs by acting as an aerial base station for them. This model is very useful in case of UAV failures or base station failures.

In this decision-based approach, the base station divides the complete area into the priority zones on the basis of the user requests from them. The upper range of the user requests is accounted by the upper limit of the requests a UAV can handle as defined in the network model. For an area A, let $A_1, A_2, , A_{|B|}$ be the demand areas generating requests for extra users. Now, the complete request areas are assigned a priority value based on the number of requests pending in the particular zone provided that the number of requests pending is always less than or equal to the number of requests supported by a single UAV. The decision of the network topology is totally based on the priority value. The part of the area which is not included as the demand zone will either be handled using the multi-layer model or will be assigned a number of UAVs for direct connectivity with the base station [85].

The problem of locating and routing UAV-UGV teams with a specific objective function can be formulated as an mixed integer linear problem with the aim of maximization of the total score collected from visited points of interest by flight routes of UAVs. These routes are originated from assigned base stations determined simultaneously along with the visit order of points of interest [100]. Ideally, in order to obtain the minimum cost mission plan, these decisions would be made via a single optimization problem. To tackle the complexity of the optimization problem, given mission parameters, a multi-stage optimization algorithm can compute a mission plan with locations for each of UAV to visit:

1. In the first stage, an optimal deployment location problem for ground

vehicles is formulated as a mixed-integer linear programming problem, with the objective to minimize the summation of the distances from each ground vehicle location to the associated aid requests.

2. In the second stage, optimal routes for ground vehicles from their initial locations to the destinations are determined such that the total travel time is minimized. Optimal trajectories for the UAVs along the locations of aid requests are determined. Once the mission plan has been determined, the way-points are sent to the respective autonomous vehicles. The way-points serve as input to the vehicle guidance and so enable automated deployment given the ability to operate autonomously for each of the vehicles.

4.2.1.2 Deployment strategies

Initial deployment The initial deployment of UAVs from the ground and the re-deployment of UAVs once an area is searched are investigated to reduce energy costs and search time. Three strategies are compared that are **scalable**, decentralized, and require low computational and communication resources. The strategies exploit environment information to reduce unnecessary motion, and reduce diminishing returns and interference between UAVs:

1. **Linear-temporal incremental deployment** (LTID): This strategy deploys UAVs one at a time with a fixed time interval between consecutive launches. Longer inter-launch intervals (λ) slow deployment, but decrease the number of concurrent UAVs. This reduces spatial interference and unnecessary flight by exploiting environmental information acquired from the expanding network. Once a sub-area of the environment has been searched, UAVs re-deploy as explorers to new unexplored areas. Before this re-deployment commences, there may be multiple explorers flying into this sub-area where they are not required, which is reduced with longer inter-launch intervals. Thus, LTID reduces energy consumption by reducing interference and unnecessary movement.

2. **Single incremental deployment** (SID): This strategy is similar to LTID and deploys one UAV at a time, but waits for the previous UAV to become a beacon before launching the next. Single incremental deployment reduces unnecessary flight time because the next UAV will only deploy once the beacon network has sensed the environment and perceived if and where a new beacon is required. Thereby, explorers always fly directly to the desired deployment location. To implement SID, the network communicates if an explorer is flying. This can be achieved by propagating local messages across the beacon network. Beacons signal to the whole team if they perceive a flying explorer and UAVs only deploy if no signal is received. To ensure only a single UAV deploys at a time, random timeouts are used. When no flying explorer signal

is present, UAVs wait a short random time period. If after this period there is no flying explorer signal, the UAV can deploy.

3. **Adaptive group size** (AGS): This strategy adapts the density of UAVs, initially rapidly deploying UAVs, every 2-3 seconds. Explorers measure the density of neighboring UAVs using their relative-positioning sensor and probabilistically land if the density is higher than a predefined threshold. This decreases the ratio of UAVs, diminishing returns and interference. UAVs which have landed launch again when there are no UAVs flying in the vicinity.

Optimal deployment at constant altitude The flying zone can be represented as a discretized parallelepiped of height $h_{min} \leq h \leq h_{max}$, width y_{max} and length x_{max}, where U is the set of available UAVs, with the coordinates (x_u, y_u, h_u) of each UAV u and T is the set of targets to be monitored, $t_i = (X_{t_i}, y_{t_i})$ with the distance $D_{t_i}^{x_u, y_u} = \sqrt{(X_{t_i} - x_u)^2 + (Y_{t_i} - y_u)^2}$. The visibility θ of each UAV u is represented by a disk with radius r^{h_u}. Two main decisions must be taken:

1. The decision variables are given by:

$$\delta_{xyh}^u = \left\{ \begin{array}{cc} 1 & \text{if the UAV} u \text{ is located at } (x, y, h) \\ 0 & \text{otherwise} \end{array} \right\} \tag{4.1}$$

2. The targets $t_i \in T$ to be monitored by a UAV u:

$$\delta_{t_i}^u = \left\{ \begin{array}{cc} 1 & \text{if the target } t_i \text{ is observed by the UAV} u \\ 0 & \text{otherwise} \end{array} \right\} \tag{4.2}$$

The objective is to monitor all the targets with at least one UAV, minimizing either the number of UAVs or the total energy consumption. Each UAV consumes the energy:

$$E = (\beta + \alpha k)t + P_{max}(k/s) \tag{4.3}$$

where α is a motor speed multiplier and β is the minimum power needed to hover over the ground, t being the operating time.

Problem 136 *The deployment problem can be stated with the objective:*

$$\min f(\delta) = \sum_{(x,y,h)} \sum_{u \in U} \delta_{xyh}^u + E \tag{4.4}$$

subject to:

$$\sum_{x,y,h} \delta_{x,y,h}^u \leq 1 \quad \forall u \in U \tag{4.5}$$

This constraint ensures that the UAV u is located at most one position.

$$\gamma_{t_i}^u \leq \sum_{x,y,h} \delta_{x,y,h}^u \left(\frac{r^{h_u}}{D_{t_i}^{uxy}} \right) \quad \forall u \in U, t_i \in T \tag{4.6}$$

This condition is used to set the value of the variable $\gamma_{t_i}^u$ which can have the value 0 or 1 depending on the radius

$$\sum_{u \in U} \gamma_{t_i}^u, \forall t_i \in T \tag{4.7}$$

Each target is observed by at least one UAV.

Heuristics can be presented to solve this non-linear mixed integer optimization problem [107].

Generalized discrete Lloyd descent A multi-agent system is composed of interconnected subsystems or agents. In control of UAVs, the aim is to obtain a coordinated behavior of the overall system through local interactions among the agents. Communication among the agents often occurs over a wireless medium with finite capacity [3]. Some definitions useful for the sequel are introduced as follows:

Definition 137 *A **landmark** is an abstraction of a point or a small area of interest that must be kept under observation, and that may be of a larger surface. A landmark is formally defined as the tuple $\ell = (q, \hat{m})$ where $q \in \mathbb{R}^3$ is the position of the landmark and $\hat{m} \in S^2$ (unit sphere) is the orientation of the landmark. More specifically, $\hat{m}$ is a direction that characterizes the orientation of the landmark.*

Definition 138 *A **mobile sensor** is a tuple $s = (p, \hat{n}, f)$ where $q \in \mathbb{R}^3$ is the position of the sensor, $\hat{n} \in S^2$ is the orientation and $f : (\mathbb{R}^3 \times S^2)^2 \to \mathbb{R}^+$ is the footprint of the sensor. The orientation of a sensor is the unit vector corresponding to the direction where the sensor is pointing. The footprint of the mobile sensor is a function that describes the sensor's perception of the surrounding environment.*

The coverage of a finite set of landmarks attained by a set of mobile sensors is defined with respect to a partition of the landmarks among the sensors, and it is given by the sum of the coverages attained by each sensor for its subset of landmarks.

Definition 139 ***Coverage of a finite set of landmarks attained by a team of mobile sensors:*** *Consider a team of mobile sensors $S = (s_1, \ldots, s_N)$, a finite set of landmarks $L = \{\ell_1, \ldots, \ell_N\}$, and a partition $P = \{P_1, \ldots, P_N\}$ so that each subset L_i is assigned to the sensor s_i, the* *coverage of the set L attained by the team S with respect to the partition P as the sum of the coverage of L_i attained by s_i for $i \in \{1, \ldots, N\}$:*

$$cov(S, P) = \sum_{i=1}^{N} cov(s_i, L_i) = \sum_{i=1}^{N} \sum_{\ell \in L} per(s_i, \ell_i) \tag{4.8}$$

where

$$per(s_i, \ell_i) = f(p_i, \hat{n}_i, q_i, \hat{m}_i) \tag{4.9}$$

Problem 140 *The objective is to find a partition $P = \{P_1, \ldots, P_N\}$ and the positions and orientations of the sensors, such that the coverage of the set L attained by a team with respect to the partition is minimized.*

Algorithm 21 aims at progressively adjusting iteratively the positions and orientations of the sensors as well as the partition P. The coverage is improved by adjusting iteratively the positions and orientations of the sensors, as well as the partition P. The partition P is improved by considering a pair of sensors s_i and s_j at each partition and rearranging the landmarks in $L_i \cup L_j$ so that each landmark is assigned to the sensor between s_i and s_j. A lower value of coverage corresponds to a better coverage.

Location problem on the plane This paragraph considers the problem of locating M facilities on the unit square so as to minimize the maximal demand faced by any facility subject to closest assignments and coverage constraints. By minimizing the demand faced by the facilities, the difference in demand rates between the busiest and the least busy facilities is the **equitable location problem** (ELP).

Definition 141 *A location vector x represents an* ***equitable facility configuration*** *(EFC) if the demand rates to all facilities are the same, i.e., if $\lambda_{max} = \Lambda/M$, where Λ is the total demand rate.*

Let x_j be a vector denoting the location of facility $j, x_j \in P$, let $I_x^j = 1$ if the jth facility is the closest one to x and $I_x^j = 0$ otherwise, and let $R(x_j) = \max_{x \in P} \|xx_j\| \, I_x^j$ be the maximum travel distance of customers' assigned to facility j. Assume that at each $x \in P$ customers' demand rate is $\lambda(x)$ such that $\int_{x \in P} \lambda(x)dx = \Lambda < \infty$, thus $\lambda_{x_j} = \int_{x \in P} I_x^j \lambda(x)dx$ is the arrival rate to the jth facility. Let r be an exogenous given distance, which is the maximum distance allowed from a customer to a facility, and let ϵ be the smallest distance allowed between distinct facilities. The Voronoi region associated with the ith facility is denoted by V_i. Then, the demand rate to this facility is

$$\lambda_{x_i} \int_{x \in V_i} dx.dy \qquad \forall i = 1, \ldots, M \tag{4.10}$$

Assuming the Voronoi diagram is given, the equitable location problem can be formulated as

Algorithm 21 Generalized Discrete Lloyd Descent

1. Assign the mobile sensors $S = \{s_1, \ldots, s_N\}$ with $s_i = (p_i, \hat{n}_i, f_i)$
2. Assign the landmarks $L = \{\ell_1, \ldots, \ell_N\}$
3. Assign a partition $P = \{P_1, \ldots, P_N\}$
4. Assign $\epsilon > 0$ and set $Z_i \ \{s_i\}$ for $i \in \{1, \ldots, N\}$
5. while $Z_i \neq \emptyset$ for some $i \in \{1, \ldots, N\}$ do
6. pick s_i such that Z_i is not empty
7. pick $s_i \in Z_i$
8. for $\ell \in L_i$ do
9. if $per(s_j, \ell) < per(s_j, \ell) - \epsilon$ then
10. transfer ℓ from L_i to L_j
11. end if
12. end for
13. if one or more landmarks have been transferred then
14. $Z_i \leftarrow S \backslash \{s_i\}$
15. $Z_j \leftarrow S \ \{s_j\}$
16. $(p_i, \hat{n}_i) \leftarrow optcov(s_i, L_i, \Omega_i)$
17. $(p_j, \hat{n}_j) \leftarrow optcov(s_j, L_j, \Omega_j)$
18. else
19. remove
20. enf if
21. end while

Problem 142 ***Equitable location problem*** *Let P be a space $P \subset \mathbb{R}^2$ equipped with some norm $\|.\|$ and $M > 0$ denote the number of facilities. Given M facilities, the closest assignment constraints with a given distance norm, divides the plane to M areas using a Voronoi diagram of this distance norm, find the equitable location configuration.*

$$min\ \lambda^{max} \tag{4.11}$$

subject to

$$\begin{array}{rl} \lambda_{x_i} \leq \lambda^{max} & \forall i = 1, \ldots, M \\ \|x_i, x_j\| \geq \epsilon & \forall i, j = 1, \ldots, M, i \neq j \\ \|x_i, x\| \leq r & \forall i = 1, \ldots, M, x \in V_i \end{array} \tag{4.12}$$

When the values of M and r are small, $ELP(M)$ may be infeasible and therefore no feasible equitable facility configuration that is feasible exists. In [13], sufficient conditions for the existence of EFC are deduced, given the closest assignment constraints.

Remark 143 *The deterministic feasible equitable facility location is a major component in the* ***stochastic capacity and facility location problem*** *(SCFLP).*

The SCFLP focuses on three sources of uncertainty: the timing, location, and actual amount of the demand generated by the customers. This problem optimizes three types of decision variables:

1. The number of facilities to be located;
2. The location of the facilities;
3. The service capacity of each facility.

The approach to solve the stochastic capacity and facility location problem is based on the following. For a given M, the equitable location problem is solved to provide optimal location for the facilities and maximal demand rate to a facility, λ_{max}. The solution of equitable location problem on the unit square is investigated, with a uniform demand using Voronoi diagrams [74].

Team-based optimization scheme The locational optimization function can be translated to maximizing the sensing performance:

$$H(P, Q) = \sum_{i=1}^{N} \int_{W_i} f\left(\|q - p_i\|\right) \Phi(q) dq \tag{4.13}$$

where for n teams, $N = \sum_{t=1}^{n} n_t$, and P is the set of all UAVs. The ith UAV is assigned to the region W_i and the cost function H is minimized by finding the optimum locations of the UAVs and their assigned regions W_i whose union is

Q. A team-based partition of the agents considering agents as a collection of multiple teams pursuing their assigned task or objective is presented in [1]. The optimization problem is broken into two interconnected functions such that the solution to each problem represents the optimum configuration of the teams and their associated agents. Let $L = (\ell_1, \ell_2, \ldots, \ell_n)$ define the set of teams where each $\ell_t, t = 1, \ldots, n$ represents the nucleus of team t function of the agent's position in the associated team $\ell_t = g(p_{t_1}, \ldots, p_{t_{n_t}})$. The polytope Q is partitioned into a set of Voronoi cells: $V(L) = \{V_1, \ldots, V_n\}$ considered as the optimal partitioning for a set of agents with fixed locations at a given space as:

$$V_t = \{q \in Q, \|q - \ell_t\| \leq \|q - \ell_s\|\} \tag{4.14}$$

The obtained Voronoi cells associated with the nuclei of the teams are then considered as the convex polytopes set to deploy their associated agents. The Voronoi partitions $V_t(P_t) = \{V_{t_1}, \ldots, V_{t_{nt}}\}$ generated by the agents $p_{t_1}, \ldots, p_{t_{n_t}}$ belonging to the ith team are defined as:

$$V_{t_m} = \{q \in V_t, \|q - p_{t_m}\| \leq \|q - p_{t_r}\|\} \tag{4.15}$$

where p_{t_m} denotes the location of the mth in the tth team such that $m \in \{1, \ldots, n_t\}$. The basic characteristics of the **Voronoi partitions** such as:

1. **Associated mass:**

$$M_{V_{tm}} = \int_{V_{tm}} \Phi(q)dq \tag{4.16}$$

2. **Centroid:**

$$C_{V_{tm}} = \frac{1}{M_{V_{tm}}} \int_{V_{tm}} q\Phi(q)dq \tag{4.17}$$

3. **Polar moment of inertia:**

$$J_{V_{tm}, p_{tm}} = \int_{V_{tm}} \|q - p_{t_m}\|^2 \Phi(q)dq \tag{4.18}$$

The characteristics of the team's Voronoi cells can be deduced as:

$$M_{V_t} = \sum_{m=1}^{n_t} M_{V_{tm}} \tag{4.19}$$

$$C_{V_t} = \frac{1}{M_{V_t}} \int_{V_t} q\Phi(q)dq \tag{4.20}$$

The **nucleus of the team** is a function of the agents' position:

$$\ell_t = \frac{\sum_{m=1}^{n_t} M_{V_{tm}} p_{tm}}{\sum_{m=1} n_t M_{V_{tm}}} \tag{4.21}$$

It is a representative of the agents' position in the team and can be considered as the collective position of the agents for drawing the Voronoi diagram of the teams V_t. The deployment task can be addressed by solving a two-level optimization problem.

1. The first function to be minimized represents the cost associated with partitioning the main space into partitions related to the teams of agents.

$$G_t(P_t, Q_t) = \sum_{i=1}^{n_t} \int_{Q_{t_m}} f\left(\|q - p_{t_m}\|\right) \Phi(q) dq \tag{4.22}$$

2. The solution of the second optimization problem results in deploying the agents in an optimum way inside the teams:

$$G(L, Q) = \sum_{i=1}^{n} \int_{Q_t} f\left(\|q - \ell_t\|\right) \Phi(q) dq \tag{4.23}$$

   where the sensing performance is $f\left(\|q - p_i\|\right)$.

The extension of Lloyd algorithm can be used to solve this problem.

Cooperative Task Allocation UAV coordination has to be reached in the presence of multiple uncertainties in the environment and in the communication channels. The specific application considered in this paragraph involves the decentralized assignment of tasks in which UAVs can only receive delayed measurements from other vehicles, and the environment disturbances are able to disrupt the planned sequence of actions [39]. UAVs are required to perform different tasks on stationary targets with known locations. The optimal and conflict-free decentralized computation of assignment of tasks in the presence of communication delays, measurement noise, and wind disturbance is considered. **Conflict-free** means correct assignment of tasks to UAV in which a given task needs to be assigned to one and only one UAV and the tasks on the same target need to be performed in a certain order. UAVs will communicate only with a sub-set of UAVs or neighbors. Each UAV will estimate the position of every other UAV and obtain a list of assignments for the group. Local estimated positions as well as the estimates of other UAV positions are affected by zero mean white sensor noise. A **coordinated assignment plan** is required in order to guarantee that every task will be performed only once. This is to prevent that the same task on the same target is carried over by two different vehicles or that conflict in the plan occurs such that a given task is never performed. The cost function to be minimized is the cumulative distance the UAVs travel in order to perform all required tasks:

$$J = \sum_{i=1}^{N_u} D_i > 0 \tag{4.24}$$

A group of N_u UAVs, a set of targets $\{1, 2, \ldots, N_t\}$ and N_m the number of tasks to be performed at each target are considered. Each task is associated with an integer value, $N_s = N_t N_m$ is the number of single assignments and $S = \{1, 2, \ldots, N_s\}$ represents the set of stages. A decision variable $g_{i,j,k} \in$

$\{0,1\}$ is defined such that it is 1 if UAV $i \in U$ performs a task on target $j \in T$ at stage $k \in S$ and it is zero otherwise. The set of assignments, up to stage k, is represented by the list $G_k = \{\bar{g}_1, \bar{g}_2, \ldots, \bar{g}_k\}$ where $\bar{g}_k = [i,j]$ such that $g_{i,j,k} = 1$. The formulation of the cooperative multiple task assignment can be expressed as follows:

$$min\left(J = \sum_{i=1}^{N_u}\sum_{j=1}^{N_t}\sum_{k=1}^{N_s} d_{i,j,k}^{G_k} g_{i,j,k}\right) \tag{4.25}$$

subject to

$$\sum_{i=1}^{N_u}\sum_{j=1}^{N_t} g_{i,j,k} = 1 \quad k \in S \tag{4.26}$$

$$\sum_{i=1}^{N_u}\sum_{lk=1}^{N_s} g_{i,j,k} = N_m \quad j \in T \tag{4.27}$$

The constraints guarantee that exactly one task is assigned at any given stage and that on each target exactly N_m tasks are performed. A specific order on the tasks performed on the same target significantly increases the complexity of this optimization problem. To run a decentralized task allocation algorithm based on distance-to-task costs, each UAV needs an estimate of current positions of teammates. It is also desired to avoid continuous or frequent inter-UAV communication. At a reasonable increased computation cost, each UAV will implement models of all UAV dynamics, including itself. The assignment at each stage is found based on a cost matrix that evaluates the expected cost of each UAV to perform each one of the current tasks. Because of communication delays and other uncertainties, UAVs may arrive at different assignment plans. An algorithm for estimation and resolution of possible conflicts is based on generating new events when entries of the cost matrix are close to the minimum at any given stage of the optimization problem. To resolve an estimated conflict, UAVs bid on their best task at that particular stage. Their bids represent their cumulative cost on the task they are bidding on. Because these are real numbers representing the expected distance to travel to perform previous tasks and the conflicted task, the probability of UAVs bidding exactly the same cost is very low. This approach results in a trade-off between reducing inter-UAV communication and achieving a conflict-free assignment plan.

4.2.2 Mobile sensor network

Various schemes have been proposed for the deployment of **mobile sensor nodes** (MSNs), with optimal utilization of resources. The following classification of sensor deployment techniques has been proposed in [84]:

1. Based on placement strategy
   (a) Random
   (b) Deterministic
2. Based on usage
   (a) Barrier
   (b) Blanket
   (c) Area
   (d) Target oriented
3. Based on deployment domain
   (a) Open area (wide regions)
   (b) Indoor

4.2.2.1 Aerial networks

In 2-D space scenarios, the maximal coverage problem can be mapped to a circle packing formulation. The problem turns into the sphere packing problem in 3-D and the strategies designed for 2-D become NP-hard in 3-D. The problem of coverage in 3-D space is often a critical part of the scenario for the observation of an environment. The number of nodes and their locations are restricted by the investigated environment and the reception range of node. Moreover, the dynamic UAV network topology and flight must be handled efficiently considering the communication constraints of the UAVs. In [9], a node positioning strategy for UAV networks is proposed with a wireless sensor and actor network structure according to different capabilities of the nodes in the network. The positioning algorithm utilizes the **Valence shell electron pair repulsion** (VSEPR) theory of chemistry, based on the correlation between molecular geometry and the number of atoms in a molecule. By using the rules of VSEPR theory, the actor nodes in the proposed approach use a lightweight and distributed algorithm to form a self-organizing network around a central UAV, which has the role of the sink.

Minimization of sensor number The objective of the **sweep coverage problem** is to minimize the number of sensors required in order to guarantee sweep coverage for a given set of points of interest on a plane. Use of both static and mobile sensors can be more effective, in terms of energy consumption.

Definition 144 ***T-sweep coverage*** *A point is said to be* ***T-sweep*** *covered if and only if at least one mobile sensor visits the point within every T time period, where T is called sweep period of the point.*

The inputs of the algorithm proposed in [42] are the graph G, speed v, sweep period T, and energy consumption per unit time for static and mobile sensors λ and μ, respectively. The output of the algorithm is the number of mobile and static sensors with the deployment locations for the static sensors and the movement schedule for the mobile sensors.

Evasion paths In **minimal sensor network** problems, one is given only local data measured by many weak sensors but tries to answer a global question. The mobile sensor network problem is considered where sensors are ball-shaped. A sensor cannot measure its location but knows when it overlaps a nearby sensor. An evasion path exists if a **moving intruder** can avoid being detected by the sensors. The evasion problem can also be described as a **pursuit-evasion problem** in which the domain is continuous and bounded, there are multiple sensors searching for intruders, and an intruder moves continuously and with arbitrary speed. The motions of the sensors are not controlled; the sensors wander continuously but arbitrarily. The locations of the sensors cannot be measured but instead know only their time-varying connectivity data. Using this information, it is important to determine whether it is possible for an intruder to avoid the sensors. Both the region covered by the sensors and the uncovered region change with time. Zigzag persistent homology provides a condition that can be computed in a streaming fashion [24]. The technical basis for zigzag persistence comes from the theory of graph representations. However, no method with time-varying connectivity data as input can give necessary and sufficient conditions for the existence of an evasion path. The existence of an evasion path depends not only on the type of the region covered by sensors but also on its embedding in space-time. Both the region covered by the sensors and the uncovered region change with time. For planar sensors that also measure weak rotation and distance information, necessary and sufficient conditions for the existence of an evasion path are provided in [4].

Blanket coverage In the category of blanket coverage problems, the main objective is to maximize the total detection area. This coverage problem is defined as how to position or deploy the sensors in a particular **region of interest** (ROI) so that coverage percentage is maximized and coverage holes are minimized. The deployment of nodes can be done either randomly or deterministically. The deterministic deployment of nodes can be considered because sensor network coverage can be improved by carefully planning position of sensors in the ROI prior to their deployment. Grid-based sensor networks divide the ROI into square cells, and sensors can be placed at the center of the square cell in order to maximize the coverage, and also the number of sensors required for placing inside the square cell is less than the number of sensors required for placing at the intersection of the grids. The sensors can be placed at the center of the square cell. In case of grid-based deployment, problem of coverage of sensor field reduces to the problem of coverage of one cell and its neighbor because of symmetry of cells [87].

Problem 145 *Given N mobile nodes with isotropic radial of range R_s and isotropic radio communication of range R_c, how should they deploy themselves so that the resulting configuration maximizes the net sensor coverage of the network with the constraint that each node has at least K neighbors?*

Definition 146 *Two nodes are considered* ***neighbors*** *if the Euclidean distance between them is less than or equal to the communication range R_c.*

Three metrics are introduced to evaluate the performance of the deployment algorithm [77]:

1. The **normalized per-node coverage** defined as:

$$cov = \frac{\text{Net area covered by the network}}{N\pi R_s^2} \tag{4.28}$$

2. The **percentage of nodes** in the network that have at least K neighbors.

3. The **average degree** of the network.

In this deployment algorithm, **virtual forces** are constructed between nodes so that each node can attract or repel its neighbors. The forces are of two kinds. The first causes the nodes to repel each other to increase their coverage, and the second constrains the degree of nodes by making them attract each other when they are on the verge of being disconnected. By using a combination of these forces, each node maximizes its coverage while maintaining a degree of at least K.

Optimal UAV deployment for stationary nodes In order to find the optimal locations of UAVs functioning as communication relays at a fixed position between stationary nodes, the performance index for network connectivity is used. As the number of nodes increases, network complexity increases. Thus, the concept of **minimum spanning tree** can be used to obtain the highest probability of a successful transmission using minimum possible links. A **spanning tree** is a sub-graph that is itself a tree connecting all the vertices of the graph together. For a successful transmission, the weight of each graph node is

$$W_{ij} = -logP_r^{ij} \tag{4.29}$$

The smaller the weight, the higher the probability of a successful transmission. If the positions of the nodes are given, the minimum spanning tree can be constructed with the weight W_{ij}. The performance index for the global message connectivity can be set as:

$$J = \sum_{i=1}^{n} \sum_{j=1}^{n} \mathbf{A}_{ij} W_{ij} \tag{4.30}$$

where $\mathbf{A} \in \mathbb{R}^{n \times n}$ represents the adjacency matrix of the minimum spanning tree for a given configuration. The implementation of this deployment optimization can be centralized for a stationary environment [53].

Task execution In a multi-UAVs system, a mission can be divided into different tasks and a number of specialized UAVs can be introduced to solve each task concurrently. These tasks may be known by the UAVs before task execution stage or may dynamically appear during task execution. In an exploration mission, the aim is to locate and visit a number of pre-determined targets in a partially unknown terrain. The challenge is then how to assign these targets to the UAVs in order to optimize an overall system objective required by the mission. Multi-UAV **task allocation** problems are often solved in a distributed manner using **market-based algorithms**, while **auction algorithms** are efficient in terms of both computation and communication. The information of the UAVs and tasks can be compressed into numerical bids and computed in parallel by each UAV. For **single-assignment problems** where each UAV can handle at most one task, the single-item auctions can be used where UAVs bid on tasks that are auctioned off individually. The UAV with the highest bid wins the task and then has to finish it. However, for **multi-assignment problems** where each UAV is able to handle several tasks, they belong to the class of combinatorial optimization problem. Strong synergies exist between the tasks for bidders. It is considered that a set of tasks have a positive synergy for a UAV if their combined cost for executing them together is less than the sum of their individual cost incurred by doing them separately, and conversely for a negative synergy. A near-optimal allocation of a set of tasks to UAVs uses single-round combinatorial auctions, calculating the bid for every UAV, based on bundles of tasks rather than individual tasks. The bid for each UAV to hold a bundle is computed through the smallest path cost that needs to visit all tasks in the bundle from the UAV's current location.

Parallel single-item auctions treat each task independent of other tasks and every UAV bids for each task in parallel. Such mechanism has its computation and communication efficiencies while it leads to highly sub-optimal solutions since it does not account for any synergies between tasks. On balance, the sequential (multi-round) single-item auctions provide the advantages of solution quality from single-round combinatorial auctions, and computation and communication efficiencies from parallel single-item auctions. It works in a multi-round manner and in each round every UAV places bid on the unallocated tasks. The bid is computed as the smallest cost increase resulted from winning the task and the UAV with the overall smallest bid is allocated the corresponding task. The process is repeated until all the tasks have been allocated [105].

4.2.2.2 Visual coverage

The visual coverage problem differs in several respects from the standard coverage control.

1. While the standard coverage assumes **isotropic sensors**, the camera sensor has **anisotropic property**.
2. The image acquisition process of a camera sensor involves a non-linear projection from the 3-D world to 2-D image plane, which is significant especially in the monitoring problem in the 3-D world.
3. A camera sensor does not provide any physical quantity, while temperature or radiation sensors trivially sample a scalar field describing importance of each point in the environment.

Thus, computer vision techniques must be integrated with control scheme to extract the meaning of the sensed data [36]. A visual coverage problem is considered under the situation where vision sensors with controllable orientations are distributed over the 3-D space to monitor 2-D environment. In this case, the control variables, i.e., the rotation matrices must be constrained on the Lie group $SO(3)$. The problem is directly formulated as an optimization on $SO(3)$ and the gradient descent algorithm is applied on matrix manifolds. A vision sensor has an image plane containing the sensing array, whose pixels provide the numbers reflecting the amount of light incident. The objective function to be minimized by sensor i is defined by a sensing performance function and a density function at a point $q \in E$. The **sensing performance function** describes the quality of the acquired data about $q \in E$ and the **sensing density function** indicates the relative importance of $q \in E$. The function is accumulated only at the center of the pixels projected onto the environment E in order to reflect the discretized nature of the vision sensors. The gradient is derived assuming that the image density describing relative importance over the image is given in the form of the mixed Gaussian function. The gradient descent approach is a standard approach to coverage control; the rotation is updated in the direction of this gradient [44].

A **visual sensor network** (VSN) consists of a number of self-configurable visual sensors with adjustable spherical sectors of limited angle: **field of view** that are meant to cover a number of targets randomly positioned over a deployment area. One of the fundamental problems of visual sensor networks is to cover a maximum number of targets using the minimum number of sensors. Visual sensor networks can be classified in two different categories:

1. **Over-provisioned systems** when the number of sensors is sufficient to cover all the targets. In this coverage task, the number of cameras must be minimized besides maximizing coverage.

2. **Under-provisioned systems** when this number is insufficient, the target coverage should be maximized regardless of the number of cameras being used.

Two approaches can be considered [104]:

1. **Sensor-oriented approach**: One can look into the cameras and determine the exact coverage count in different field-of-views of each camera. While counting the coverage, the overlapping regions of the neighboring cameras must also be considered to exclude the possibility of redundant coverage.

2. **Target-oriented approach**: One can look into the targets first. Some targets might be located in a difficult corner of the deployment area and could be covered only by a single camera. In order to maximize the target coverage, those targets need to be covered first. The targets must be prioritized based on their coverage vulnerability and then one must select a minimal set of cameras that can cover targets in their order of priorities.

The distance at which a UAV camera sensor should be positioned from the target is

$$h = H\frac{f}{d} \tag{4.31}$$

where f is the focal length, h is the camera sensor height, d is the distance from the sensor to the far plane, and H is the height of the far plane of the frustum. A **view frustum** is a 3-D volume that defines how models are projected from camera space to projection space. Objects must be positioned within the 3-D volume to be visible. Points beyond this distance are considered invisible to the camera.

Definition 147 *The **functional coverage** is defined as the ratio of the area covered by the set of cameras to the area to be reconstructed.*

In order to determine the area covered by a given set of cameras, a viewing space for each camera is defined according to geographical position and the distance from the camera to the far plane. The terrain points that are visible to each individual camera are determined by an **occlusion test**, involving reflecting the point cloud in the frustum onto a spherical surface away from the camera. Any points that are not reflected are not included in the frustum; however, the entire set of visible points is included. The distance from the sensor to the object of interest determines the resolution of the final model. Several variables and sets of constraints are applied simultaneously in the process of adjusting each camera to find the optimal coordinates, position, altitude, and optimal flight path [80].

Problem 148 ***Schedule of the active periods of sensors:*** *Given*

1. *a set of targets* $T = \{t_1, t_2, \ldots, t_m\}$ *and their corresponding weights* $W = \{w_1, w_2, \ldots, w_m\}$,

2. *a set of homogeneous cameras* $S = \{s_1, s_2, \ldots, s_n\}$ *randomly deployed in a 2-D plane,*

3. *a sub-set* $F = \{S_{i,j}, 1 \leq i \leq n, 1 \leq j \leq q\}$ *sub-sets* $S_{i,j}s \subseteq T$ *computed by an identifiability test,*

4. *the required coverage-level* C_L, *specific to a given mission where*

$$\max_{k \in \{1,\ldots,m\}} w_k \leq C_L \leq \sum_{k \in \{1,\ldots,m\}} w_k \tag{4.32}$$

The problem is to schedule the active periods of each camera such that the sum of weights of all targets which are effectively covered is at least C_L *at any time and the network lifetime is maximized.*

This problem can be divided into two sub-problems: determining the direction (active sensor) of each node and assigning the sleep-wake-up schedule to it. A heuristic has been proposed in [47] to solve this problem.

Voronoi approach for the visibility landmark The problem in this paragraph is to consider a team of UAVs $a_i \in \mathbb{A}$, where $i = 1, \ldots, m$ and $m \in \mathbb{N}$, with a set of poses $\mathbf{A} = (A_1, A_2, \ldots, A_m)$. A set of m disjoint partitions of the set of landmarks $\mathbf{S}$ is $\mathbf{P} = (\mathbf{P}_1, \mathbf{P}_2, \ldots, \mathbf{P}_m)$. In this scenario, each UAV a_i is responsible of a sub-set $\mathbf{P}_j$ for $j = 1, \ldots, m$ with $\mathbf{P}_i \bigcap \mathbf{P}_j = \emptyset$. The visibility of a landmark is calculated with respect to the pose $\mathbf{A}_i$ of the UAV a_i that is responsible for this landmark [89]. The coverage score of a_i is given by:

$$C(\mathbf{A}_i, \mathbf{P}_i) = \sum_{s_k \in \mathbf{P}_i} vis(\mathbf{A}_i, s_k) \tag{4.33}$$

where the visibility of a point $s \in \mathbb{R}^2$ with respect to an agent a with generic pose $\mathbf{A}$ can be derived as

$$vis(A, s) = vis_{\mathbf{I}_3}\left(\left(\mathbf{A}^{-1}\tilde{s}\right)_{xy}\right) \tag{4.34}$$

where the homogeneous coordinates of s are $\tilde{s} = [s_x, s_y, 1]^T$ and

$$vis_{\mathbf{I}_3}(s) = \left\{ \begin{array}{cc} 0 & s \in \mathbf{R}_1 \\ s_x & s \in \mathbf{R}_2 \\ s_x / \|s\|^2 & s \in \mathbf{R}_3 \end{array} \right\}$$

while the partition of $\mathbb{R}^2$ is
$\mathbf{R}_1 = \left\{s \in \mathbb{R}^2 : s_x \leq 0\right\}$,
$\mathbf{R}_2 = \left\{s \in \mathbb{R}^2 : s_x > 0 \text{ and } \|s\| \leq 1\right\}$,
$\mathbf{R}_3 = \left\{s \in \mathbb{R}^2 : s_x > 0 \text{ and } \|s\| > 1\right\}$.

The coverage score of the whole team of UAVs is calculated as the coverage score of each UAV as:

$$C(\mathbf{A}, \mathbf{P}) = \sum_{i=1}^{m} C(\mathbf{A}_i, \mathbf{P}_i) \tag{4.35}$$

In order for the algorithm to take the landmarks partitions to a Voronoi configuration, while improving the visibility of each landmark and thus the overall coverage score, it has to be designed for handling non-trivial interaction between the agents, as well as simultaneous communications between different pairs of agents. After the initialization procedure has finished, all the agents first calculate the visibility of each landmark in their own set $\mathbf{P}_i$ from their current positions, then optimize the pose in order to maximize the coverage score on that set. The optimization process takes place every time the set of landmarks $\mathbf{P}$ of an agent changes, i.e., every time two agents trade some landmark successfully as in Algorithm 22.

Algorithm 22 Pose Optimization Procedure a_i

---

1. $old_{score} \leftarrow$ Old coverage score
2. $new_{score} \leftarrow$ New coverage score
3. $p \leftarrow$ current position of the agent
4. $\psi \leftarrow$ current orientation of the agent
5. $p_n \leftarrow$ optimized position of the agent
6. $\psi_n \leftarrow$ optimized orientation of the agent
7. $(p_n, \psi_n) \leftarrow$ optimization routine on (p, ψ)
8. for $s_k \in \mathbf{P}$ do
9. calculate visibility of s_k from (p_n, ψ_n)
10. calculate the coverage score of the agent from (p_n, ψ_n)
11. if $new_{score} > old_{score}$ then
12. new way-point $\leftarrow (p_n, \psi_n)$
13. else
14. new way-point $\leftarrow (p, \psi)$
15. end if
16. end for

The trading procedure is the main part of the coverage task, presented in Algorithm 23. This part involves actual communication between agents, with exchange of information about current pose of each agent and on the partition of landmarks currently owned by each UAV. A UAV can start its trading routine only if it has reached the last way-point generated by the optimization algorithm. The way-point is considered reached if the distance between the way-point and the current position of the UAV is below a certain threshold. In order to maintain consistence in the information about the ongoing coverage mission, a trading procedure can involve only two UAVs at a time.

Algorithm 23 Trading Algorithm for Client Agent a_i

$o_w \leftarrow$ last way-point generated by the optimization routine 22

$q_c \leftarrow$ client agent

$q_s \leftarrow$ server agent

$A_c \leftarrow$ the pose of q_c

$\mathbf{P}_c \leftarrow$ the landmark partition of q_c

$\mathbf{Q}_{c_{in}} \leftarrow$ initial $\mathbf{Q}_c$

$state_c \leftarrow (A_c, \mathbf{P}_c)$

if o_w is reached then

pick an item $q_i \in \mathbf{Q}_c$

$q_i \leftarrow q_s$

send to q_s state

if q_s is available and $n \neq 0$ landmarks can be traded then

call the optimization routine 22

$\mathbf{Q}_c = \mathbf{Q}_c\ q_s$

else if $\mathbf{Q}_i = \emptyset$ then

all possible trading completed

end if

end if

Visibility coverage in polygonal environments with hole A visual coverage problem can be considered under the situation where vision sensors with controllable orientations are distributed over the 3-D space to monitor a 2-D environment. In the case, the control variables, i.e., the rotation matrices must be constrained on the **Lie group** $SO(3)$. The problem is directly formulated as an optimization on $SO(3)$ and the gradient descent algorithm is applied on matrix manifolds. A vision sensor has an image plane containing the sensing array, whose pixels provide the numbers reflecting the amount of light incident. The objective function to be minimized by sensor i is defined by a **sensing performance function** and a **density function** at a point $q \in E$. The sensing performance function describes the quality of the acquired data about $q \in E$ and the density function indicates the relative importance of $q \in E$. The function is accumulated only at the center of the pixels projected onto the environment E in order to reflect the discretized nature of the vision sensors. The gradient is derived assuming that the image density describing relative importance over the image is given in the form of the mixed Gaussian function [44].

In another scenario, UAVs begin deployment from a common point, possess no prior knowledge of the environment, and operate only under **line-of-sight sensing and communication**. The objective of the deployment is for the agents to achieve **full visibility coverage** of the environment while maintaining line-of-sight connectivity with each other. This is achieved by incrementally partitioning the environment into distinct regions, each completely visible from some UAV. Approaches to visibility coverage problems can be divided into two categories:

1. Those where the environment is known a priori, in the **art gallery problem**, one seeks the smallest set of guards such that every point in a polygon is visible to some guard.

2. Those where the environment must be discovered, **simultaneous localization and mapping** (SLAM) techniques explore and build a map of the entire environment, then use a centralized procedure to decide where to send agents. For example, deployment locations can be chosen by a human user after an initial map has been built. Waiting for a complete map of the entire environment to be built before placing agents may not be desirable.

Problem 149 *The **Distributed Visibility-Based Deployment Problem with Connectivity** is stated as: Design a distributed algorithm for a network of UAVs to deploy into an unmapped environment such that from their final positions every point in the environment is visible from some UAV. The UAVs begin deployment from a common point, their visibility graph $G_{vis,E}(P)$ is to remain connected, and they are to operate using only information from local sensing and line-of-sight communication. Each UAV is able to sense its visibility gaps and relative positions of objects within line-of-sight. Additionally, the following main assumptions are made:*

1. *The environment* **E** *is static and consists of a simple polygonal outer boundary together with disjoint simple polygonal holes. Simple means that each polygon has a single boundary component, its boundary does not intersect itself, and the number of edges is finite.*
2. *UAVs are identical except for their unique identifiers* $(0, \ldots, N-1)$.
3. *UAVs do not obstruct visibility or movement of other UAVs.*
4. *UAVs are able to locally establish a common reference frame.*
5. *There are no communication errors nor packet losses.*

A centralized algorithm to incrementally partition the environment E into a finite set of openly disjoint star-convex polygonal cells is used. The algorithm operates by choosing at each step a new vantage point on the frontier of the uncovered region of the environment, then computing a cell to be covered by that **vantage point** (each vantage point is in the kernel of its corresponding cell). The frontier is pushed as more and more vantage point-cell pairs are added until eventually the entire environment is covered. The vantage point-cell pairs form a directed rooted tree structure called the partition tree [73].

4.2.2.3 Wireless sensor network

A **wireless sensor network** (WSN) is a distributed system of sensor nodes interconnected over wireless links. Sensors gather data about the physical world and transmit these data to a central through single-hop or multi-hop communications. Wireless sensor nodes have integrated batteries with limited energy resources. Sensor nodes can be either thrown in as a mass or placed one by one during deployment. **Deployment planning** requires consideration of several objectives such as energy consumption, sensing coverage, network lifetime, network connectivity, etc. Often these objectives conflict with one another, and operational trade-offs must be established during network design. In pre-determined deployment, the locations of the nodes are specified. This type of deployment is used when sensors are expensive or their operation is meaningfully affected by their position. Self-deployment is proposed as a technique assuming the sensors' own mobility. **Potential fields** or **virtual force** based approaches are used to spread sensors out from a compact or random initial configuration to cover an unknown area [2].

Wireless sensor network deployment strategy A major design step is to selectively decide the locations of the sensors in order to maximize the covered area of the targeted region. There are different sensor network deployment strategies: sensor capabilities, base station options, initial energy assignment, sensor locations, and traffic generation pattern are the key parameters used to describe WSN deployment strategies. Most strategies with fixed sinks try to optimize routing/transmission power. They try to maximize coverage while keeping the number of sensors at a minimum. Another strategy is the use of redundant sensors at some points for redundancy. Deployment strategies can be introduced as:

1. Coverage maximizing
2. Connectivity maximizing
3. Energy efficiency and lifetime optimization
4. Multi-objective deployment.

A common method is as follows [91]:

1. During deployment, UAVs can measure the average of the **received signal strength indicator** (RSSI) values, and then they can stop and place wireless sensor nodes where the average of measured RSSI values falls below a pre-determined threshold.
2. Determining a fixed distance between each sensor node.

The important contribution for implementing autonomous deployment on the multi-UAV system is saving time. In [8], the meaning of the time is duration of the complication deployment task.

Service optimization in a convex field Let Q be a convex field in $\mathbb{R}^2$ with a group of n agents randomly distributed in Q where the position of the ith agent is denoted by p_i. Let $\Phi : Q \to \mathbb{R}^+$ be a priority function representing the likelihood of an event taking place at any arbitrary point $q \in Q$. This function can reflect a measure of the relative importance of different points in the field. Let the strictly increasing convex function be $f_i : \mathbb{R} \to \mathbb{R}^+, f_i(q) = \alpha_i \left\| p_i - q \right\|^2, i = 1, \ldots, n$ denoting the cost of serving an event taking place at point q by the ith agent, α_i are pre-specified strictly positive coefficients. The cost function can encode the travel time or the energy consumption required to serve any point in the field. Let S denote a set of n distinct weighted nodes $\{(S_1, w_1), \ldots, (S_n, w_n)\}$ in a 2-D field where $w_i > 0$ is the weighting factor associated with the node $(S_i, w_i), i = 1, \ldots, n$.

Definition 150 *The* ***weighted distance*** *of a point q from the node $(S_i, w_i), i = 1, \ldots, n$ is defined as:*

$$d_w(q, S_i) = \frac{d(q, S_i)}{w_i} \tag{4.36}$$

where $d(q, S_i)$ denotes the Euclidean distance between q and S_i.

Problem 151 ***Service optimization problem:*** *The objective is to develop* *an appropriate agent-deployment algorithm such that the following cost function is minimized:*

$$H(P, W) = \sum_{i=1}^{n} \int_{W_i} f_i(q)\Phi(q)dq \tag{4.37}$$

where the set $W = \{W_1, W_2, \ldots, W_n\}$ represents a partition of the field Q into n regions, where the ith agent is in charge of serving all points in region W_i.

Minimizing the above cost function implies maximizing the overall efficiency of the network. When f_i are agent-dependent, the conventional Voronoi partitioning is no longer optimal. The **multiplicatively weighted (MW)-Voronoi diagram partition** is such that each region contains only one node, which is the nearest node, in the sense of weighted distance to any point inside the region. It is described as follows:

$$\Pi_i = \{q \in \mathbb{R}^2; d_w(q, S_i) \leq d_w(q, S_j), j = 1, \ldots, i-1, i+1, \ldots, n\} \tag{4.38}$$

Definition 152 *Given two points A and B in a 2-D plane and a positive constant k, the **Apollonian circle** $\Omega_{S_i, S_j, w_i, w_j}$ is the locus of any point E such that $AE/BE = k$.*

The smallest region generated by the Apollonian circles and containing the ith node is the ith MW-Voronoi diagram.

Definition 153 *The mass and center of mass of a region W_i with respect to the priority function Φ are respectively defined as follows:*

$$M_{W_i} = \int_{W_i} \Phi(q)dq \tag{4.39}$$

$$C_{W_i} = \frac{1}{M_{W_i}} \int_{W_i} q\Phi(q)dq \tag{4.40}$$

A distributed coverage control law for double-integrator dynamics is proposed in [77, 96].

4.3 PATROLLING

Patrolling is a standard way of addressing security threats. Securing citizens, infrastructures, forests, as well as wild animals is a critical concern around the world.

Definition 154 ***Patrolling*** *is the activity of going around or through an area at regular intervals for security purposes. The goal of patrolling is to repeatedly visit a set of positions and often to minimize the downtime between visits.*

However, security resources are often limited or expensive preventing full security coverage at all the times. Instead, these limited security resources can be allocated and scheduled efficiently while simultaneously taking into account adversary responses to the security coverage and potential uncertainty over such preferences and capabilities. It involves one or multiple decision-makers reasoning about the course of actions to achieve so as to cover an environment as quickly as possible [20]. The environment to patrol is commonly abstracted through a navigation graph and numerous works explore spanning trees or graph partitioning to compute minimal-cost cycles that assign efficient routes for each UAV in the patrolling mission. **Auctions and market-based coordination**, **task allocation**, **artificial forces**, **Gaussian processes** theory, **evolutionary algorithms**, **linear programming** modeling are also popular. Agents may bid to exchange vertices of the patrol graph to increase overall patrol performance. Based on a topological representation of the environment and using global/centralized information, optimal patrolling could eventually be obtained for the single UAV case by solving the traveling salesperson problem. For the multi-UAVs situation, optimal performance depends on the environment topology and the number of UAVs. Theoretically superior performance can be obtained either by optimal **k-way** graph partitioning, especially for high number of agents or graphs with unbalanced edges; or having all UAVs following the same traveling salesperson problem cycle, equally distributed in time and space, especially for low number of agents or balanced graphs. Both the traveling salesperson problem and the graph partitioning problem are NP-hard. Therefore, solving these problems is non-trivial, particularly in sparse topologies, which is the case of most real world environments [78].

For applications of multi-UAVs **persistent monitoring**, the goal can be to patrol the whole mission domain while driving the uncertainty of all targets in the mission domain to zero [88]. The uncertainty at each target point can be assumed to evolve non-linearly in time. Given a closed path, multi-UAVs persistent monitoring with the minimum patrol period can be achieved by optimizing the UAV's moving speed and initial locations on the path [98].

Recent work in patrolling can be classified as [91]:

1. **Offline versus online**: offline algorithm computes patrols before sensors are deployed, while online algorithm controls the sensor's motion during operation and is able to revise patrols after the environment has changed.

2. **Finite versus infinite**: finite planning horizon algorithm computes patrols that maximize reward over finite horizon, while infinite horizon maximize an expected sum of rewards over an infinite horizon.

3. **Controlling patrolling versus single traversal**: this is dynamic environment monitoring versus a one snapshot of an environment.

4. **Strategic versus non-strategic patrolling**.

5. **Spatial or spatio-temporal dynamics**.

4.3.1 Perimeter patrol

The application of UAVs in perimeter surveillance missions can be assumed as a patrolling mission along defined paths: the perimeter [82]. The main challenges involved with target localization include maintaining tracks of all potential targets within sensor coverage region and developing platform trajectories, such that the target localization error is minimized. Maintaining the sensor's field of view over the target is extremely challenging due to certain factors like limited UAV maneuverability, payload range constraints, environmental disturbances like wind or turbulence, etc. The function of tracking multiple targets in an aerial surveillance mission is further affected due to much more complex sources of uncertainties such as false alarms, ambiguity in data association, lower detection probability, the target sudden maneuvers, closed spaced target trajectories with multiple miss detections, etc.

The patrolling problem has a graphical structure: V is the vertex set of that graph and E the edges set. Let L be an $|V| \times |V|$ matrix in which L_{ij} is a real number that represents the time required to go travel from i to j if $[i,j] \in E$ and is infinite otherwise. Each vertex i has a non-negative importance weight w_i. **Idleness** can be used as a performance measure. The idleness of vertex i, noted τ_i represents the time since the last visit of a UAV to that vertex. The idleness is 0 if and only if a UAV is currently at vertex i and $\tau_{i+1} = \tau_i + \Delta t$ if there are no visits to i in the time interval $(t, t+\Delta t)$. As idleness is an unbounded quantity, exponential idleness is used $k_i^t = b\tau_i^t$ with $0 < b < 1$. It can be seen as the expected value of a **Bernoulli random variable** and k_i^t is the probability that this random variable is 1 at time t. b is the rate at which k_i decays over time. The probability evolves as $k_i^{t+\Delta t} = k_i^t b \Delta t$ if there are no visits to i during time interval $(t, t+\Delta t)$. If a UAV with noisy observations visits i at time t, idleness becomes 0 with probability $b < (1-a) \leq 1$, where a is the probability that the idleness does not become 0 when a UAV visits a vertex. If n UAVs visit vertex i at time $t + \Delta t$ and there were no visits since time t, then

$$k_i^{t+\Delta t} = k_i^t b \Delta t + 1 - a^n \tag{4.41}$$

An instance of the patrolling problem is a tuple $\langle L, w, a, b \rangle$ consisting respectively of the matrix L of edge lengths, the vector w of importance weights and parameters a, b.

In patrol missions, the status of some sites must be monitored for events [66]. If a UAV must be close to a location to monitor it correctly and the number of UAVs does not allow covering each site simultaneously, a path planning problem arises.

Problem 155 *How should the UAVs visit the locations in order to make sure that the information about all locations is as accurate as possible?*

One patrolling algorithm is based on a graph patrolling formulation on which agents use reinforcement learning on a particular MDP. The MDP is defined aver a countably infinite state space, assuming that as UAVs communicate by leaving messages on the nodes of the graph, leading to unrealistic communication model. Reactive algorithms, such as the **ant colony** approach, have been shown to perform well; however, this approach relies on the simplistic communication models relying on pheromones. When all locations are equally important, the shortest **Hamiltonian circuit** is an optimal solution for a single UAV. Multi-agent strategies using a unique cycle are the best whatever the graph is. However, as some locations may be more important than others, not visiting the less important ones from time to time may be advantageous.

In [50, 51], the following base perimeter **patrol problem** is addressed:

Problem 156 *A UAV (or more) and a remotely located operator cooperatively perform the task of perimeter patrol.* ***Alert stations*** *consisting of* ***unattended ground sensors*** *(UGS) are located at key locations along the perimeter. Upon detection of an incursion in its sector, an alert is flagged by the UGSs. The statistics of the alert arrival process are assumed known. A camera-equipped UAV is on continuous patrol along the perimeter and is tasked with inspecting unattended ground sensors with alerts. Once the UAV reaches a triggered unattended ground sensor, it captures imagery of the vicinity until the controller dictates it to move on.*

There are m alerts stations/sites on the perimeter where a nearby breaching of the perimeter by an intruder is flagged by an UGS. Upon detection of an incursion in its sector, an alert is flagged by the UGS. Camera-equipped UAVs are on a continuous patrol along the perimeter and their task is to inspect UGS with alerts. To determine whether an incursion flagged by a UGS is a false alarm or a real threat, a patrolling UAV flies to the alert site to investigate the alert. The longer a UAV loiters at an alert site, the more information it gathers; however, this also increases the delay in responding to other alerts. The decision problem for the UAV is to determine the **dwell time** so as to maximize the expected payoff. This perimeter patrol problem falls in the domain of discrete time controlled queuing systems. A **stochastic dynamic programming** approach may be employed to obtain optimal policies for the patrolling UAV. The customers are the flagged UGS/alerts waiting to be serviced and the UAVs are the servers. Only unit/single buffer queuing is considered, for the UGS either flags an alert or it does not. Once it flags an alert, its state does not change even if additional triggering events were to occur, until the flag is reset by a loitering UAV. Thus, this perimeter patrol problem constitutes a **multi-queue multi-server**, unit buffer queuing system with deterministic interstate travel and service times. Because the UAVs are constantly on patrol or are servicing a triggered unattended ground sensor, the framework considered here is analogous to a cyclic polling system. The basic model of a **cyclic polling system** consists of separate queues with independent Poisson arrivals served by a single server in cyclic order. A related problem is the **dynamic traveling repairmen problem**, where the stations are not restricted to being on a line segment or a closed path. One is interested from a patrolling perspective in the optimal service time in addition to the dynamic scheduling of the server's movement. The basic question then would be to decide how long the server/UAV should dwell at a triggered alert station/UGS as well as in which direction is a bi-directional server.

One objective may be to maximize the information gained and at the same time reduce the expected response time to alerts elsewhere [67]. The problem is simplified by considering discrete time evolution equations for a finite fixed number m of UGS locations. The UAV has access to real-time information about the status of alerts at each alert station. Because the UAV is constantly on patrol and is servicing a triggered UGS, the problem is a **cyclic polling system** in the domain of discrete time controlled queuing system. The patrolled perimeter is a simple closed curve with $N \geq m$ nodes that are spatially uniformly separated of which m are the alert stations (UGS locations). The objective is to find a suitable policy that simultaneously minimizes the service delay and maximizes the information gained upon loitering. A stochastic optimal control problem is thus considered. A **Markov decision process** is solved in order to determine the optimal control policy [21]. However, its large size renders exact dynamic programming methods intractable. Therefore, a state aggregation based approximate linear programming method is used instead, to construct provably good sub-optimal patrol policies. The state space is partitioned and the optimal cost to go or value function is restricted to be a constant over each partition. The resulting restricted system of linear inequalities embeds a family of Markov chains of lower dimension, one of which can be used to construct a lower bound on the optimal value function. The perimeter patrol problem exhibits a special structure that enables tractable linear programming formulation for the lower bound [11].

Remark 157 *Scalable swarm robotics [41] and game-theoretic [94] approaches allow taking into account adversary responses to the security strategy. The issue of designing security patrolling strategies using UAVs while providing strong performance guarantees requires guarantees on the level of security.* ***Partially observable stochastic patrolling game*** *(POSG) considers a general framework for modeling security patrolling problems with UAVs [43]. The goal of solving POSG is to find a patrol strategy that achieves the best security performance. In this game-theoretic framework, two teams are considered: defenders and attackers. The security game proceeds by alternating between defender and attacker decisions [76], [101], [102]. However, this security game repeats until the attackers get arrested. Other salient features include uncertain action effects and partial observability about both the state of the world and what each team does or plans to do.*

In multi-UAV Markov decision process, the problem is assumed to be fully observable, i.e., each UAV has the same complete information to make its decision. In the patrolling problem, however, the actions of each UAV have a concurrent effect on the environment. These actions have also different durations. Concurrency in decision processes is modeled with a **generalized MDP**. Such decision processes generalize **multi-UAV MDP** to continuous time with asynchronous events. The state variables for this problem describe the position of each UAV and the idleness of each vertex. If the total number of UAVs is N, the state space is

$$S = V^N \times [0, 1]^{|V|} \tag{4.42}$$

Given some states $s = (v, k) \in S$, v_i is the position of the ith UAV and k_i the idleness of the ith vertex. At various time points, called **decision epochs**, the UAV must choose an action. The actions from which a UAV can choose depend on the structure of the graph and on its position: if a UAV is at vertex v, it can choose its action from $A_v = \{u : [v, u] \in E\}$. If a UAV occurs at time $t^{i+1} = t^i + L_{vu}$ and $v^t = v$ while $t \in \left[t^i, t^{i+1}\right]$ and $v^t = u$ as soon as $t = t^{i+1}$. The problem is concurrent because the decision epochs of all UAVs can be interleaved arbitrarily, each component k_i of k and the number of UAVs n. Let $\left\{y^j\right\}_j$ be the non-decreasing sequence of decision epochs and n_i^j is the number of UAVs arriving at vertex i at time t^j, $\Delta t^j = \left\{t^{j+1} - t^j\right\}$:

$$k_i^{t^{j+1}} = k_i^{t^j} a^{n_i^{j+1}} b^{\Delta t^j} + 1 - a^{n_i^{j+1}} \tag{4.43}$$

The reward R is defined in terms of k, the rate at which reward is gained is given by:

$$dR = w^T k^t dt \tag{4.44}$$

The discounted value function for a generalized Markov decision process is defined as:

$$\begin{aligned} V^{\pi}(s) = E\left[\int_0^{\infty} \gamma^t dR\right] &= E\left[\gamma^{t^j} \int_0^{\Delta t^j} \gamma^t w^T k^t dt\right] \\ &= E\left[\gamma^{t^j} w^T k^{t^j} \frac{(b\gamma)^{\Delta t^j} - 1}{ln(b\gamma)}\right] \end{aligned} \tag{4.45}$$

where $\gamma \in (0, 1]$ is the discount factor.

On-line planning has the advantage that it solves Equation (4.45) only for the current state, in contrast with offline algorithms that do so for all states. The patrolling problem is simpler to solve online than offline. **Anytime error minimization search** (AEMS) performs a heuristic search in the state space. The search procedure proceeds using a typical **branch-and-bound** scheme. Since the exact long-term value of any state is not exactly known, it is approximated using lower and upper bounds, guiding the expansion of the search tree by greedily reducing the error on the estimated value of the root node. In the patrolling problem, actions have the same interpretation as in a partially observable setting, whereas observations are the travel durations. In anytime error minimization search, the error is defined using the upper bound and the lower bound on the value of some state. Let $s \in S$ be a state,

$L(s) \leq V(s) \leq U(s)$, where $L(s), U(s)$ are, respectively, the lower and upper bounds and $V(s)$ the actual value of s. Given some search tree T, whose set of leaf nodes is noted $F(T)$, the bounds for the root node are estimated recursively. When a UAV must choose an action, the action of maximum lower bound is chosen. A lower bound for the value of any state is the value of following any policy from that state. An upper bound is usually obtained by relaxing problem constraints, assuming that UAVs are ubiquitous (they can be in more than one location at the same time). Whenever a UAV reaches a vertex, it instantaneously multiplies itself and starts heading to adjacent unvisited locations. This bound estimates the shortest time that a team of UAVs would take to cover the entire graph and estimates through the discount factor and upper bound on the maximum reward obtainable. This bound implicitly assumes that the optimal policy does not require having more than one UAV at any vertex.

Extending anytime error minimization search to asynchronous multi-UAV is simple: whenever a node is expanded, there is a branch for every joint action and observation. **Asynchronicity** is handled with state augmentation. The state is now (s, η) where η_i is the time remaining before the next decision epoch of UAV i. At anytime t, the next decision epoch happens at time $t + min_i \{\eta_i\}$. The expand operation adds actions and observations for any UAV for which $\eta = 0$. Whenever UAV i performs an action of duration $\Delta t, \eta_i$ is assigned Δt. Otherwise, η_i is updated according to its depth in the search tree. Anytime error minimization search can be used to perform online planning for any sub-set of UAVs. The complexity being exponential in the number of UAVs, these UAVs are thus coordinated locally and a partial order is defined amongst UAVs. A UAV is said to be greater than (respectively less than) another UAV if it must choose its policy before (respectively after). The UAVs compute their policy according to that order. Once a UAV knows the policies of all greater UAVs, it proceeds to compute its policy and then communicates it to the lesser UAVs. Whenever a UAV selects its policy, it chooses the best policy given the policy of greater UAVs. A useful property of this coordination algorithm is that if the UAVs use an online anytime planner, then it is also anytime and online. A **fallback strategy** is to ignore the presence of the greater UAVs until their policy has been received.

4.3.2 Area cooperative patrolling

This section considers a problem where the area of interest has to be patrolled by a team of multiple UAVs from a frequency-based approach. These approaches assume that the probability of event appearing is equal along the whole area and are useful when there are not a priori information about the problem. The objective is to cover any position in the area with the maximum frequency, which is equivalent to minimize the elapsed time between each two consecutive visits to any position or refresh time. A solution where each position in the path is visited with the same period is the optimal solution to cover a path with a set of UAVs. Different cooperative patrolling strategies may be analyzed and compared from the elapsed time (frequency-based) and latency time criteria. The area partitioning strategy allows the system to obtain theoretically the optimal performance from frequency-based approach (**elapsed time criterion**) if the sub-areas are sized correctly. Surveillance scenarios usually involves large areas where the communications among the UAVs and with the control stations can not be guaranteed. In order to improve the robustness and scalability of the whole system, a distributed coordination control is the most efficient option to approach the surveillance missions with multiple UAVs. The distributed coordination method should allow the multi-UAV system to converge to a common cooperative patrolling strategy from local decisions and asynchronous communications among pair of them. Some authors propose algorithms based on peer-to-peer (or one-to-one) coordination. The algorithms based on the one-to-one coordination assume a different problem to be solved by each pair of contacting UAVs. Each pair of contacting UAVs addresses a reduced version of the whole problem considering just their own information. These algorithms require low information storage capabilities for the UAVs because they just have to store their own local information. Moreover, this technique has proved to converge to the desired solution, but its convergence time complexity is increased quadratically with the total number of UAVs. On the other hand, the algorithms based on the coordination variables assume that the problem can be totally described by a limited set of variables (the coordination variables) and that, using these variables, each UAV can solve independently the whole problem [2].

4.3.2.1 Multiple depot multi–TSP

The multi-UAV patrolling can be cast as a multiple depot **multiple traveling salesmen problem** (MTSP), where a set of m UAVs, located at different initial positions, must visit a set of n target locations and come back to their depots. The main objective is to find an efficient assignment of the target locations to the team of UAVs such that all the targets are covered by exactly one UAV and the cost is minimal. A distributed solution based on a market-based approach is presented in [58], involving the cooperation of the UAVs to incrementally allocate targets and remove possible overlap. In each step, a UAV moves and attempts to improve its solution while communicating with its neighbors. The approach consists of four main phases: an initial target allocation followed by a tour construction, then negotiation of conflicting targets and finally solution improvement.

The MTSP arises when several Hamiltonian tours are considered. It can in general be defined as follows:

Problem 158 ***Multiple traveling salesmen problem:*** *Given a set of nodes, let there be m salesmen located at a single depot node. The remaining nodes that are to be visited are called intermediate nodes. Then, the MTSP consists of finding tours for all m salesmen, who all start and end at the depot,* *such that each intermediate node is visited exactly once and the total cost of visiting all nodes is minimized.*

The cost metric can be defined in terms of distance and time [14]:

1. **Single versus multiple depots**: In the single-depot case, all UAVs start from and end their tours at a single point. On the other hand, if there exist multiple depots with a number of UAVs located at each, the UAVs can either return to their original depot after completing their tour or return to any depot with the restriction that the initial number of UAVs at each depot remains the same after all the travel. The former is referred as the fixed destination case, whereas the latter is named as the non-fixed destination case.

2. **Number of UAVs**: The number of UAVs in the problem may be a bounded variable or fixed a priori.

3. **Fixed charges**: When the number of UAVs in the problem is not fixed, then each UAV usually has an associated fixed cost incurring whenever this UAV is used in the solution.

4. **Time windows**: In this variation, certain nodes need to be visited in specific time periods, named as time windows. This is the **multiple traveling salesperson problem with time windows** (MTSPTW).

5. **Other special restrictions**: These restrictions may consist of bounds on the number of nodes each UAV visits, the maximum or minimum distance a UAV travels or other special constraints.

The MTSP can be defined by a complete graph $G(V, A)$ where $V\{0\}$ is the set of nodes to be visited, node 0 is the depot, and A is the set of arcs connecting the n nodes to be visited. A c_{ijk} value is associated with each arc (i, j) which is traversed by the salesperson k. The matrix C is symmetric when an undirected graph is considered and asymmetric otherwise. The traveled distance of a tour is calculated as the sum of the lengths of the arcs belonging to the tour. In addition, the travel time of a salesperson for each arc (i, j) of A is given by t_{ijk}. The travel time of a tour is calculated as the sum of the lengths of the arcs belonging to the tour. Given an integer m, the problem consists of finding m tours passing through the depot node. Each node must be visited by a tour only once so that the total traveled distance is minimized and the working times of the traveling salesmen are similar. The multi-objective version of the MTSP can be formulated as a multi-objective mixed integer linear model with two objective functions:

1. Minimization of the distance traveled by all the salesmen, and

2. Similar working times of the salesmen.

A multi-objective optimization problem can be formulated through a mathematical model defined by a set of h objective functions which must be minimized or maximized subject to a set of m inequality constraints, a set of ℓ equality constraints and lower and upper limits in the k decision variables. With each arc (i, j) of G, is associated a binary variable which takes the value of 1 if arc (i, j) was traversed by the salesperson k into the final solution and 0 otherwise [19].

Problem 159 *The optimization problem can be formulated as:*

$$\min Z_1 = \sum_{k=1}^{m} \sum_{i=1}^{n} \sum_{j=1}^{n} c_{ijk} x_{ijk} \tag{4.46}$$

$$\min Z_2 = \sum_{k=1}^{m} |t_{avg} - t_k| \tag{4.47}$$

where t_{avg} is the average travel times of the tours defined as:

$$t_{avg} = \frac{1}{m} \sum_{k=1}^{m} \sum_{i=1}^{n} \sum_{j=1}^{n} t_{ijk} x_{ijk} \tag{4.48}$$

and t_k is the time spent on each tour $k, k = 1, \dots, m$:

$$t_k = \sum_{i=1}^{n} \sum_{j=1}^{n} t_{ijk} x_{ijk} \tag{4.49}$$

The following set of constraints must be also considered, with the binary decision variables $x_{ijk} \in \{0, 1\}$:

$$\sum_{k=1}^{m} \sum_{i=1}^{n} x_{ijk} = 1, \forall j = 1, \dots, n \tag{4.50}$$

$$\sum_{k=1}^{m} \sum_{j=1}^{n} x_{ijk} = 1, \forall i = 1, \dots, n \tag{4.51}$$

These two constraints are related to the degree restrictions and ensure that each node is visited only once by a single UAV, except the depot node.

$$\sum_{k=1}^{m} \sum_{i=1}^{n} x_{i0k} = m \tag{4.52}$$

$$\sum_{k=1}^{m} \sum_{j=1}^{n} x_{0jk} = m \tag{4.53}$$

These two constraints ensure that each UAV must leave and return to the depot node.

$$\sum_{i \in S} \sum_{j \notin S} x_{ijk} \geq 1, \forall k = 1, \ldots, m, \forall S \subseteq V; 0 \in S \tag{4.54}$$

This constraint is the connectivity constraint, avoiding sub-tours generation on the final solution.

Many approaches can be proposed to solve this problem such as the computational intelligence with **genetic algorithm** [54], **artificial bee colony** algorithm [99], and **a set covering approach** [12].

4.3.2.2 Exploration

There are different strategies proposed for robotic exploration [95]:

1. **Frontier-based exploration**: A group of UAVs are directed to regions on the boundary between unexplored space and the space known to be open. In this strategy, each UAV builds and maintains its global map. During the mission, each UAV shares perceptual information with the others.

2. **Role-based exploration**: UAVs are categorized in two groups:

   (a) **Relays**: They maintain the connection between explorers and a base station responsible for controlling the mission. If relays discover information about the zone while relaying, this information is added to the exploration team knowledge.

   (b) **Explorers**: They explore an unknown zone and communicate their findings to a relay in previously agreed rendez-vous points. They use frontier-based exploration strategy.

3. **Market-driven exploration**: The exploration task is divided into subtasks. In this strategy, UAVs place bids on the sub-tasks and when communication strength is factored into the bids, UAVs avoid going beyond communication range.

The design of a multi-target exploration/visiting strategy for a team of UAVs in a cluttered environment is able to

1. Allow visiting multiple targets at once (for increasing the efficiency of the exploration), while

2. Always guaranteeing connectivity maintenance of the group despite some typical sensing/communication constraints representative of real-world situations,

3. Without requiring presence of central nodes or processing units, and

4. Without requiring that all the targets are known at the beginning of the task.

Designing a decentralized strategy that combines multi-target exploration and **continuous connectivity** maintenance imposes often antithetical constraints. A fixed-topology and centralized method which, using a virtual chain of mobile antennas, may be able to maintain the communication link between a ground station and a single UAV visiting a given sequence of target points. A similar problem is addressed by resorting to a partially centralized method where a linear programming problem is solved at every step of motion in order to mix the derivative of the second smallest eigenvalue of a weighted **Laplacian** or algebraic connectivity, or **Fiedler eigenvalue** and the **k-connectivity** of the system. A line-of-sight communication model is considered where a centralized approach, based on polygonal decomposition of the known environment, is used to address the problem of deploying a group of UAVs while achieving periodical connectivity. The case of periodical connectivity optimally solves the problem of patrolling a set of points to be visited as often as possible [72]. Triangulations in [60] provide complete coverage with only basic local geometry. The underlying topological structure of a triangulation allows to exploit its dual graph for mapping and routing, with performance guarantees for these purposes. The maximum edge length of a triangle is bounded by the communications range of the UAVs. If the number of available UAVs is not bounded a priori, the problem of minimizing their number for covering all of the region is known as the **minimum relay triangulation problem** (MRTP); if their number is fixed, the objective is to maximize the covered area, which is known as the **maximum area triangulation problem** (MATP). Both problems have been studied both for the offline scenario, in which the region is fully known, and the online scenario, where the region is not known in advance.

Covering in minimal time For the exploration and rescue application, an important goal is covering the whole environment to find possible victims in a minimal time. Compared with single-UAV systems, multi-UAV systems not only speed up the exploration process but also have advantages of robustness and redundancy. In addition, multi-UAV systems can finish tasks that the single UAV systems cannot do. In order to utilize the advantages of multi-UAV systems, it is essential to have a coordinated algorithm which realizes **moving target selection** for each UAV while considering the environment status and the moving targets of other UAVs. A primary method for multi-UAV coordination is **market economy based approach**. All UAVs in the team trade tasks and resources with each other to maximize individual profits. In the **threshold-based approach**, each UAV has to handle the closest event, without duplicating another UAV's work, the stimulus $\sigma(r, e)$ produced by an event e for UAV r can be proposed as:

$$\sigma(r,e) = \frac{1}{d(r,e)} \quad \theta_e = \frac{1}{D_r} \quad p_e = \frac{\sigma(r,e)^n}{\sigma(r,e)^n + \theta_e^n} \tag{4.55}$$

In [47], the best performance was found when the threshold θ_e for every event e is equal to the inverse of expected distance D_r between the UAVs. This threshold value is consistent with the best reserve price for the market.

Coordination in a unknown environment :
A **distributed bidding algorithm** (DBA) can be proposed to coordinate the multi-UAV systems exploring unknown environments where a nearness measure was introduced in bidding to maintain all UAVs close to each other to overcome the shortage of communication range. On the other hand, a decision theoretic approach dispersed UAVs over an unknown environment through a utility calculation scheme that a UAV utility reduction was inversely related to distances to targets of other UAVs. For a known environment, task allocation is realized by repeating the auction algorithm when a UAV finished its task. More recently, an **energy consumption based bidding scheme** was proposed. In the proposed approach, a hierarchical coordinated architecture for multi-UAV exploring an unknown environment is proposed where the lower level is developed by maximizing two kinds of new utility, respectively. The other kind of improvement is to deal with heterogeneity in multi-UAV systems. The method developed is that each UAV offers three bids from the aspects of the shortest path length, moving speed and battery life, respectively. A back-propagation neural network based approach was proposed to decide which UAV was the winner. The method used vectors to express capabilities of UAV and task, respectively. The auctioneer broadcasts a task and its capability vector, and each UAV bids for the task based on its capability. The task allocation was realized within the framework of **contract net protocol**. Finally, an index for describing exploration performance of each UAV is proposed for a heterogeneous multi-UAV system. However, the approaches mentioned above depend on a perfect communication among all teammate UAVs. Different to the existing results, the proposed algorithm in [30] improves the market economy based approach for the purpose of dealing with both target selection and heterogeneity. **Colored Petri nets** were used to realize the complex and concurrent conversations between agents. According to whether there is an installed task detection device, all UAVs are categorized into initiators or helpers, respectively. The latter assisted the former to finish task when it matched the task requirements. The UAVs can change their teammate selection strategies to adapt to a dynamic environment.

Communication Early exploration strategies were based on the principle of keeping the UAVs within communication range. Some authors propose market-driven exploration strategies, in which an exploration task is divided into sub-tasks and UAVs place bids on these sub-tasks. In these strategies, goal point selection is an important decision and aims to select unexplored regions for UAVs to visit. Here, bids are based on specific values such as traveling cost to a particular location and expected information gain. Though a market-driven exploration approach does not rely on perfect communication, and is still functional with zero communication, when communication strength is factored into these bids, UAVs avoid going beyond the communication range. Goal point selection strategies are

1. **Random goal point** selection is the simplest strategy, in which goal points are chosen randomly. If the area that surrounds the goal point has already been visited, it is discarded. This strategy has been effectively used in practice.
2. In the **greedy exploration strategy**, a goal point in the closest unexplored region is selected as a candidate and this strategy is effective for single UAV exploration.
3. **Frontier-based exploration** is an exploration strategy with a key idea of gaining the most new information by moving to the boundary between open space and uncharted territory. In this strategy, UAVs share perceptual information while maintaining separate global maps. Even if UAVs are dispersed to the maximum extent which their communication ranges allow, unexplored spaces may remain.

A solution to deal with the limited communication during multiple UAV exploration is rendez-vous points in which shared knowledge is communicated. **Clustering approach** was proposed by some authors, in which groups of UAVs stay close together while exploring the zone. Another strategy using the same principle is **role-based exploration**. In this strategy, UAVs are categorized into two groups: explorers and relays. While explorers explore an unknown zone by using frontier-based exploration strategy and communicate their findings to a relay in rendez-vous points, relays maintain the connection between the base station responsible for the mission and the explorers. Role-based exploration strategy offers a solution to connectivity related issues in large environments at the expense of additional UAVs responsible for messaging. Though most exploration strategies are successful in maintaining connectivity during an exploration mission, their performances are limited due to the constraint of having to keep the UAVs within communication range [95].

Ergodic trajectories In the context of exploration, **ergodic trajectory optimization** computes control laws that drive a dynamic system along trajectories such that the amount of time spent in regions of the state space is proportional to the expected information gain in those regions. Using **ergodicity** as a metric encodes both exploration and exploitation, both needed for non-myopic search when variance is high and convexity is lost, as well as myopic search when variance is low and the problem is convex. By encoding these needs into a metric, generalization to non-linear dynamics is possible using tools from optimal control. Different dynamical systems can achieve nearly identical estimation performance using **ergodic exploration of distributed information**. The ability to actively explore and respond to uncertain scenarios is critical in enabling UAVs to function autonomously. Active sensing or sensor path planning refers to control of sensor parameters, such as position, to acquire information or reduce uncertainty. Planning for search/exploration is challenging as the planning step necessarily depends not only on the sensor being used but also on the quantity being estimated, such as target location versus target size. Methods for representing and updating the estimate and associated uncertainty, and the belief state to determine expected information are, therefore, required. The best choice for representing and updating the belief state for a given application depends on UAV dynamics, sensor physics, and the estimation task (modeling a field versus target localization). Ergodic theory relates the time-averaged behavior of a system to the space of all possible states of the system. Ergodicity can be used to compare the statistics of a search trajectory to a map of **expected information density** (EID). An efficient exploration strategy should spend more time exploring regions of space with higher expected information, where useful measurements are most likely to be found. The UAV should not, however, only visit the highest information region, but distribute the amount of time spent searching proportional to the overall expected information density. This is the key distinction between using **ergodicity** as an objective and previous work in **active sensing** (e.g., information maximization); the ergodic metric encodes the idea that measurements should be distributed among regions of high expected information. Information maximizing strategies otherwise require heuristics in order to force subsequent measurements away from previously sampled regions so as not to only sample the information maxima [69].

Human–UAV collaboration To realize effective **human-UAV collaboration** (HAC) under uncertainty, in the search task, a UAV is capable of searching the mission domain autonomously by utilizing its cameras, thus allowing the exploration of a potentially dangerous environment that would otherwise be unsafe for a human to enter. However, due to the limitations in sensing capabilities and image processing, it is difficult for the UAV to confirm a found object of interest (e.g., to distinguish between human, animal, and mechanical movement). This might lead to errors (missed detections or false alarms) and hence low overall task performance. It is therefore necessary to have timely and more effective manual processing of these camera images. When situation demands, the UAV will send an intervention request to the human operator through a human-machine interface. However, keeping human monitoring for a long duration will probably make the operator cognitively overloaded due to decrements in vigilance, which consequently degrades the overall performance. This is especially true in the case when the human operator is required to collaborate with multiple UAVs in a large-scale task domain. Hence, a reasonable solution is to switch between the manual and autonomous operations to balance human workload while guaranteeing a desired level of task efficiency and accuracy. This type of mixed autonomous operation and tele-operation is called tele-autonomous operation. Existing decision making approaches such as Bayesian methods and data-driven techniques for process monitoring and control almost always seek optimal solutions based on measurements in the presence of uncertainty [96].

Uncertain terrain UAVs often fly in a complicated environment. Many threats such as hills, trees, and other aircraft can be fatal in causing the UAV to crash. These threats can only in general be detected within a limited range from a single UAV. However, by sharing information with other UAVs, these threats can be detected over a longer distance. Furthermore, an effective path for navigation should be smooth, provide an escape route and must be computationally efficient. In previous work on path planning for a single UAV, Voronoi graph search and visibility graph search have been proven to be effective only in a simple environment. They are not real-time and also lead to fatal failure when the map information is not entirely available, such as when obstacles are not detected. Path planning of multiple UAVs concentrates on the **collaborative framework**, **collaborative strategies** and **consistency**. The Voronoi graph search and the A^* or Dijkstra algorithms plan a global path for multiple UAVs to simultaneously reach the target in an exhaustive procedure [7, 8, 65]. Path planning of multiple UAVs can also be addressed from the perspective of reinforcement learning. Q-learning is a way to solve the path planning problem. The basic idea of Q-learning is to obtain the optimal control strategy from the delayed rewards according to the observed state of the environment in a learning map, and to make a control strategy to select the action to achieve the purpose.

Modeling of probabilistic risk exposure to obstacles can be presented as follows. It is necessary for a UAV to keep a certain distance from regions of high risk to ensure safe flying. So the measure of probabilistic risk exposure to obstacles can be seen as a continuous distribution function. For example, consider the case where an obstacle is at position (x_i, y_i, z_i) the measure of the risk is denoted by F_i, in which the parameters are related to the dimension of the planning space. In the 3-D space,

$$F_i(x, y, z) = \frac{1}{\sqrt{2\pi}\sigma_i} \exp\left(-\frac{d_i^2}{2\sigma_i}\right) \tag{4.56}$$

where

$$d_i = \sqrt{(x - x_i)^2 + (y - y_i)^2 + (z - z_i)^2} \tag{4.57}$$

σ_i is an adjustable parameter. The probabilistic risk of the area where UAVs could not fly over can be represented as a very big value. Furthermore, when more than one obstacle exists on the map, the probabilistic risk at position (x, y, z) can be calculated as:

$$F(x, y, z) = 1 - \prod_{i=1}^{M} [1 - f_i(x, y, z)] \tag{4.58}$$

The key idea of cooperative and geometric learning algorithm is to calculate the cost matrix $\mathbf{G}$ which can be used to find the optimal path from a starting point to a target point in terms of distance and integral risk. Each element in the matrix $\mathbf{G}$ is defined to be the sum of cost from its position to a target point. The process underlying path planning in an unknown environment can be modeled as a controlled Markov chain or a Markov decision process. In the case of an unknown environment, the transition probabilities of the underlying process are not known and have to be estimated while planning paths in the environment. This corresponds to **adaptive control of a Markov decision process**. The state space is defined by the ordered pair (s, q) where s represents the system state and q denotes the environment state. The system state is assumed to be known perfectly, while the environment state may be noise corrupted. It is assumed that the terrain does not change throughout the duration of the path planning maneuvers. If the environment at any system state can be modeled as a stationary random process, the method developed in [25] can be applied. The system state is assumed to be finite (for example, in 2-D, 8 states: North, East, South, West, North-East, North-West, South-East, South-West; in 3-D, 26 possible states), N is the number of system states, D the number of environment states, M the number of controls, and U is the set of control actions.

Problem 160 *Let* $\mathbb{F}^T = \{(s_0, q_0), u_0, \dots, (s_{t-1}, q_{t-1}), u_{t-1}\}$ *represent the history of the process till t then*

$$Prob\left((s_t, q_t)|\mathbb{F}^T\right) = Prob\left((s_t, q_t)|(s_{t-1}, q_{t-1}), u_{t-1}\right) \tag{4.59}$$

The dynamics of the system can be represented probabilistically through the probability density function $Prob\left((s_t, q_t)|(s_{t-1}, q_{t-1}), u_{t-1}\right)$ *where the function represents the probability of the system making a transition to state* (s_t, q_t) *given that the current state is* (s_{t-1}, q_{t-1}) *and the current control is* u_{t-1}.

The current system state s_t is assumed independent of the past environment state q_{t-1}:

$$Prob\left(s_t|(s_{t-1}, q_{t-1}), u_{t-1}\right) = Prob\left(s_t|s_{t-1}, u_{t-1}\right) \tag{4.60}$$

The above assumption is satisfied if the UAV is controllable. The current environment state q_t is dependent only on the current system state:

$$Prob\left((s_t, q_t)|(s_{t-1}, q_{t-1}), u_{t-1}\right) = Prob\left(s_t|s_{t-1}, u_{t-1}\right) Prob\left(, q_t|(s_t)\right. \tag{4.61}$$

The probability distribution function quantifies the localization and the control uncertainties inherent in the system and is assumed to be known beforehand. The terrain/environment uncertainty $p(q_t, s_t)$ is unknown and successive estimates $p_t(q_t, s_t)$ are made of the uncertainty as the planning proceeds to completion. The question is then how to use this increasing information of the terrain in order to plan better paths. In fact, the goal of path planning can be framed as an **infinite horizon discounted stochastic optimization problem**. The above assumption defines the goal of path planning in an uncertain terrain, i.e., the average value of the total discounted cost incurred by the system needs to be minimized. Adaptive control involves controlling an uncertain system while simultaneously obtaining better estimates of the system parameters. The problem of path planning can be envisaged as one of adaptive control of an uncertain Markov decision process, i.e., the transition probabilities are not known. In such a scenario, the strategy of adaptive control is to use the policy that is optimal with respect to the current estimate of the system, since it corresponds to the current knowledge of the system that is being controlled and referred to as the certainty equivalence principle in adaptive control. This is the equivalent of the **persistent excitation condition** in adaptive control which seeks to tackle the exploration/exploitation tradeoff inherent in every adaptive control problem. More details on this implementation can be found in [25].

Rescue in Uncertain Adversarial Environment The rescue path planning problem is a variant of a resource and risk constrained shortest path planning. Non-linearity characterizing problem complexity introduces a composite measure of performance coupling rescue UAV **survivability**, and path length subject to multiple side constraints makes problem solving very challenging [15]. Examples may be hurricane or earthquake rescue tasks [6]. The simplest rescue path planning problem in an uncertain adversarial environment involves a UAV planning a trajectory over a certain 2-D area avoiding obstacles and threats in order to rescue stranded individuals while minimizing a combination of traveled distance and threat level exposure of an equivalent survivability measure subject to a variety of temporal, itinerary and/or survivability constraints such as deadline or survivability threshold. The UAV is assumed to move from a base station (source site s) to a rescue service point. The adversarial environment is pictured as a grid or lattice composed of N cells. Cells are colored using a spectrum of shades to depict various threat levels (or survivability) ranging from white for safe cells to red for physical obstruction or zero survivability cells. Modeled as a directed graph $G(V, A)$, the grid specifies a set of vertices or nodes representing individual cells connected to one another (center to center) through a set A of arcs (i, j). Arcs define neighborhood relationship or possible moves between adjacent cells i, j. Typically, in a 2-D environment, a non-boundary node i has 8 outgoing arcs and 8 incoming arcs. The rescuing UAV is assumed to move one step at a time. Map representation uses a discretized probabilistic threat risk exposure of survivability. The grid is initally built from the probabilistic threat presence $p_T(\ell)$ estimated from a previous search task. Threat risk exposure, UAV destruction or detection risk contribution $p_{dest}(\ell, j)$ in cell j from threat presence in cell ℓ for a UAV traveling over a given cell is then derived from the map as:

$$risk_j = 1 - \underbrace{1 - p_{dest}(\ell, j)p_T(\ell)}_{survivability_j} \tag{4.62}$$

or

$$risk_j = 1 - p_{s_j} \tag{4.63}$$

where p_{s_j} is the UAV probability to survive cell j visit.

The basic rescuing path planning problem consists in minimizing traveled distance and threat risk exposure measures. Given user preferences or threat risk exposure over traveled distance, a UAV must move from a source node s to a destination site d over an $N-$cells grid environment characterized by a survivability distribution P_s, progressively constructing a path by making decisions x_{ij}, binary decision variables on visiting arcs (i, j). The ideal solution would be to find the shortest and safest paths separating source s and destination d sites subject to various constraints. The problem can be formulated as follows:

Problem 161

$$\min_{x_{ij}} (F = (1 - \alpha)L + \alpha(1 - S)) \tag{4.64}$$

where

$$0 \leq L \leq 1 \text{ and } 0 \leq \alpha \leq 1 \tag{4.65}$$

with

$$L = \frac{\sum_{(i,j)\in A} d_{ij}x_{ij} - L_{min}}{L_{max} - L_{min}} \tag{4.66}$$

and

$$S = \prod_{(i,j)\in A} p_{sj} \tag{4.67}$$

subject to

$$\sum_{j\in V} x_{sj} = 1 \quad \text{path starts at source node, } s \tag{4.68}$$

$$\sum_{j\in V} x_{jd} = 1 \quad \text{path ends at destination node, } d \tag{4.69}$$

$$\sum_{i\in V} x_{ij} \leq 1 \quad \text{At most one visit per node} \tag{4.70}$$

$$t_i^s + d_{ij} - K(1 - x_{ij}) \leq t_j^s \quad (i,j) \in A, k \in \mathbb{N} \quad \text{Disjoint sub-tours elimination} \tag{4.71}$$

$$\sum_{j\subset V} x_{ij} - \sum_{j\in V} x_{ji} = 0, \forall i \in V/\{s,d\}, (i,j) \in A \quad \text{Flow constraint/continuity} \tag{4.72}$$

$$\sum_{(i,j)\in A} d_{ij}x_{ij} \leq L_{max}, \forall i \in V/s,d, (i,j) \in A \quad \text{Itinerary constraint/deadline} \tag{4.73}$$

$$- \sum_{(i,j)\in A} x_{ij} \log(p_{sj} \leq \log(S_m), S_m \in [0,1] \quad \textit{Survivability threshold} \tag{4.74}$$

$$x_{ij} = 0 \forall (i,j) \notin A \tag{4.75}$$

$$x_{ij} \in \{0,1\}, t_i^s \geq 0 fori \in V \tag{4.76}$$

Where F is the objective function, α is a user-defined threat exposure, S_m is the minimum survivability probability, P_s survival probability distribution over the N cells grid, x_{ij} is a binary decision variable representing the UAV positive ($x_{ij} = 1$) or negative ($x_{ij} = 0$) decision to travel along arc (i, j), d_{ij} is the distance separating centroids from cells i and j, L_{max} is user defined maximum distance or $= 2\sqrt{N}$ and $L_{min} = \sqrt{2N}$. They are respectively the maximal and minimal possible traveled distance from s to d, and S is the overall UAV survivability on path completion.

The **survivability biased risk free path** $\alpha = 1$ attempts to minimize any threat exposure at the expense of extra travel cost, whereas the strictly distance biased route $\alpha = 0$ slightly move around the deadly core of the threat to satisfy minimal survivability requirements. This problem is then simplified to a quadratic objective function approximation through logarithmic survivability function transformation. It consists in replacing S by an approximate function $\bar{S}$ having the property to make the resulting approximate quadratic program formulation capture key elements of the original problem [15].

One computationally low cost and efficient on-line complete coverage path planning approach for an unknown environment is presented in [52]. The coverage task is performed by following an online **boustrophedon** motion along with an efficient backtracking technique called **two-way proximity search** (TWPS) to reduce total coverage time. The proposed algorithm generates the shortest possible path for backtracking. For the complete coverage path planning, it is assumed that the environment must be closed, where all the regions are connected so that any accessible position within the environment is reachable by UAV.

4.4 FORAGING

Foraging robots are mobile robots capable of searching for and, when found, transporting objects to one or more collection points. Foraging robots may be single robots operating individually, or multiple robots operating collectively. Single foraging robots may be remotely tele-operated or semi-autonomous; multiple foraging robots are more likely to be fully autonomous systems. The study of foraging strategies has led to the stochastic optimization methods such as: **ant colony optimization** (ACO), **particle swarm optimization** (PSO), **bacterial foraging optimization algorithm** (BFOA), **artificial bee colony** (ABC), and **information foraging theory** (IFT), among others [26].

4.4.1 Problem formulation

4.4.1.1 Abstract model

A **finite state machine** (FSM) can represent a foraging UAV. In the model, the UAV is always in one of four states [97]:

1. **Searching**: In this state, the UAV is physically moving through the search space using its sensors to locate and recognize the target items. At this level of abstraction, it could, for instance, wander at random, or it could employ a systematic strategy such as moving alternately left and right in a search pattern. The fact that the UAV has to search follows from the real-world assumptions that either the UAV's sensors are of short range and/or the items are hidden (behind occluding obstacles for instance); in either event, the UAV cannot find items simply by staying in one place and scanning the whole environment with its sensors. Object identification or recognition could require one of a wide range of sensors and techniques. When the UAV finds an item, it changes state from searching to grabbing. If the UAV fails to find the target item, then it remains in the searching state forever; searching is therefore the default state.

2. **Grabbing**: In this state, the UAV physically captures and grabs the item ready to transport it back to the home region. Here a single UAV is capable of grabbing and conveying an item. As soon as the item has been grabbed, the UAV will change state to homing.

3. **Homing**: In this state, the UAV must move, with its collected object, to a home or nest region. Homing clearly requires a number of stages, firstly, determination of the position of the home region relative to where the UAV is now, secondly, orientation toward that position and, thirdly, navigation to the home region. Again there are a number of strategies for homing: one would be to re-trace the UAV's path back to the home region; another would be to home in on a beacon with a long-range beacon sensor. When the UAV has successfully reached the home region it will change state to depositing.

4. **Depositing**: In this state, the UAV deposits or delivers the item in the home region, and then immediately changes state to searching and hence resumes its search.

Successful object collection and retrieval can be formulated as follows:

$$F(O_i, t) = \left\{ \begin{array}{cc} 1 & \text{Object } O_i \text{ is in a sink at time } t \\ 0 & \text{Otherwise} \end{array} \right\} \tag{4.77}$$

If the foraging task is performance time limited and the objective is to maximize the number of objects foraged within fixed time T, then a performance metric for the number of objects collected in time T, can be defined:

$$P = \sum_{i=1}^{N_O} F(O_i, t_0 + T) \tag{4.78}$$

where N_o is the number of objects available for collection and t_0 is the start time. A metric for the number of objects foraged per second $P_t = P/T$ is independent of the number of UAVs. In order to measure the performance improvement of multi-UAV foraging, the **normalized performance** P_m of a multi-UAV system,

$$P_m = \frac{P}{N_r} \tag{4.79}$$

where N_r is the total number of UAVs. The **efficiency** of multi-UAV foraging is then the ratio P_m/P_s.

4.4.1.2 Continuous foraging

In continuous foraging, UAVs visit locations in the environment to forage resources and deliver them to a home location. The resources replenish over time, and three models of resource replenishment can be considered: The first two are the **Bernoulli** and **Poisson** models where resources replenish probabilistically, which is suitable for scenarios with independently occurring resources. The third model is the **stochastic logistic model** where the rate of resource replenishment depends on the number of existing resources. The logistic model is suitable when the resources are populations of living things. The problem can be formulated as follows [63]:

Problem 162 *Maximize the rate of resources foraged to l_0 the UAVs' home location after T time-steps, i.e., maximize $\frac{v_{0,T}}{T}$ where $A = \{a_1, \dots, a_n\}$ be the foraging UAVs, each UAV a_i has an associated speed V_i, maximum capacity c_i, and payload $y_i \leq c_i$, i.e., the number of resources a_i is currently carrying. Let R be the reconnaissance UAV that performs information gathering and $L = \{l_0, \dots, l_m\}$ be the set of locations, where $v_{j,t}$ be the number of resources available at location l_j at time-step t. The number of resources at each location changes over time, in a Markovian way, i.e., the number of resources $v_{j,t}$ at a location l_j at time-step t depends only on $v_{j,t-1}$. Let $\hat{v}_{j,t}^{(i)}$ be a_i's estimate of $v_{j,t}$. When a foraging UAV a_i arrives at a location $l_j (j > 0)$, $max(v_{j,t}, c_i - y_i)$ resources at l_j (i.e., all resources at l_j subject to the remaining capacity of a_i) are transferred to a_i, and a_i makes an observation $\hat{v}_{j,t}^{(i)}$ of the number of resources remaining.*

When a_i arrives at l_0, all y_i resources carried by a_i are transferred to l_0. Let $D : L \times L \rightarrow \mathbb{R}^+$ be the distance function of the locations. Thus, a UAV a_i takes $t(a_i, l_j, l_k) = \frac{D(l_j, l_k)}{V_i}$ time-steps to move from location l_j to location l_k.

Every time-step, the reconnaissance UAV R observes the number of resources at $M \leq m$ locations. This corresponds to the UAV being launched from l_0, instantaneously visiting M locations, observing the number of resources, and sharing the information with all $a_i \in A$.

The important features of the system for online decentralized information gathering cited in [31] are

1. Tasks are distributed randomly within the environment. The spatial distribution of tasks is not known a priori by the UAVs. The UAVs perform a distributed search in the environment to discover the tasks. A UAV needs to move to the vicinity of a task to be able to discover it by perceiving/observing it on its sensors.

2. A single UAV is only capable of discovering and partially performing tasks but lacks the computational resources required to complete a task on its own. A task can be completed only if multiple UAVs collaborate to share their computational resources towards executing the task. Loosely coupled tasks are considered and different UAVs collaborating with each other to perform the same task can execute those actions asynchronously and independently of each other.

3. To enlist the cooperation of other UAVs required to complete a task, a UAV that discovers a task communicates the task's information to other UAVs.

4. On completing its portion of execution of a task, a UAV communicates the progress of the task after its execution to other UAVs within its communication range. Those UAVs then consider their own commitments and selectively choose to visit the task to perform it. After completing its portion of a task, a UAV either continues to perform any other tasks it is already committed to, or reverts to individually searching the environment for tasks if it does not have any other committed tasks.

Different multi-UAV aggregation strategies can be used. Each strategy is implemented using a heuristic function that calculates the suitability or priority of a task in the UAV's allocated task list. The different heuristic functions that have been used for this purpose are described below:

1. **Closest task first**: Each UAV selects a task from its task list that is closest to it. Distances are normalized over the sum of distances to all tasks in a UAV's task list to enable comparison between task distances. The value of this heuristic is very simple to calculate. However, it can lead all UAVs to prefer tasks that are close to them and consequently give a lower preference to tasks further away from them. Consequently, tasks that are further away from most of the UAVs can remain incomplete for long times and more than the required number of UAVs can get allocated to tasks that are closer to most of the UAVs. Overall, the closest task first heuristic is not very efficient in spatially distributing the task-load across the UAVs.

2. **Most starved task first**: To address the drawbacks of the closest task first heuristic and to balance the task load across the environment, a higher priority is given to tasks that have the least number of UAVs in their vicinity and are likely to be requiring more UAVs to complete them. The most starved task first heuristic does this by enabling each UAV to select a task from its task list that has the least number of UAVs in its vicinity. A potential drawback of the most starved task first heuristic is that UAVs are attracted towards a recently discovered task which is likely to have few UAVs near it. This can result in almost complete tasks being left incomplete because UAVs prefer more starved and possibly less complete tasks over a task that is almost nearing completion but has more UAVs (which possibly already visited the task) near it.

3. **Most starved, most complete task first**: the first heuristic extends the most starved task first heuristic by considering the number of UAVs still required to complete a task. While selecting a task, a UAV using this heuristic considers a product of the number of UAVs in the vicinity of the task as well as the progress of the task from the pheromone value associated with the task. Tasks that are nearing completion are given higher preference.

4. **Most proximal task first**: In the previous two heuristics, the task allocation algorithm only considers the effect of other UAVs on tasks in its allocated task list, but it does not consider the UAV's own relative position to the other UAVs. This can result in UAVs that are further away from the task allocating the task and moving towards it, only to be informed en-route that other UAVs have completed the task before it. To alleviate this problem, the most proximal task first heuristic first determines how many other UAVs are closer to the task being considered than the UAV itself. It then selects the task that has the least number of UAVs closer to the task than itself and also is the nearest towards being completed.

4.4.1.3 Foraging algorithms

Levy flight An observation made about foraging animals is that foragers when in no or limited prior knowledge of food show searching patterns that have special characteristics. The patterns are different from what can be seen in Brownian motion, the random walk by particles diffusing in a liquid. Foragers sometimes take long paths in just one direction. This strategy is found to be the key to the foragers success in finding food rapidly in an unknown environment. In a target searching context, flight lengths, ℓ, are said to have the power-law distribution when all these lengths are drawn from a probability distribution of the form:

$$p(\ell) = C\ell^{-\mu} \tag{4.80}$$

where $p(\ell)$ is a probability density, μ is the scaling parameter, and C is the normalization constant. A more general Levy distribution can be classified as a power-law distribution where the scaling parameter, μ, lies in the range $1 < \mu \leq 3$. Thus, a **Levy flight** consists of flight lengths that obey the power-law distribution with $1 < \mu \leq 3$. The following form for the continuous case can be given:

$$p(\ell) = \frac{\mu - 1}{\ell_{min}} \left(\frac{\ell}{\ell_{min}} \right)^{-\mu} \tag{4.81}$$

Assuming a 2-D random search, the forager follows these rules:

1. The UAV moves from one point to another until it locates a target, i.e, this UAV keeps drawing lengths to follow from a power-law distribution with a fixed μ.

2. If the UAV locates a target in its vicinity, the sensor range of radius r_v, the UAV sends out the information to an operator outside.

3. Without stopping, the UAV moves on to its search, i.e., the next flight length is drawn from the power-law distribution with a pre-determined value of scale factor μ. The angle UAV takes is drawn from a uniformly distributed set of range $[0, 2\pi]$.

4. The mapping of the region has not been specified; the UAV can re-visit any region.

One flight length can be defined as the distance traveled by the forager between one point to another without stopping and changing the angle of its path. The forager may take several of such flight lengths to find one target [86].

When the target sites are sparsely and randomly distributed, the optimum strategy is a Levy flight. By performing Levy flight, a forager optimizes the number of targets encountered versus the traveled distance. Combining Levy flight algorithm with artificial potential field method allows to perform an optimum bio-inspired foraging strategy. The **Levy flight** algorithm generates the length of the movement, while the artificial potential method improves the dispersion of the deployed UAVs by generating repulsion forces among them [93].

1. The **Levy probability distribution** has the following form:

$$p_\alpha(\ell) = \frac{1}{\ell} \int_0^{10} \exp\left(-\gamma q^\alpha\right) \cos(q\ell) dq \tag{4.82}$$

where γ is the scaling factor and α determines the shape of the distribution such that different shapes of probability distribution in the tail region can be obtained. The distribution is symmetrical with respect to $\ell = 0$. The length of the Levy flight of the UAV during the foraging phase can be calculated as the limit of the following sequence (or n

around 100):

$$z_n = \frac{1}{n^{1/\alpha}} \sum_{k=1}^{n} v_k \tag{4.83}$$

requiring two independent random variables a, b having a normal Gaussian distribution from this non-linear transformation $v = \frac{a}{|b|^{1/\alpha}}$

2. The basic idea of the potential field method is that the UAV is attracted towards searching target while being repulsed by the known obstacles. To improve the dispersion process during deployment, the repulsion forces among UAVs are used. A repulsive field can be proposed as:

$$U_{rep}(q) = \left\{ \begin{array}{cc} \frac{1}{2} k_{rep} \left(\frac{1}{\rho(q)} - \frac{1}{\rho_0} \right) & \text{if } \rho(q) \geq \rho_0 \\ 0 & \text{elsewhere} \end{array} \right\} \tag{4.84}$$

where k_{rep} is a scaling factor, $\rho(q)$ is the minimal distance from q to the adjacent UAV and ρ_0 the threshold value of the distance. This leads to the repulsive force:

$$F_{rep}(q) = \left\{ \begin{array}{cc} k_{rep} \frac{1}{\rho^2(q)} \left(\frac{1}{\rho(q)} - \frac{1}{\rho_0} \right) & \text{if } \rho(q) \geq \rho_0 \\ 0 & \text{elsewhere} \end{array} \right\} \tag{4.85}$$

This approach has the advantage that it does not require a centralized control or localization system, and therefore scales the possibility to apply very large number of UAVs. An alternative macroscopic model to the Levy flight model is the intermittent search model that views foraging in two alternating phases. In the first phase, the animal performs a local Brownian search, and in the second phase the animal performs a ballistic relocation. In both the Levy flight and intermittent search models, the key macroscopic observation is that the animal performs a local exploration for some time and then moves to a far-off location. In animal foraging, the energy aggregated from a patch can be thought of as the reward from the patch, and the animal's objective is to maximize intake energy rate, while minimizing expenditure in time and energy. In robotic foraging, the UAV searches an area, and the reward is the aggregated evidence. Analogous to the animal, the UAV's objective is typically to maximize evidence collected, while minimizing expenditure of time and energy. The **multi-armed bandit problem** models the foraging objective in optimal foraging theory well and the associated block allocation strategy captures the key features of popular macroscopic search models [90].

Fish school The UAVs, just like fish, forage when in search of a target. The team of UAVs can behave/respond like a school of fish in search of food, thus the **fish prey algorithm** can be used. It uses the concept of adaptive networks and adds mobility as another attribute. The adaptive network is strongly connected and examined from the viewpoint of a group of nodes capable of learning and interacting with one another locally to accomplish distributed inference and processing challenges in real time. The nodes relate locally with their neighbors within neighborhoods which are constantly changing due to the mobility of nodes. This leads to a network topology adaptive in nature. Each UAV can be represented as a fish in this school [32].

Greedy rate The Greedy rate algorithm actively re-plans the UAV a_i's destination based on the estimates $\hat{v}_{j,t}^{(i)}$. The Greedy rate algorithm is inspired by continuous area sweeping approaches and its pseudo-code is given by Algorithm 24.

Algorithm 24 Next Destination of UAV a_i at Location l_α

1. if $c_i = y_i$ then
2. return l_0
3. end if
4. Compute the rate if a_i heads home
5. $(r_{best}, l_{best}) \leftarrow \left(\frac{y_i}{t(a_i, l_\alpha, l_0)}, l_0\right)$
6. Compute the rate if a_i visits l_j then heads home
7. for all $l_j \in L$ such that $j > 0$ do
8. $e_j \leftarrow \sum_{a_k \in A \text{ heading to } l_j} (c_k - l_k)$
9. $y_i' \leftarrow max\left(c_i, y_i + max(0, \hat{v}_{j,t+t(a_i,l_\alpha,l_j)} - e_j)\right)$
10. $r' \leftarrow \frac{y_i'}{t(a_i,l_\alpha,l_j)+t(a_i,l_j,l_0)}$
11. if $r' > r_{best}$ then
12. $(r_{best}, l_{best}) \leftarrow (r', l_j)$
13. enf if
14. end for
15. return l_{best}

Earmarking can be used to minimize communication among the foraging UAVs. Foraging UAVs tend to have limited computational power and communication bandwidth, to minimize the computational and communication requirements. If communication among foraging UAVs was impossible, it would still be feasible for a foraging UAV to infer another UAV's destination by observing its direction of travel.

Adaptive sleep The adaptive sleep algorithm is adapted from sustainable foraging. In Algorithm 25, each UAV a_i chooses a location l_α, and the UAV forages from the location when it has $K_\alpha/2$ resources, where K_α is the maximum number of resources at l_α.

Algorithm 25 Compute if UAV a_i that is Assigned to location l_α Should Sleep Further

---

1. if a_i is not at l_0 then
2. return false
3. end if
4. if $\hat{v}_{\alpha,t+t(a_i,l_l,l_\alpha)} < K_\alpha/2$
5. return true
6. else
7. return false
8. end if

The UAV sleeps until the estimated number of resources is at least $K_\alpha/2$ and the foraging UAVs coordinate on the locations, subject to communication range limits.

4.4.1.4 Anchoring

Autonomous systems situated in the real world often need to recognize, track, and reason about various types of physical objects. In order to allow reasoning at a symbolic level, one must create and continuously maintain a correlation between symbols denoting physical objects and sensor data being collected about them, a process called anchoring. Anchoring must necessarily take place in any robotic system that comprises a symbolic reasoning component.

Definition 163 ***Anchoring*** *is the process of creating and maintaining the correspondence between symbols and sensor data that refer to the same physical objects.*

The anchoring problem is the problem of how to perform anchoring in an artificial system. Anchoring can be performed top-down, bottom-up or in both directions simultaneously [28]. The anchoring process must take the temporal dimension into account with the flow of continuously changing sensor input.

An anchor can be seen as an internal model of a physical object that links together the symbol-level and sensor-level representations of that object. Uncertainty and ambiguity arise when anchoring is performed using real sensors which have intrinsic limitations and in an environment which cannot be optimized in order to reduce these limitations. Moreover, at the symbolic level, there are several ways to refer to objects. An anchor is a unique internal representation of an object o in the environment. At every moment t, $\alpha(t)$ contains a symbol meant to denote o; a percept generated by observing o; and a signature meant to provide the current (best) estimate of the values of the observable properties of o [27]. The anchoring process is performed in an intelligent embedded system that comprises a **symbol system** σ and a **perceptual system** π. The symbol system manipulates individual symbols meant to denote physical objects. It also associates each individual symbol with a set of symbolic predicates that assert properties of the corresponding object. The perceptual system generates percepts from the observation of the physical objects. A **percept** is a structured collection of measurements assumed to originate from the same physical object.

For a UAV performing surveillance tasks of a road network, the symbol system consists of the planner, individual symbols denote cars and elements of the road network. The perceptual system is a reconfigurable active vision system able to extract information about car-like objects in aerial images, and they have attributes like position, width and color. The predicate grounding relation is a table that associates each predicate symbol with a fuzzy set of admissible values for the corresponding attribute. An anchor stores an individual symbol, the index of a region, and an association list recording the current estimates of the object's attributes. Symbolic formalisms, such as chronicle recognition, require a consistent assignment of symbols, or identities, to the physical objects being reasoned about and the sensor data received about those objects. Image analysis provides a partial solution, with vision percepts having symbolic identities that persist over short intervals of time. However, changing visual conditions or objects temporarily being out of view leads to problems that image analysis often cannot handle. This is the task of the anchoring system, which also assists in object classification and in the extraction of higher level attributes of an object. A geographic information system is used to determine whether an object is currently on a road. Concrete events corresponding to changes in such attributes and predicates provide sufficient information for the chronicle recognition system to determine when higher-level events such as reckless overtakes occur.

In the case study, anchoring links vision percepts from an object tracker to world objects, which are then linked to on-road objects. Link conditions are intended to demonstrate key concepts and could be elaborated to take more complex conditions into account. The temporal unit in the formulas is milliseconds, which is the temporal granularity used in the real system. When a traffic monitoring scenario is considered, traffic violations and other events to be detected should be represented formally and declaratively, which is a specific example of a general classification task which is common in robotic applications. This can be done using chronicle recognition, where each chronicle defines a parameterized class of complex events as a simple temporal network whose nodes correspond to occurrences of high-level qualitative events, and edges correspond to metric temporal constraints between event occurrences. Creating these high-level representations from low-level sensor data, such as video streams from the color and thermal cameras on-board the UAVs, involves extensive information processing within each sensor platform. Anchoring is a central process making symbolic reasoning about the external world possible by creating symbols referring to objects in the world based on processing of sensor data. The anchoring process is actually a set of concurrent link processes [45].

4.4.2 Aerial manipulation

Recently, UAVs have been employed in tasks such as grasping and manipulation as well as in cooperative transportation. An aerial manipulator, a multi-rotor UAV equipped with n degrees of freedom robotic arm, merges the versatility of multi-rotor UAV with the precision of a robotic arm. Multi-rotor UAVs have simple mechanical structures; they have the desirable capabilities of hovering and vertical take-off and landing. In particular, several aggressive maneuvers have been demonstrated by utilizing their high thrust-to-weight ratio. These properties are also particularly useful for autonomous load transportation. Small-size single or multiple autonomous vehicles can be considered for load transportation and deployment. Load transportation with multiple multi-rotors is useful when the load is heavy compared with the maximum thrust of a single multi-rotor, or when additional redundancy is required for safety. But, this is challenging since dynamically coupled quad-rotors should cooperate safely to transport load [49]. These are challenging issues in foraging since the system is characterized by unstable dynamics and the presence of the object causes non-trivial coupling effects where the dynamic model of the whole system, UAV plus manipulator, is devised. A Cartesian impedance control can be developed in such a way to cope with contact forces and external disturbances. The problem of motion control of the end-effector of a manipulator mounted on a UAV can be tackled through a hierarchical control architecture. Namely, in the top layer, an inverse kinematics algorithm computes the motion references for the actuated variables, i.e., position and yaw angle of the UAV and joint variables for the manipulator, while in the bottom layer, a motion control algorithm is in charge of tracking the motion references. The previous scheme can be extended by adding, at the motion control level, an adaptive term, in charge of taking into account modeling uncertainties due to the under-actuation of the system [22]. A gripper on a UAV can serve for object retrieval, sensor installation, courier services, logistics chains, etc. The aerial grasping process can be linked to alignment and grasp, coupled to ground and partial ground contact. However, UAVs have limited positioning accuracy, unstable dynamics, are sensitive to load imbalances, and are susceptible to aerodynamic effects. Any disturbance causes drift, and GPS has limited spatial/temporal resolution. It is difficult to have a high-gain position control.

4.4.2.1 Aerial transportation

The single-lift configuration, where a long rope couples one helicopter and one load, is the only configuration commercially utilized for the transportation of slung loads. However, the manual maneuvering of a helicopter with an attached slung load is very difficult and requires a skillful and experienced pilot. In particular the active damping of load oscillations is a difficult task, which most pilots avoid. Instead, the pilots stabilize only the helicopter and wait for the load oscillation to die down. The motivation for using two or more small helicopters instead of one with bigger load capacity are

1. In the case of real manned transport helicopters, the costs for two small helicopters are often less than for one with double load capacity.
2. Independent from the load capacity of the most advanced helicopters, there is always a task, which requires more load capacity than provided by a single helicopter.

In this case the control software should allow the coupling of the existing helicopters, in order to form a system with sufficient load capacity. The coordinated control of the motion of the UAVs needs to consider the involved forces. Thus, each UAV could be controlled around a common compliance center attached to the transported object. Assuming that each UAV holds the object firmly, the trajectories of all UAVs determine the trajectory of the object. Both centralized and decentralized compliant motion control algorithms have been proposed, including the consideration of non-holonomic constraints [16]. Experimental results with a team of UAVs to manipulate and transport a payload in three dimensions via cables has been recently presented [16]. The transportation of loads using only one UAV is strongly limited by the payload capacity of the UAV itself. Assuming the use of small sized UAVs, this constraint may prevent the transportation and deployment of loads required for a particular application. The system designed allows the transportation of a single load by means of several helicopters. The number of helicopters is configurable, depending on the capabilities of the helicopters and the load to be transported. The platform comprises self-deployment capabilities using one or several small sized helicopters. The systems integrated in the platform include UAVs, wireless sensor networks, ground fixed cameras and ground vehicles with actuation capabilities. This load can be a variety of things, such as cargo transportation or sensor deployment.

Model for the interaction The most substantial challenges of this generalized object retrieval and transport of unstructured objects from UAVs can be classified into three categories: **aerial approach and descent**, **object capture**, and **UAV stability** after object contact and acquisition. Instability and fragility of hovering UAVs encourages avoiding at all costs contact with surroundings. Landing and take-off generally involve rapidly transitioning through partial contact conditions, with minimal time in intermediate states between static and dynamic stability, where the danger of ground collision is high. The ground effect should also be considered. For grasping and manipulating external objects, operation in these regimes is required, both when grasping objects and in the process of lifting a target clear of the ground. In the case of a UAV interacting with the environment and/or other UAVs, a UAV can perform k different tasks $\Omega = \{\tau^1, \tau^2, \dots, \tau^k\}$ with n logical interactions requiring a change of task in the current plan. Let $E = \{\epsilon^1, \epsilon^2, \dots, \epsilon^n\}$ be a set of discrete events associated with n logical conditions requiring a change of task during the execution. The ith UAV's current task has a discrete dynamics $\delta : \Omega \times E \rightarrow \Omega$, $\tau_i^+ = \delta(\tau^i, e^i)$ where $e^i \in E$ is an event requiring a change of task from τ^i to τ_i^+. For example, the following sequence of tasks should be executed in an aerial manipulation task:

1. τ^1: approach from home,
2. τ^2: precise positioning for manipulation,
3. τ^3: grasping and manipulation,
4. τ^4: departure,
5. τ^5: return to home.

The stability of the UAV in coupled and partial contact with ground, and once airborne with payload, must be analyzed and assured. The grasping task can be divided into four phases:

1. Approach band alignment to the target,
2. Grasping hover while coupled to an object resting on the ground,
3. Partial coupling during liftoff,
4. Departure.

Each phase poses specific challenges. Disruptive aerodynamic surface effects make near-ground position-keeping outdoors more difficult than hovering in free air. The wake of the rotor is contained by the surface underneath it, creating a repelling cushion of air referred to as **ground effect**. As a rotorcraft moves laterally through ground effect, the deflected wake is pushed ahead of the UAV and can be entrained and recirculated by the rotor, causing a ground vortex. When the vortex enters the rotor, the thrust decreases rapidly; together these create an instability that causes the UAV to bounce on descent and then drift and plunge from wake interactions. In practice, UAV hovering is not yet sufficiently precise to enable grasping with a rigid manipulator. When the vehicle is in position and has a secure grasp on the target, its flight dynamics become coupled to the ground through forces transmitted by the gripper. Certain ratios of lateral and angular coupling stiffness can destabilize the UAV. As thrust increases and the surface normal force decreases, this coupling must remain well-conditioned. Once the object is lifted clear of the ground, the added load must not destabilize the UAV. The added mass changes physical parameters of the system such as the net mass, moment of inertia, and location of the center of gravity of the vehicle are all altered [79].

Layers The interactions between UAVs are not only information exchanges but also physical couplings required to cooperate in the joint transportation of a single load, it requires the consideration of physical interactions between the UAVs. The coordinated control of the motion of the UAVs should consider the involved forces. Thus, each UAV could be controlled around a common compliance center attached to the transported object. Under the assumption that each UAV holds the object firmly, the real trajectories of all of the UAVs are equal to the real trajectory of the object. Both centralized and decentralized compliant motion control algorithms can be proposed. Basically, those vehicles should be able to move to a given location and activate their payload when required. In each UAV, there are two main layers:

1. The **on-board deliberative layer** for high-level distributed decision making, and

2. The **proprietary executive layer** for the execution of the tasks.

In the interface between both layers, the on-board deliberative layer sends task requests and receives the execution state of each task and the UAV state. For distributed decision making purposes, interactions among the on-board deliberative layers of different UAVs are required. Finally, the human-machine interface software allows the user to specify the missions and tasks to be executed by the platform, and also to monitor the execution state of the tasks and the status of the different sub-systems. The different modules shown in the on-board deliberative layer support the distributed decision making process involving cooperation and coordination. The vehicles to be integrated in the platform should be able to receive elementary tasks, report their associated execution events, and execute them. The task manager module receives the planned tasks from the plan builder module. Those tasks can have preconditions and/or post-conditions and the task model assumed is based on elementary events processing, which are expected to occur whenever the states of the tasks and the environment evolve [68].

Assembling bar structure Initially, low-capacity UAVs incorporated fixed claws under the platform, allowing the system to carry lightweight and small size objects. However, the uncertainty in the positioning maneuver during hovering, inherent to rotary wing platforms, and the reduced motion of claws made an autonomous accurate grasping difficult. To face this problem, solutions with either **magnetic devices** or by using poorly articulated claws were proposed. Although this last option had the additional advantage of extending the range of applications of the robotic system, it also implied a significant increase in on-board weight, which was difficult to afford by conventional UAVs. The three basic functions required for an aerial manipulator intended for assembling bar structures are

1. The capture, including the manipulator approach to the bar and the grasping by the end effector;
2. The transport that involves the displacement of the load from the storage place to the construction site; and
3. The assembling of the bar, whose purpose is to install it at the assigned location within the structure.

These functions will be executed in a cyclic manner during the construction process, so that once a bar has been captured, transported, and finally assembled, the aerial manipulator returns to the storage place to capture a new one and to repeat the process. The successful achievement of these capabilities is conditioned by the fulfillment of several design requirements. Some of these requirements are generic, typical of any manipulator, and others are specific, due to the particularity of being on-board a rotary-wing UAV. When the rotary-wing UAV hovers, it is not completely stable, but there are small oscillations around the control reference of all degrees of freedom. These oscillations, mostly caused by electromechanical asymmetries and turbulence in the air stream entering the platform rotors, may cause difficulties in the end effector positioning. Manipulator dynamics is usually faster than the aerial platform and therefore it is able to compensate quicker this kind of disturbances. However, the manipulator dynamic is strongly coupled with the aerial platform and it would be necessary to find a compromise between the manipulator speed and the reaction forces at the platform. There are also disturbances caused by displacements of the manipulator center of gravity. When the manipulator configuration changes, the position of its center of gravity varies, generating a reaction torque at the platform. In rotary wing UAVs, this torque produces an inclination of the propeller's plane, which also induces a displacement of the entire aerial platform. This disturbance is particularly evident in those cases in which combined mass of the manipulator and its load is significant in comparison with the mass of the UAV. On heavy platforms with lightweight manipulators and loads, this disturbance is negligible. The UAV control system can be used to compensate the effect of this perturbation, being not strictly necessary to develop any other additional mechanism [23].

4.4.2.2 *Coupled dynamics*

This system is composed of the main components: a helicopter, a manipulator mounted on the fuselage, and the object itself. The task of this system is to fly closely enough to the object, to activate the sensor for object tracking, to go closer to the object, and to perform the manipulation task. The manipulation task can range from picking up objects to performing force interaction with an object for assembly operations. The control can be realized using one of the following approaches or their combinations [55]:

1. Completely decoupled control,
2. Coupling on the kinematical level,
3. Coupling on the dynamical level.

Grasping task A pick and place or insertion task requires a brief moment of loose coupling with the ground during the grasp or insertion. Gripper compliance reduces coupling forces/torques transmitted to the airframe. When grasping in hover, elastic gripper forces can be added to the helicopter flight dynamics. Another technique can be used such as in [83]. Considering a world-fixed inertial frame $\{O, X, , Y, Z\}$ and a body-fixed frame $\{O_b, X_b, Y_b, Z_b\}$ centered at the multi-rotor center of gravity, the dynamic model of a multi-rotor is given by:

$$\begin{aligned} m\ddot{p}_b + mg_b &= \mathbf{R}_b(\eta_b)\left(f_b^b + f_v^b\right) \\ \mathbf{I}_b\dot{\omega}_b^b + Sk(\omega_b^b)\mathbf{I}_b\omega_b^b &= \tau_b^b + \tau_v^b \end{aligned} \tag{4.86}$$

where $p_b \in \mathbb{R}^3$ is the position of the multi-rotor with respect to the inertial frame, $\mathbf{R}_b(\eta_b) \in SO(3)$ is the rotation matrix, $\eta = (\phi, \theta, \psi)$ Euler angles, m is the multi-rotor mass, $\mathbf{I}_b \in \mathbb{R}^{3\times3}$ is its constant diagonal inertia matrix expressed with respect to the body-frame, $\omega_b^b \in \mathbb{R}^3, \dot{\omega}_b^b \in \mathbb{R}^3$ are respectively the angular velocity and acceleration of the multi-rotor expressed in the body frame, $g_b = (0, 0, g)^T$ is the gravity vector, $f_b^b \in \mathbb{R}^3$ and $\tau_b^b \in \mathbb{R}^3$ are respectively the external forces and torques input vectors acting on the aerial vehicle and expressed in the body-frame (effects of the manipulator on the helicopter, neglected aerodynamic effects, physical interaction of the system with the environment). In the multi-rotor, $f_b^b = (0, 0, u)^T$ and $\tau_b^b = (\tau_\phi, \tau_\theta, \tau_\psi)^T$ where u represents the thrust perpendicular to the propellers' plane. Generally, an aerial manipulator consists of three components: a base (fixed to the landing gear of the multi-rotor), a multi-joint arm, and an end-effector. The center of gravity of the whole system is as close as possible to the multi-rotor geometric center. The instantaneous center of gravity position of each link i, referred to as the arm fixed axis system $\{O_0, X_0, Y_0, Z_0\}$, is given by:

$$[x_{A_i}^0, y_{A_i}^0, z_{A_i}^0, 1]^T = \mathbf{T}_i^0[x_{A_i}^i, y_{A_i}^i, z_{A_i}^i, 1]^T \tag{4.87}$$

for $i = 1, \ldots, n$ and $\mathbf{T}_i^0 \in \mathbb{R}^{4\times4}$ is the homogeneous transformation matrix of link i. The robotic arm center of gravity position vector $p_A^b \in \mathbb{R}^3$ referred to the body frame is given by:

$$p_A^b = \frac{1}{m_A}\mathbf{E}_3\mathbf{T}_0^b\left(\sum_{i=1}^{n}[x_{A_i}^0, y_{A_i}^0, z_{A_i}^0, 1]^T\right) \tag{4.88}$$

where m_i is the ith link mass, $m_A = \sum_{i=11}^{n} m_i$, $\mathbf{T}_0^b \in SE(3)$ is the homogeneous transformation matrix from arm to body frame, $\mathbf{E}_0 \in \mathbb{R}^{4\times4}$ selects the first three components.

The dynamic model of the system can be written as:

$$\mathbf{M}(\zeta)\ddot{\zeta} + \mathbf{C}(\zeta, \ddot{\zeta}) + g(\mathbf{M}(\zeta) + C(\zeta, \ddot{\zeta}) = u \tag{4.89}$$

where $\zeta = (x_b^T, q^T)^T \in \mathbb{R}^{6+n\times1}$, $\mathbf{M}$ represents the symmetric and positive definite inertia matrix of the system:

$$\mathbf{M} = \begin{pmatrix} \mathbf{M}_{pp} & \mathbf{M}_{p\phi} & \mathbf{M}_{pq} \\ \mathbf{M}_{p\phi}^T & \mathbf{M}_{\phi\phi} & \mathbf{M}_{\phi q} \\ \mathbf{M}_{pq}^T & \mathbf{M}_{\phi q}^T & \mathbf{M}_{qq} \end{pmatrix} \tag{4.90}$$

where $\mathbf{M}_{pp}, \mathbf{M}_{p\phi} \in \mathbb{R}^{3\times3}$, $\mathbf{M}_{pq}, \mathbf{M}_{\phi q} \in \mathbb{R}^{3\times n}$. The inertia matrix can be viewed as a block matrix.

Similarly, matrix $\mathbf{C}$ and vector g can be written as:

$$\mathbf{C} = (\mathbf{C}_p, \mathbf{C}_\phi, \mathbf{C}_q)^T \quad g = (g_p, g_\phi, g_q)^T \tag{4.91}$$

with $\mathbf{C}_p \in \mathbb{R}^{3\times(6+n)}, \mathbf{C}_\phi \in \mathbb{R}^{3\times(6+n)}, \mathbf{C}_q \in \mathbb{R}^{n\times(6+n)}$ and $g_p \in \mathbb{R}^3, g_\phi \in \mathbb{R}^3, g_q \in \mathbb{R}^n$.

The physical interaction between UAVs and their surrounding environment is currently investigated. The goal is to explore the potentialities of systems that are not only able to fly autonomously, but also to safely interact with remote objects to accomplish tasks such as data acquisition by contact, sample picking, and repairing and assembling objects. This feature affects the design of the control law, particularly because stability has to be preserved even in the presence of disturbances derived from physical interaction. The airframe of a generic quad-rotor helicopter has been adapted to carry a miniature robotic manipulator, which has been specifically designed for aerial inspection of industrial plants [38]. By analyzing the interaction between the quad-rotor, the manipulator, and the environment, the dynamical properties of the aerial manipulator are investigated near hovering both during free flight and during a docking maneuver with a vertical surface, i.e., when the end effector is in contact with the environment. Building upon this analysis, an energy-based control strategy is proposed. The main idea is to make the position dynamics of the vehicle passive by relying upon a cascade control strategy in which the attitude is considered a virtual available control input. The closed-loop passive system can then be controlled as a standard robotic manipulator, implementing impedance control strategies, suitable to handle both contact and no-contact cases.

Valve turning Valve, knob, and handle turning has been widely studied for use with industrial robots, mobile manipulators, and personal assistance robots. A typical requirement involves a grasp and turn of an object that remains fixed to the environment but allowed to rotate. Various techniques such as compliant motion, learning, passive compliance, hybrid position and force control, and impedance control have been implemented. All of these solutions deal with the challenges in the dynamic interaction of the manipulator with its environment. However, the strong coupling required during valve or knob turning greatly influences the dynamics of an aerial manipulator. Rigidity in the manipulator and the propagation of contact forces when interacting with the environment can cause crashes. The direct coupling between the manipulators and valve can cause sudden unexpected changes in the flight dynamics. The aerial manipulator must constantly adjust to compensate for the vehicle movement and further have adequate compliance to prevent a crash, particularly during the manipulator-environment coupling after grasping and while turning. Handles come in different shapes and sizes. The hand-wheel shape is ideally designed for the envisioned scenario: the aerial manipulator grabs onto the valve handle and twists it using its own degrees of freedom. The main approach used for detecting circular shapes of known radius R in 3-D environments by observing their elliptic perspective projection based on projective linear transformation, namely collineation of a circle, by observing the camera with a pinhole model approach [56].

In order for a 2-arm manipulator A, B to grab the valve and remain symmetric with respect to the UAS ZY plane, certain kinematic constraints on joint movements need to be applied. In order to maintain symmetry, both manipulators need to move in exactly the same way (i.e., $q_1^A = -q_1^B$ and $q_2^A = -q_2^B$). Lengths L_1 and L_2 denote link sizes from Denavit-Hartenberg parameters. Given a desired vertical distance $H(q_1, q_2)$ along the z-axis, and horizontal distance $X(q_1, q_2)$ along the x axis of the body frame, the following constraints on joint movements can be written [75]:

$$q_1 = \pm \arccos \left[\frac{L_2^2 - H(q_1, q_2)^2 - X(q_1, q_2)^2 - L_1^2}{2L_1\sqrt{H(q_1, q_2)^2 + X(q_1, q_2)^2}} \right] \tag{4.92}$$

$$q_2 = \pm \arccos \left[\frac{H(q_1, q_2)^2 + X(q_1, q_2)^2 - L_1^2 - L_2^2}{2L_1 H(q_1, q_2)^2 X(q_1, q_2)} \right] \tag{4.93}$$

By keeping the manipulator motion slow with respect to the quad-rotor attitude motion, the dynamic coupling between UAS body and arm torques is minimized. Next, model simplification comes from the fact that the payload limits joint actuator choice, so the manipulators need to be constructed using lightweight servo-motors. Once the UAV-arm system has taken position over the valve and the geometric center of the valve is aligned with the aerial system centroid, the valve is constrained on a plane parallel to the bottom plate of the quad-rotor. In this situation the quad-rotor center of mass is positioned directly above the pivot of the valve. The valve is balanced, that is the pivot of the valve represents the center of mass in the plane on which the valve turns. As the arms are symmetric and articulate equally and opposite of each other, the combined arms' center of mass shifts only along the z axis of the quad-rotor geometric center.

Pick-up and delivery problem The objects with a-priori known masses are located in a bounded 2-D space, where the UAV is capable of localizing itself using a state-of-the-art simultaneous localization and mapping (SLAM) system. A challenge follows from the requirement to collect a finite number of objects and drop them at a particular spot, both leading to autonomous switchings of the UAV's continuous dynamics. Deterministic versions of this problem can be handled efficiently, e.g., by two-stage optimization or relaxation. Since the UAV has to reach the corresponding locations of the objects or the depot with minimal overall cost, the overall problem also contains an instance of the traveling salesperson problem. Further, optimal exploration of a limited space is a difficult problem by itself. Minimizing the expected time for detecting a target located on a real line can be addressed with a known probability distribution by a searcher that can change its motion direction instantaneously, has a bounded maximal velocity and starts at the origin. Different versions of this problem are a **pursuit-evasion game** or a **coverage problem**, but its solution for a general probability distribution or a general geometry of the region is still an open question [71].

UAV tele-operation Tele-operation of a group of UAVs in a bilateral way, providing a force feedback to the human operator, is an emerging topic which combines the field of autonomous multi-robot systems and the studies on human-robot interaction. As an alternative to unilateral tele-operation, the use of suitable sensorial feedback has also been proven to improve the telepresence of the human operator, in particular by exploiting haptic cues via force-feedback. It is then interesting to study the possibility of establishing a bilateral tele-operation channel interfacing a human operator with a remote group of UAVs possessing some local autonomy, but still bound to follow the high-level human motion commands [35]. The feedback upon which operators in tele-operation tasks base their control actions differs substantially from the feedback to the pilot of a UAV. There is often a lack of sensory information and there is an additional status information presented via the visual channel. Haptic feedback can be used to unload the visual channel and to compensate for the lack of feedback in other modalities. For collision avoidance, haptic feedback could provide repulsive forces via the control inceptor [34]. Haptic feedback allows operators to interpret the repulsive forces as impedance to their control deflections when a potential for collision exists. Tele-operation performance and efficiency may be improved by providing additional information. Haptic feedback allows the operator to directly perceive the information about the environment through the sense of touch. Using haptic feedback for a collision avoidance system requires an algorithm to generate artificial forces on the control inceptor in order for the operator to perceive through the haptic channel, information about the environment before actual contact with an obstacle occurs. Both the magnitude of the repulsive forces and also the mapping algorithm affect operator performance and workload. Haptic information can be generated from an artificial force field that maps environment constraints to repulsive forces [59]. The parametric risk field allows adjustments of the size, shape, and force gradient by means of parameter settings, which determine the sensitivity of the field. Because of its smaller size, the field yields lower repulsive forces, results in less force cancellation effects, and allows for larger UAV velocities. This indicates less operator control demand and more effective UAV operations, both expected to lead to lower operator workload, while increasing safety.

The key features of the direct-manipulation interface can be described in:

1. **Perception**: The first task of the interface is to help the user perceive the current relationship between the UAV and the world, while not overwhelming the user with unnecessary information.

2. **Comprehension**: The next level of situational awareness is obtained by combining perceptual data to comprehend how these data relate to the overall goal.

3. **Projection**: The highest level of situational awareness is projection: the ability to predict what will happen to the system in the near future.

By using direct manipulation and visual/overlays, the behavior of the UAV can be predictable. The interface maintains constant flight parameters unless it receives a new command, in which case it will seek to match the new parameters.

4.5 CONCLUSION

The aim is to develop a new generation of service robots capable of supporting human beings in all those activities that require the ability to interact actively and safely with environments not constrained on the ground but airborne. Deployment problem is considered first, using location approaches as well as optimal control methods. The application to homogeneous point UAVs and heterogeneous groups of UAVs with various sensori-motor capabilities is considered from different points of view. Patrolling is a very active field of research. UAVs equipped with on-board radar or high resolution imaging payloads like electro-optic infrared sensors are used to localize and track the targets in perimeter surveillance missions. Foraging combines in fact all the previously presented robotic generic problems. A natural progression is to advance beyond simple motion and observation to interaction with objects and the fixed environment. Of specific interest is grasping and retrieving objects while hovering, combining robotic manipulation capabilities with the range, speed and vertical workspace of UAVs. This last section illustrates how a team of UAVs can cooperatively transport a single payload to distribute and minimize the load on each UAV.

Bibliography

[1] Abbasi, F.; Mesbahi, A.; Mohammadpour, J.: *Team-based coverage control of moving sensor networks*, In American Control Conference, pp. 5691-5696, DOI: 10.1109/ACC.2016.7526561, 2016.

[2] Acevedo, J. J.; Arrue, B. C.; Maza, I.; Ollero, A.: *A distributed algorithm for area partitioning in grid-shape and vector-shape configurations with multiple aerial robots*, Journal of Intelligent and Robotic Systems, vol. **84**, pp. 543-557, DOI 10.1007/s10846-015-0272-5, 2015.

[3] Adaldo, A.: *Event-triggered control of multi-agent systems: pinning control, cloud coordination and sensor coverage*, PhD thesis, Royal institute of Technology, School of Electrical Engineering, Dept. of Automatic Control, 2016.

[4] Adams, H.; Carlsson, G.: *Evasion paths in mobile sensor networks*, The International Journal of Robotics Research, vol. **34**, pp. 90-104, 2015.

[5] Alitappeh, R. J.; Jeddisaravi, K.; Guimaraes, F. G.: *Multi-objective multi-robot deployment in a dynamic environment*, Soft Computing, vol. **21**, pp. 6481-6497, 2017.

[6] Althoff, D.; Kuffner, J.; Wollherr, D.; Buss, M.: *Safety assessment of robot trajectory for navigation in uncertain and dynamic environment*, Autonomous Robots, vol. **32**, pp. 285-302, 2010.

[7] Angelov, P., Filev, D. P., Kasabov, N.: *Evolving intelligent systems*, IEEE Press, 2010

[8] Angelov, P.: *Sense and Avoid in UAS - research and applications*, Wiley aerospace series, 2012.

[9] Akba, M. .; Solmaz, G.; Turgut, D.: *Molecular geometry inspired positioning for aerial networks*, Computer Networks, vol. **98**, pp. 72-88, 2016.

[10] Arslan, O.; Koditschek, D. E.: *Voronoi-based coverage control of heterogeneous disk-shaped robots*, In IEEE International Conference on Robotics and Automation (ICRA), pp. 4259-4266, DOI: 10.1109/ICRA.2016.7487622, 2016.

[11] Atkins, E.; Moylan, G.; Hoskins, A.: *Space based assembly with symbolic and continuous planning experts*, IEEE Aerospace Conference, 2006.

[12] Barbato, M.; Grappe, R.; Lacroix, M.; Calvo, R. W.: *A set covering approach for the double traveling salesman problem with multiple stacks*, In International Symposium on Combinatorial Optimization, pp. 260-272, doi/10.1007/978-3-319-45587-7-23, 2016.

[13] Baron, O.; Berman, O.; Krass, D.; Wang, Q.: *The equitable location problem on the plane*, European Journal of Operational Research, vol. **183**, pp. 578-590, 2007.

[14] Bektas, T.: *The multiple traveling salesman problem: an overview of formulations and solution procedures*, Omega, vol. **34**, pp. 209-219, 2006.

[15] Berger, J.; Boukhtouta, A.; Benmoussa, A.; Kettani, O.: *A new mixed integer linear programming model for rescue path planning in uncertain adversarial environments*, Computers and Operations Research, vol. **39**, pp. 3420–3430, 2012.

[16] Bernard, M.; Kondak, K.; Maza, I.; Ollero, A.: *Autonomous transportation and deployment with aerial robots for search and rescue missions*, Journal of Field Robotics, vol. **28**, pp. 914-931, 2011.

[17] Besada-Portas, E.; De La Torre, L.; de la Cruz, J. M.; de Andrs-Toro, B.: *Evolutionary trajectory planner for multiple UAVs in realistic scenarios*, IEEE Transactions on Robotics, vol. **26**, pp. 619-634, 2010.

[18] Bestaoui Sebbane, Y.: *Planning and Decision Making for Aerial Robots*, Springer Switzeland, 2014.

[19] Bolanos, R.; Echeverry, M.; Escobar, J.: *A multi-objective non-dominated sorting genetic algorithm (NSGA-II) for the Multiple Traveling Salesman Problem*, Decision Science Letters, vol. **4**, pp. 559-568, 2015.

[20] Bryngelsson, E.: *Multi-robot distributed coverage in realistic environments*, MS Thesis in Computer science, Chalmers University of Technology, Gothenburg, Sweeden, pp. 1-46, 2008.

[21] Brooks, A.; Makarenko, A.; Williams, S.; Durrant-Whyte, H.: *Parametric POMDP for planning in continuous state spaces*, Robotics and autonomous systems, vol. **54**, pp. 887-897, 2006.

[22] Caccavale, F.; Giglio, G.; Muscio, G.; Pierri, F.: *Adaptive control for UAVs equipped with a robotic arm*, In Preprints of the 19th World Congress, The International Federation of Automatic Control, Cape Town, South Africa, pp. 11049-11054, 2014.

[23] Cano, R.; Perez, C.; Pruano, F.; Ollero, A.; Heredia, G.: *Mechanical design of a 6-DOF aerial manipulator for assembling bar structures using UAVs*, In 2nd RED-UAS Workshop on Research, Education and Development of Unmanned Aerial Systems, 2013.

[24] Carlsson, G.; De Silva, V.; Morozov, D.: *Zigzag persistent homology and real-valued functions*, Proceedings of the Annual Symposium on Computational Geometry, pp. 247-256, 2009.

[25] Chakravorty, S.; Junkins, J.: *A methodology for intelligent path planning*, IEEE International Symposium on Mediterranean Control, Cyprus, pp. 592-597, DOI: 10.1109/.2005.1467081, 2005.

[26] Chaumont, N.; Adami, C.: *Evolution of sustained foraging in three-dimensional environments with physics*, Genetic Programming and Evolvable Machines, vol. **17**, pp. 359-390, 2016.

[27] Coradeschi, S.; Saffiotti, A.: *Anchoring symbols to sensor data: preliminary report*, In AAAI/IAAI American Association for Artificial Intelligence, pp. 129-135, 2000.

[28] Coradeschi, S.; Saffiotti, A.: *An introduction to the anchoring problem*, Robotics and Autonomous Systems, vol. **43**, pp. 85-96, 2003.

[29] Couceiro, M. S.; Figueiredo, C. M.; Rocha, R. P.; Ferreira, N. M.: *Darwinian swarm exploration under communication constraints: Initial deployment and fault-tolerance assessment*, Robotics and Autonomous Systems, vol. **62**, pp. 528-544, 2014.

[30] Dai, X.; Jiang, L.; Zhao, Y.: *Cooperative exploration based on supervisory control of multi-robot systems*, Applied Intelligence, pp. 1-12, 2016.

[31] Dasgupta, P.: *Multi-robot task allocation for performing cooperative foraging tasks in an initially unknown environment*, In Innovations in Defence Support Systems, pp. 5-20, Springer Berlin Heidelberg, 2011.

[32] El Ferik, S., Thompson, O. R.: *Biologically inspired control of a fleet of UAVs With Threat Evasion Strategy*, Asian Journal of Control, vol. **18**, pp. 2283-2300, 2016.

[33] Ergezer, H.; Leblebiciolu, K.: *3-D path planning for multiple UAVs for maximum information collection*, Journal of Intelligent and Robotic Systems, vol. **73**, pp. 737-762, 2014.

[34] Field, E.; Harris, D.: *A comparative survey of the utility of cross-cockpit linkages and autoflight systems' backfeed to the control inceptors of commercial aircraft*, Ergonomics, vol. **41**, pp. 1462-1477, 1998.

[35] Franchi, A.; Bulthoff, H. H.; Giordano, P. R.: *Distributed online leader selection in the bilateral teleoperation of multiple UAVs*, In 50th IEEE Conference on Decision and Control and European Control Conference, pp. 3559-3565, DOI: 10.1109/CDC.2011.6160944, 2011.

[36] Forstenhaeusler, M.; Funada, R.; Hatanaka, T.; Fujita, M.: *Experimental study of gradient-based visual coverage control on SO (3) toward moving object/human monitoring*, In American Control Conference, pp. 2125-2130, DOI: 10.1109/ACC.2015.7171047, 2015.

[37] Frost, J. R.; Stone, L. D.: *Review of search theory: advances and application to search and rescue decision support*, U.S. Coast Guard Research and Development Center, report CG-D-15-01, 2001.

[38] Fumagalli, M.; Naldi, R.; Macchelli, A.; Forte, F.; Keemink, A. Q.; Stramigioli, S.; Carloni, R.; Marconi, L.: *Developing an aerial manipulator prototype: Physical interaction with the environment*, IEEE Robotics and Automation Magazine, vol. **21**, pp. 41-50, 2014.

[39] Garcia, E.; Casbeer, D. W.: *Cooperative task allocation for unmanned vehicles with communication delays and conflict resolution*, AIAA Journal of Aerospace Information Systems, vol. **13**, pp. 1-13, 2016.

[40] Gaynor, P.; Coore, D.: *Towards distributed wilderness search using a reliable distributed storage device built from a swarm of miniature UAVs*, In International Conference on Unmanned Aircraft Systems (ICUAS), pp. 596-601, DOI: 10.1109/ICUAS.2014.6842302, 2014.

[41] Glad, A.; Buffet, O.; Simonin, O.; Charpillet, F.: *Self-organization of patrolling-ant algorithms*, IEEE 7th International Conference on Self-Adaptive and Self-Organizing Systems, pp. 61-70, 2009.

[42] Gorain, B.; Mandal, P. S.: *Solving energy issues for sweep coverage in wireless sensor networks*, Discrete Applied Mathematics, http://dx.doi.org/10.1016/j.dam.2016.09.028, 2016.

[43] Hansen, E. A.; Bernstein, D. S.; Zilberstein, S.: *Dynamic programming for partially observable stochastic games*, In AAAI Conference on Artifical Intelligence, vol. **4**, pp. 709-715, 2004.

[44] Hatanaka, T.; Funada, R.; Fujita, M.: *3-D visual coverage based on gradient descent algorithm on matrix manifolds and its application to moving objects monitoring*, In 2014 American Control Conference, pp. 110-116, DOI: 10.1109/ACC.2014.6858663, 2014.

[45] Heintz, F.; Kvarnstrm, J.; Doherty, P.: *Stream-based hierarchical anchoring*, KI-Knstliche Intelligenz, vol. **27**, pp. 119-128, 2013.

[46] Holzapfel, F., Theil, S.(eds): *Advances in Aerospace Guidance, Navigation and Control*, Springer, 2011.

[47] Hong, Y.; Kim, D.; Li, D.; Xu, B.; Chen, W.; Tokuta A. O.: *Maximum lifetime effective-sensing partial target-coverage in camera sensor networks*, In 11th IEEE International Symposium on Modeling and Optimization in Mobile, Ad Hoc and Wireless Networks (WiOpt), pp. 619-626, 2013.

[48] Hunt, S.; Meng, Q.; Hinde, C.; Huang, T.: *A consensus-based grouping algorithm for multi-agent cooperative task allocation with complex requirements*, Cognitive Computation, vol. **6**, pp. 338-350, 2014.

[49] Ibarguren, A.; Molina, J.; Susperregi, L.; Maurtua, I.: *Thermal tracking in mobile robots for leak inspection activities*, Sensors, vol. **13**, pp. 13560-13574, 2013.

[50] Kalyanam, K.; Chandler, P.; Pachter, M.; Darbha, S.: *Optimization of perimeter patrol operations using unmanned aerial vehicles*, AIAA Journal of Guidance, Control, and Dynamics, vol. **35**, pp. 434-441, 2012.

[51] Kalyanam, K.; Park, M.; Darbha, S.; Casbeer, D.; Chandler, P.; Pachter, M.: *Lower bounding linear program for the perimeter patrol optimization problem*, AIAA Journal of Guidance, Control, and Dynamics, vol. **37**, pp. 558-565, 2014.

[52] Khan, A.; Noreen, I.; Ryu, H.; Doh, N. L.; Habib, Z.: *Online complete coverage path planning using two-way proximity search*, Intelligent Service Robotics, DOI 10.1007/s11370-017-0223-z, pp. 1-12, 2017.

[53] Kim, S.; Oh, H.; Suk, J.; Tsourdos, A.: *Coordinated trajectory planning for efficient communication relay using multiple UAVs*, Control Engineering Practice, vol. **29**, pp. 42-49, 2014.

[54] Kiraly, A.,; Christidou, M.; Chovan, T.; Karlopoulos, E.; Abonyi, J.: *Minimization of off-grade production in multi-site multi-product plants by solving multiple traveling salesman problem*, Journal of Cleaner Production, vol. **111**, pp. 253-261, 2016.

[55] Kondak, K.; Ollero, A.; Maza, I.; Krieger, K.; Albu-Schaeffer, A.; Schwarzbach, M.; Laiacker, M.: *Unmanned Aerial Systems Physically Interacting with the Environment: Load Transportation, Deployment, and Aerial Manipulation*, In Handbook of Unmanned Aerial Vehicles, pp. 2755-2785, Springer Netherlands, 2015.

[56] Korpela, C., Orsag, M.; Oh, P.: *Towards valve turning using a dual-arm aerial manipulator*, In IEEE/RSJ International Conference on Intelligent Robots and Systems, pp. 3411-3416, DOI: 10.1109/IROS.2014.6943037, 2014.

[57] Korsah, G. A.; Stentz, A.; Dias, M. B.: *A comprehensive taxonomy for multi-robot task allocation*, The International Journal of Robotics Research, vol. **32**, pp. 1495-1512, 2013.

[58] Koubaa, A.; Cheikhrouhou, O.; Bennaceur, H.; Sriti, M. F.; Javed, Y.; Ammar, A.: *Move and mmprove: a market-based mechanism for the multiple depot multiple travelling salesmen problem*, Journal of Intelligent and Robotic Systems, vol. **85**, pp. 307-330, 2017.

[59] Lam, T. M.; Boschloo, H. W.; Mulder, M.; Van Paassen, M. M. : *Artificial force field for haptic feedback in UAV teleoperation*, IEEE Transactions on Systems, Man and Cybernetics, Part A: Systems and Humans, vol. **39**, pp. 1316-1330, 2009.

[60] Lee, S. K.; Becker, A.; Fekete, S. P.; Kroller, A.; McLurkin, J.: *Exploration via structured triangulation by a multi-robot system with bearing-only low-resolution sensors*, In IEEE International Conference on Robotics and Automation, pp. 2150-2157, 2014.

[61] Levine, D.; Luders, B.; How, J. P.: *Information-theoretic motion planning for constrained sensor networks*, AIAA Journal of Aerospace Information Systems, vol. **10**, pp. 476-496, 2013.

[62] Li, W.; Wu, Y.: *Tree-based coverage hole detection and healing method in wireless sensor networks*, Computer networks, vol. **103**, pp. 33-43, 2016.

[63] Liemhetcharat, S.; Yan, R.; Tee, K. P.: *Continuous foraging and information gathering in a multi-agent team*, In Proceedings of the International Conference on Autonomous Agents and Multiagent Systems, pp. 1325-1333, 2015.

[64] Maftuleac, D.; Lee, S. K.; Fekete, S. P.; Akash, A. K.; Lpez-Ortiz, A.; McLurkin, J.: *Local policies for efficiently patrolling a triangulated region by a robot swarm*, In IEEE International Conference on Robotics and Automation (ICRA), pp. 1809-1815, 2015.

[65] Maravall, D.; De Lope, J.; Martin, J.A.: *Hybridizing evolutionary computation and reinforcement learning for the design of almost universal controllers for autonomous robots*, Neurocomputing, vol. **72**, pp. 887-894, 2009.

[66] Marier, J. S.; Besse, C.; Chaib-Draa, B.: *A Markov model for multiagent patrolling in continuous time*, In Neural Information Processing, pp. 648-656, Springer Berlin Heidelberg, 2009.

[67] Matveev, A.S.; Teimoori, H.; Savkin, A.: *Navigation of a uni-cycle like mobile robot for environmental extremum seeking*, Automatica, vol. **47**, pp. 85-91, 2011.

[68] Maza, I.; Kondak, K., Bernard, M.; Ollero, A.: *Multi-UAV cooperation and control for load transportation and deployment*, Journal of Intelligent and Robotic Systems, vol. **57**, pp. 417-449, 2010.

[69] Miller, L. M.; Silverman, Y.; MacIver, M. A.; Murphey, T. D.: *Ergodic exploration of distributed information*, IEEE Transactions on Robotics, vol. **32**, pp. 36-52, 2016.

[70] Mladenovic, N.; Brimberg, J.; Hansen, P.; Moreno-Perez, J. A.: *The p-median problem: A survey of metaheuristic approaches*, European Journal of Operational Research, vol. **179**, pp. 927-939, 2007.

[71] Nenchev, V.; Cassandras, C.G.; Raisch, J.: *Optimal control for a robotic exploration, pick-up and delivery problem*, arXiv preprint arXiv:1607.01202., 2016.

[72] Nestmeyer, T.; Giordano, P. R.; Bulthoff, H. H.; Franchi, A.: *Decentralized simultaneous multi-target exploration using a connected network of multiple robots*, Autonomous Robots, pp. 1-23, DOI: 10.1007/s10514-016-9578-9, 2016.

[73] Obermeyer, K. J., Ganguli, A.; Bullo, F.: *Multiagent deployment for visibility coverage in polygonal environments with holes*, International Journal of Robust and Nonlinear Control, vol. **21**, pp. 1467-1492, 2011.

[74] Okabe, A.; Boots, B.; Sugihara, K.; Chiu, S. N.: *Spatial tessellations: concepts and applications of Voronoi diagrams*, John Wiley, 2009.

[75] Orsag, M.; Korpela, C.; Bogdan, S., Oh, P.: *Valve turning using a dual-arm aerial manipulator*, In IEEE International Conference on Unmanned Aircraft Systems, pp. 836-841, DOI: 10.1109/ICUAS.2014.6842330, 2014.

[76] Pita, J.; Jain, M.; Tambe, M.; Ordonez, F.; Kraus, S.: *Robust solutions to stackelberg games: Addressing bounded rationality and limited observations in human cognition*, Artificial Intelligence, vol. **174**, pp. 1142 1171, 2010.

[77] Poduri, S., Sukhatme, G. S.: *Constrained coverage for mobile sensor networks*, In IEEE International Conference on Robotics and Automation, vol. **1**, pp. 165-171, 2004.

[78] Portugal, D.; Rocha, R. P.: *Cooperative multi-robot patrol with Bayesian learning*, Autonomous Robots, **40**, pp. 929-953, 2016.

[79] Pounds, P. E.; Bersak, D. R.; Dollar, A. M.: *Grasping from the air: Hovering capture and load stability*, In IEEE International Conference on Robotics and Automation, pp. 2491-2498, 2011.

[80] Rojas, I. Y.: *Optimized Photogrammetric Network Design with Flight Path Planner for UAV-based Terrain Surveillance*, MS Thesis, Brigham Young university, 2014.

[81] Rout, M.; Roy, R.: *Dynamic deployment of randomly deployed mobile sensor nodes in the presence of obstacles*, Ad Hoc Networks, vol. **46**, pp. 12-22, 2016.

[82] Roy, A.; Mitra, D.: *Unscented Kalman filter based multi-target tracking algorithms for airborne surveillance applications*, AIAA Journal of Guidance, Control and Dynamics, vol. **39**, pp. 1949-1966, 2016.

[83] Ruggiero, F.; Trujillo, M. A.; Cano, R. Ascorbe, H., Viguria, A., Perz, C., Lippiello, V.; Ollero, A.; Siciliano, B. : *A multilayer control for multirotor UAVs equipped with a servo robot arm* In IEEE International Conference on Robotics and Automation (ICRA), pp. 4014-4020, 2015.

[84] Sharma, V.; Patel, R. B.; Bhadauria, H. S.; Prasad, D.: *Deployment schemes in wireless sensor network to achieve blanket coverage in large-scale open area: A review*, Egyptian Informatics Journal, vol. **17**, pp. 45-56, 2016.

[85] Sharma, V.; Srinivasan, K.; Chao, H.C.; Hua, K.L.: *Intelligent deployment of UAVs in 5G heterogeneous communication environment for improved coverage*, Journal of Network and Computer Applications, http://dx.doi.org/10.1016/j.jnca.2016.12.012, 2017.

[86] Singh, M. K.: *Evaluating Levy flight parameters for random searches in a 2D space*, Doctoral dissertation, Massachusetts Institute of Technology, 2013.

[87] Sivaram Kumar, M. P.; Rajasekaran, S.: *Path planning algorithm for extinguishing forest fires*, Journal of Computing, vol. **4**, pp. 108-113, 2012.

[88] Song, C.; Liu, L.; Feng, G.; Xu, S.: *Optimal control for multi-agent persistent monitoring*, Automatica, vol. **50**, pp. 1663-1668, 2014.

[89] Sposato, M.: *Multiagent cooperative coverage control*, MS thesis, KTH, Royal Institute of Technology, Stockholm, Sweden, pp. 67, 2016.

[90] Srivastava, V.; Reverdy, P.; Leonard, N. E.: *On optimal foraging and multi-armed bandits*, In 51st IEEE Annual Allerton Conference on Communication, Control, and Computing (Allerton), pp. 494-499, 2013.

[91] Stranders, R.; Munoz, E.; Rogers, A.; Jenning N.R.: *Near-optimal continuous patrolling with teams of mobile information gathering agents*, Artificial Intelligence, vol. **195.**, pp. 63-105, 2013.

[92] Sun, S.; Sun, L.; Chen, S.: *Research on the target coverage algorithms for 3D curved surface*, Chaos, Solitons and Fractals, vol. **89**, pp. 397-404, 2016.

[93] Sutantyo, D. K.; Kernbach, S.; Nepomnyashchikh, V. A.; Levi, P.: *Multi-robot searching algorithm using Levy flight and artificial potential field*, IEEE International Workshop on Safety Security and Rescue Robotics (SSRR), 2010.

[94] Tambe, M.: *Security and Game Theory: Algorithms, Deployed Systems, Lessons Learned*, Cambridge University Press, 2012.

[95] Tuna, G., Gulez, K.; Gungor, V. C.: *The effects of exploration strategies and communication models on the performance of cooperative exploration*, Ad Hoc Networks, vol. **11**, pp. 1931-1941, 2013.

[96] Wang, Y.P. : *Regret-Based Automated Decision-Making Aids for Domain Search Tasks Using Human-Agent Collaborative Teams*, IEEE transactions on control systems technology, vol. **24**, pp. 1680-1695, 2016.

[97] Winfield, A. F.: *Towards an engineering science of robot foraging*, In Distributed Autonomous Robotic Systems (Springer Berlin Heidelberg), vol. **8**, pp. 185-192, 2009.

[98] Wilkins, D. E.; Smith, S. F.; Kramer, L. A.; Lee, T.; Rauenbusch, T.: *Airlift mission monitoring and dynamic rescheduling*, Engineering Application of Artificial Intelligence, vol. **21**, pp. 141-155, 2008.

[99] Xue, M. H.; Wang, T. Z.; Mao, S.: *Double evolutsional artificial bee colony algorithm for multiple traveling salesman problem* In MATEC Web of Conferences, vol. **44**, EDP Sciences, 2016, DOI:10.1051/mateconf/20164402025.

[100] Yakici, E.: *Solving location and routing problem for UAVs*, Computers and Industrial Engineering, vol. **102**, pp. 294-301, 2016.

[101] Yang, R.; Kiekintvled, C.; Ordonez, R.; Tambe, M.; John, R.: *Improving resource allocation strategies against human adversaries in security games: An extended study*, Artificial Intelligence Journal, vol. **195**, pp. 440-469, 2013.

[102] Yin, Z.; Xin Jiang, A.; Tambe, M.; Kiekintveld, C.; Leyton-Brown, K.; Sandholm, T.; Sullivan, J. P.: *Trusts: Scheduling randomized patrols for fare inspection in transit systems using game theory*, AI Magazine, vol. **33**, pp. 5972, 2012.

[103] Younis, M.; Akkaya, K.: *Strategies and techniques for node placement in wireless sensor networks: A survey*, Ad Hoc Networks, vol. **6**, pp. 621-655, 2008.

[104] Zannat, H.; Akter, T.; Tasnim, M.; Rahman, A.: *The coverage problem in visual sensor networks: A target oriented approach*, Journal of Network and Computer Applications, vol. **75**, pp. 1-15, 2016.

[105] Zhao, W.; Meng, Q.; Chung, P. W.: *A Heuristic distributed task allocation method for multi-vehicle multi-task problems and its application to search and rescue scenario*, IEEE Transactions on Cybernetics, vol. **46**, pp. 902-915, 2016.

[106] Zhu, C.; Zheng, C.; Shu, L.; Han, G.: *A survey on coverage and connectivity issues in wireless sensor networks*, Journal of Network and Computer Applications, vol. **35** pp. 619-632, 2012.

[107] Zorbas, D.; Pugliese, L. D. P.; Razafindralambo, T.; Guerriero, F.: *Optimal drone placement and cost-efficient target coverage*, Journal of Network and Computer Applications, vol. **75**, pp. 16-31, 2016.

CHAPTER 5

Search, Tracking, and Surveillance

5.1 INTRODUCTION

Target search and tracking is often tackled by the use of multi-robot systems, making use of the advantage mobile robots have with being capable of dynamically adapting to the target movements by changing their spatial distribution accordingly. Target tracking is a complicated problem involving multi-sensor information fusion, image processing, control technology, etc. Various problem setups vary in certain parameters and assumptions used [51]:

1. **Number of targets**: The problem of target search and tracking may be divided into two main scenarios depending on the number of targets to be searched or tracked:

   (a) **Single target**: When tracking a single target with a team of UAVs, the main focus would be on sensor data fusion from multiple trackers in order to improve the target state estimation accuracy.

   (b) **Multiple targets**: The multiple targets scenario can be viewed as an extension of the single target case, where many other uncertainties come into play. For example, the number of targets may be unknown, or may even vary with time. But even when the number of targets is known and constant, there is still uncertainty in sensor measurements. This is the problem of data association. UAVs need to spread themselves appropriately among the multiple targets, with a task allocation method. The ratio between the number of targets and trackers is another important characteristic that influences the solution approach. Multiple target tracking algorithms are divided into two categories:

      i. **Classical data association** based approach works on the measurement to target association principle. In these methods, initial measurement to track association is based on track gate formulation. Gate size is selected such that the probability of a target oriented measurement falling with the gate is maximized.

ii. **Random finite set** based approach propagates the posterior intensity or probability hypothesis density in time and it updates based on a set of target observations.

2. **Mobility of targets**: The problem becomes either searching for stationary targets or tracking moving targets. In the case of stationary targets, the only uncertainty is of noisy observations, i.e., there might be false alarms or missing measurements. But for moving targets, there is additional uncertainty in target motion.

3. **Mobility of the trackers**: The mobility mode of the UAVs highly influences the problem solution. It governs the trackers view of the world as well as speed and agility of motion. The trackers may be the same as the targets. Alternatively, the mobility mode of the targets and trackers might be different.

4. **Complexity of environment**: Complexity of the environment is an important factor governing the design of a **multi-robot system** (MRS), because a robot's interactions with other robots and with the environment play crucial roles. In the case of an open space, the only interactions to take into account are those among the trackers and targets. In structured environments, the structure of the environment can be exploited for target detection or the motion planning of the robots. However, in unstructured, cluttered environments, occlusion caused by the environment structure should be taken into account as uncertainties in sensor measurements. The environment may also affect the mobility of the trackers and targets, due to uneven terrains, obstacles and in dynamic environments, environmental changes such as wind forces.

Remark 164 *When the targets significantly outnumber the trackers, it may not be possible to track all the targets all the time, and the objective may be maximizing the average number of targets that are being observed by at least one robot throughout the mission. Another possible approach is to group targets into clusters and track those clusters instead of tracking them individually. On the other hand, when the number of targets is significantly less than the trackers, it would be possible to track all the targets all the time, and small sub-groups of robots may be formed and assigned to each target.*

Surveillance missions give rise to many challenges:

1. The **management of uncertainty** in an unpredictable environment,

2. The handling of **restricted resources**,

3. The right balance between **reactivity and deliberation** and the **communication of requests and commitments** between multiple heterogeneous observers, including human operators in mixed-initiative scenarios.

Planners must propose effective strategies for the observers to achieve their mission goals in the face of all relevant constraints, such as restricted resources, tight deadlines and uncertainty. A search and track mission aims to follow the target to its destination and proceeds in two phases, which constantly interleave:

1. **Tracking**: the observer flies over the target observing its progress; and
2. **Search**: the observer loses the target and flies a series of maneuvers to rediscover it.

Once the target is spotted, the observer switches back to tracking. The following procedure is employed to manage the search phase of a search and track mission, where each step is fully automated:

1. Determine the optimal area where the search effort should be deployed, which is an area where the target is most likely to be found;
2. Divide this area into appropriate sub-areas for assignment to individual search patterns, which are sets of maneuvers for surveying specified regions;
3. Generate a set of search patterns to optimally cover each sub-area and choose their orientations;
4. Select a sub-set of the generated search patterns for execution and sequence them over time; and
5. Execute the chosen sequence of patterns, switching back to tracking if the target is rediscovered.

In real-world search and track operations, many biasing factors exist: the target **last known position** (LKP), its intentions if predictable, its size and motion characteristics, possible hazards, results of previous searches, the terrain characteristics, the road network structure and the weather conditions. These features are used to make predictions on where the target might be over time and to construct a probability detection for the target location. The outcome of both steps consists of [14]:

1. A confined search area, usually a circular sector that is centered on the last known position of the target;
2. A probability density for the target position defined across this sector and constructed considering the above-mentioned factors; and

3. A number of points within the sector that present the highest probability of rediscovering the target and on which candidate search patterns are deployed.

Civilian UAVs have an important impact in the evolution of the security professions. The use of UAVs for security purposes has technical, legal and ethical implications which should be resolved through the collaboration and complementarity of both communities of UAVs and security. This partnership will help the evolution of the regulations inherent to civilian UAVs and facilitate the integration of this new technology in the activity of the security domain.

5.2 BASICS OF SEARCH THEORY AND DECISION SUPPORT

The initial work on search was carried out by simplifying the search problem as an **area coverage problem** and applying area coverage techniques accordingly. The search study later has expanded with the introduction of the probability of detection as a probabilistic measure. A Bayesian approach for the searching problem of lost targets found the efficient search plans that maximize the probability of finding the target given a fixed time limit by maintaining an accurate target location probability density function, and by explicitly modeling the target's process model. An approach to search for a target, based on modeling the predicted position of the target and then controlling the UAV to maximize the expectation of rediscovery, was also used. This approach generates local control instructions for the UAV, responding to relatively fine-grained predicted behavior of the target. The probabilistic model of the predicted behavior of the target may be constructed within a confined area. Another approach is to exploit standard search patterns and use these as the building blocks for a search plan that attempts to maximize the expectation of rediscovering the target. Some problems of search may impose using more than one searcher for a valuable target or a serious target [29].

Finding the allocation (time, place, and amount) over some subset(s) of the possibility area for the limited amount of available search effort that maximizes the probability of success is called the **optimal search problem**. The solution to this problem tells the search planner the sub-area where search effort should be placed and how much effort should be placed in each [37]. The basic elements in the optimal search problem include a prior distribution on target location, a function associated with search resource and detection probability, and a constrained amount of search resource. The exponential function is a common assumption to describe the probability of target detection. The elements of the basic problem of optimal search are

1. A prior probability density distribution on search object location, so the **probability of containment** (POC) for any subset of the possibility area can be estimated,

2. A detection function relating search effort density (or coverage, C) and

the **probability of detection** (POD) of the object if it is in a searched area,

3. A constrained amount of search effort, and

4. An optimization criterion of maximizing probability of finding the object: **probability of success** (POS) subject to the constraint on effort. POS is the likelihood that the target will be found when accounting for all reasonable known variables. In the random search case, the following equation can be proposed:

$$POS = 1 - \exp\left(-W.L/A\right) \tag{5.1}$$

where W is the sweep width (the largest distance where a target can reasonably be detected from a sensor), L is the total length of the random search and A is the total area to be searched.

There are two quality metrics for the probability-maximizing, path-planning problem:

1. Find the path that maximizes the **cumulated detection probability** (CDP) after a specific flight time.

2. Find the path that achieves a desired cumulated detection probability in the shortest amount of time.

5.2.1 Types of search problems

Some basic search concepts are defined as:

1. **Effective search** (or sweep) width

$$W = \frac{\text{Number of objects detected per unit time}}{\text{Number of objects per unit area} \times \text{searcher speed}} \tag{5.2}$$

It depends on three classes of factors:

(a) The search **object's characteristics** affecting detection by the sensors in use (object size, color, reflectivity/emission properties, etc.),

(b) The **capabilities of the sensors** in use (visual acuity, a lidar's or radar's ability to reliably detect a standard test object at various ranges, the effect of speed on performance, etc.), and

(c) The **environmental conditions** at the place and time of the search that affect the performance of the sensors in use (visibility, weather, sea state, vegetation (ground cover), etc.).

2. **Effective search (or sweep) rate** = W x search speed;

3. **Lateral range (detection) function**: probability of detecting as a function of distance off a searcher's single track;

4. **Effort**: total length of the searchers' tracks while searching, $L = v \times t$;

5. **Search effort**: area effectively swept, $Z = W \times L$;

6. **Search effort density**: coverage, $C = Z/\text{area searched}$

7. **Detection function**: a probability of detecting versus coverage function, e.g., $POD = 1 - \exp^{-C}$. In a uniform search through a region A with the effectiveness of random search, the probability of detecting the search object by time t given it is located in the region is

$$P(t) = 1 - \exp^{-Wvt/A} \tag{5.3}$$

   the random search $Wvty/A$ is the density of search effort or coverage function.

It may be also convenient to categorize search theory according to the type of search problem involved:

1. **one-sided search problems**

   (a) **Stationary search objects**: for a search object location distribution, the density function is given by:

$$p(x_1, x_2) = \frac{1}{2\pi\sigma_1\sigma_2} \exp^{\left(-\frac{1}{2}\left(\frac{x_1^2}{\sigma_1^2} + \frac{x_2^2}{\sigma_2^2}\right)\right)} \tag{5.4}$$

   a bi-variate normal probability distribution with its mean center at $(0, 0)$.

   (b) **Moving search objects**: most moving search object problems were approached by freezing the search object motion over some increment of time, allocating effort as though the search object were stationary during that time increment, and then repeating the process for the next time increment.

2. **Two-sided search problems**: although drifting search objects move very slowly when compared to the speeds of searching aircraft, it is not unusual for them to drift several miles while the UAVs are on scene searching.

   (a) **Cooperative search**: teams of UAVs are deployed to perform search and track-related tasks. At first, an operator marks out the boundary of an area for search. There exist multiple static targets (may change to be mobile targets) placed in the **area of operation** (AO). Upon reaching the area, the UAVs collaboratively search for the target in the AO. The UAVs will head for the high uncertainty locations, exchange search results, update the target location probability map to be maintained on each UAV. The UAVs ensure that they do not overlap in their respective search areas to ensure the best search effectiveness. When a target is detected, the founder UAV will assign itself as a manager to assign a target monitoring (tracking) task. Task assignment will be done based on contract net protocol. The cooperative search pseudo-code is presented in Algorithm 26 [38].

(b) **Non-cooperative search** can be formulated as the problem of uncertainty reduction by using multiple searchers and obtaining strategies for search route determination using a game theoretical framework.

Algorithm 26 Cooperative Search

---

1. **Initialization**: Initialize way-point (entry area of the area of operation), detection probability, and false alarm, then divide the area into cells and set the initial probability of each cell to the largest uncertainty of target existence.

2. **Sampling step**: The UAV takes measurement and updates its probability map, then it shares the probability map with its neighbors.

   (a) **If** the target is found **Then** the UAV first checks whether the target is tracked by the other UAVs

   (b) **Else** the UAV will assign itself as the leader, prepare and send out an announcement to ask for task assignement proposal. The leader will propose where task assignment is based on contract net protocol. The designated UAVs will switch to the tracking mode.

3. **Way-point planning**: If no target is found, the UAV is kept in search mode. The way-point generation is based on the Lloyd-like gradient descent control law taking into account the probability map, the UAV flight constraints, and multi-UAV collision avoidance.

4. **Way-point based path planning** according to the real UAV model.

5. **Return to Step 2**.

The search mission is subject to a time or distance constraint. The practical interpretation of this constraint could be the limited battery capacity or the presence of time window to perform the mission. Due to this constraint, optimal search paths may not visit every location. Indeed, in some scenarios it is possible that some locations are visited more than once, while other locations are never visited at all. Another task is to find the most likely location of a single object or declare that the object is absent. The goal is either to maximize the probability of detecting the object or to minimize the expected time until a decision about presence or absence of the object is made. In a realistic search problem, there are certain limitations on successfully locating the objects such as imperfect sensor measurements or uncertain knowledge of the search environment. Noisy sensor measurements often include missed detections, i.e., failing to detect an object that is present, and false alarms, i.e., detection of an object that is not present. Local environmental conditions may also affect the number of false alarms and missed detections the sensor observes. When using a UAV's on-board camera to assist wilderness search and rescue operations, factors such as dense vegetation, lighting conditions, shadows, or distance between the camera and the ground can lower the quality of the UAV's aerial view and decrease the probability of detection. This can be attributed to both sensor and human limitations (such as limited attention span and cognitive workload). **Partial detection** can be represented in the form of a task difficulty map, where a more difficult sub-region on the map has lower probability of detection. Using a **task difficulty map** enables to integrate geo-referenced and spatial-related sensor constraints into the problem formulation, which supplements traditional sensor modeling methods and potentially improves search performance in real-world search scenarios. Rather than flying long distances or loitering, a fleet of aircraft will maintain a permanent presence at a remote location. Two questions arise:

1. At what distance can this presence be maintained?
2. At what aerodynamic efficiency shall the aircraft fly during the transits?

The **Breguet equations** for range and endurance provide a preliminary formula for the **range for continuous coverage**, in which the aerodynamic coefficients for both transit phases appear. The optimization of these coefficients provides good results: for a given propulsion type and number of aircraft, the lift coefficient for the optimal range for continuous coverage is always a constant multiple of the lift coefficient for the best range [56].

Rather than the range or endurance, however, the performance parameter to be optimized may be the range at which a fleet of N UAVs can maintain continuous coverage, continuously replacing an UAV that is ending its mission with another one arriving from a base. The **fleet factor** is the ratio between the **mission active time** or **time on station** and the **mission idle time**, which is the sum of the transit times and maintenance time. The system performance will be evaluated through the value of the range (denoted by D_p) at which continuous coverage can be maintained for a given number of UAVs in the fleet. The problem is the optimization of the distance one UAV can fly such that the ratio $(Transit_{Time} + Turn_{around_{Time}} / Time_{On_{Station}}$ remains below a certain value, assuming that the transits are performed at the speed corresponding to the best range, so that fuel burn can be minimized for the distance covered.

Remark 165 *The following assumptions can be made:*

1. *The UAVs in the multi-UAV swarm are isomorphic and every UAV can execute search.*

2. *All UAVs fly in a constant altitude with the same constant velocity. Therefore, the mission space can be simplified to be 2-D space.*

Given the UAV activity area in 2-D space, the mission area is discretized to a grid map with size of $L \times W$ [37].

The search and rescue mission planning in large wilderness areas can be formulated as a **mixed integer linear programming** optimization problem. The objective is to jointly define, for all agents in a heterogeneous search and rescue team, the search trajectories and the activity scheduling that maximize the coverage of the area. Resulting trajectories consist of sequences of adjacent environment sectors, while activity scheduling specifies the amount of time that each agent should spend searching inside each sector along its trajectory. In practice, the maximization of the overall area coverage might not be the only representing factor of efficiency for mission performance. Other important aspects, common to search and rescue missions, need to be accounted for in order to be able to deal with the complexity of the real-world and to address the needs and the possibilities offered by the presence of a heterogeneous team. Among these aspects, the management of agents' interactions in terms of spatio-temporal relations represents a challenging issue to the mission planner:

1. **Provisioning of topological connectivity** to support wireless communications (to stream mission updates to a control center);

2. **Boosting cooperation between the agents** (if a human rescuer and a UAV are concurrently exploring closeby or overlapping areas, the UAV could augment the rescuer's view by serving requests for the real-time video streaming of areas selected by the human);

3. **Increasing safety of agents** (at night time, the human agents performing a search in the wilderness might be required to stay relatively close to each other for personal safety reasons); and

4. **Minimizing negative interferences** (different air-scent dogs should be sent out over different, far away areas to avoid disturbing each other).

The search area is discretized into a set of squared environment cells. A uniform cell grid decomposition is considered, and the cells' set is indicated by $C = \{c_1; \ldots; c_n\}$. Environment cells also serve as a means of evaluating mission status in terms of coverage. The coverage map $C_m : C \rightarrow [0; 1]$ relates cells to numerical values representing the amount of coverage currently required, or the residual need of exploration of each cell. Based on the above cell definitions, the area is further decomposed into sectors, that is, clusters of contiguous environment cells (i.e., a subset of C). Sectors are needed to account for the different sensory-motor capabilities of the different agents, and for the purpose of efficient search, since performing mission planning at the resolution of individual environment cells may become both impractical and computationally infeasible if cells are very small. Defining the sector boundaries requires a careful analysis, and ideally, knowledge of the region. In case of heterogeneous team of agents, the definition of possible sectors must be done taking into account the agents' capabilities and terrain conditions in order to properly pair them. Once the sectors have been defined, then the allocation of the resources inside the sectors is made, deciding when, from whom, and how much effort each sector will receive. This is accomplished by the specification of **agent plans**, which are defined in terms of **search tasks**: dispatching the agents to sectors with the objective of carrying out exploration activities for a certain amount of time.

A global mission plan consists of sequences of search tasks to be executed one after the other by each one of the agents. Search tasks are represented by $(L; t_{start}; t_{end})$, where $L \subset C$ is a sector, and t_{start} and t_{end} are start and end times for a search task inside sector L. The whole mission time is discretized into mission intervals of length equal to Δ_t seconds. For planning purposes, the traveling times between tasks are implicitly taken into account within the time allocated to perform the tasks. Clearly, traveling across a sector is performed while searching. The set of sectors of each agent must be congruent to the definition of the mission interval length Δ_t. Given that this interval represents the minimum duration of any task assigned to any agent, its length must be enough to enable each agent to traverse the sector corresponding to its current task and ensure the timely arrival to the sector assigned for the next task. Therefore, mission plans should be designed taking into account that consecutive tasks belonging to a plan must involve sectors that are physically close, if not adjacent. To this end, a **traversability graph** $G_k = (\Gamma_k; E_k)$ is defined for each agent $k \in A$, where E_k contains an edge $(i; j)$ if a task at sector j can be scheduled right after a task at sector i. In order to compute efficient joint mission plans, the planner must explicitly take into account the fact that different agents may show different levels of performance accomplishing a task in the same portion of environment, due to their heterogeneous skills as well as to the effect of local conditions. Relevant environment properties are extracted from spatial data provided by **geographic information systems** and procedures are defined for estimating the expected search performance or **search efficiency**, for each single agent and for each different portion of the environment. The **search efficiency** is specified by the coverage rate, by tracking the movements of each agent and adding up the amount of time spent inside each cell [31].

Definition 166 ***Surveillance coverage rate*** *is the ratio of grids which have* *been searched to all the grids in mission area as:*

$$P = \sum_{x=1}^{L} node_{x,y}/L \times W \qquad node_{x,y} \in \{0, 1\} \tag{5.5}$$

Problem 167 *The goal of the searching mission is to maximize the* ***surveillance coverage rate*** *in order to find the most targets. Each UAV searches the mission area with the goal of maximizing the surveillance coverage rate in order to find the most targets.*

When the targets are clustered, that is, located in groups of linked sites, each group is to be inspected non-stop from one site to the other. The problem is to efficiently detect the targets. The choice of a search strategy strongly influences the losses incurred by the search as well as the probability of finding the target within a time limit. In many cases, two types of errors in search-and-detection tests may occur:

1. A **false-negative detection** wherein a target is overlooked; and
2. A **false-positive detection** (also called a false alarm) which wrongly classifies a good object as malicious.

Unfortunately, the problem of selecting the best search strategy is fundamentally hard due to its stochastic nature and the non-linearity induced by the detection probability [52].

Search efficiency is very important in wilderness search and rescue because, as time progresses, the survivability of the missing person decreases and the effective search radius increases by approximately 3 km/h for a pedestrian. Therefore, a good flight path should rapidly maximize the probability of finding the missing person to make efficient use of the limited flying time. Each UAV path accumulates information over time as the UAV's sensors scan the ground. Various paths do so in different ways depending on how information is distributed in the environment. The goal is to maximize the total **probability of detection** which is the likelihood of detecting a target.

Reconnaissance problems can be defined as the **vehicle routing problem**. The cooperative system of UAVs can considerably improve information superiority in this problem. Road search problems can be generally classified into two categories: one is the traveling salesperson problem and the other the postperson problem. The TSP using multiple UAV can be considered as **task assignment problem** to minimize the cost (time or energy) by assigning each target to the UAV [46]. The problem of managing uncertainty and complexity of planning and executing **intelligence, surveillance, reconnaissance** (ISR) mission uses a network of UAV sensor resources. In such applications, it is important to design uniform coverage dynamics such that there is little overlap of the sensor footprints and little space left between the sensor footprints. The **sensor footprints** must be uniformly distributed so that it becomes difficult for a target to evade detection. For the search of a **stationary target**, the uncertainty in the position of the target can be specified in terms of a fixed probability distribution. The spectral multi-scale coverage algorithm makes the sensors move so that points on the sensor trajectories uniformly sample this stationary probability distribution. Uniform coverage dynamics, coupled with sensor observations, helps to reduce the uncertainty in the position of the target [7].

5.2.2 Camera properties

Regarding the surveillance problem, the **sensor coverage range and resolution** must be defined. It can be modeled as a linear function of altitude: the sensor has a constant field of view and its coverage range increases as the UAV gains altitude [27]. The sensor resolution diminishes at high altitude, which is modeled as:

$$c(z) = \left\{ \begin{array}{cc} z \tan\theta & \text{If } 0 \leq z \leq h_{max} \\ h_{max} \tan\theta - \frac{h_{max} \tan\theta}{h_0 - h_{max}} (z - h_{max}) & \text{If } h_{max} \leq z \leq h_0 \end{array} \right\} \quad (5.6)$$

where c and z are, respectively, the coverage range and altitude. The sensor constant **field of view** is denoted by θ, h_{max} is the altitude at which the sensor has maximum coverage range, h_0 represents the altitude at which the sensor resolution is not appropriate for the aerial surveillance. A **probability of object detection, recognition, or identification** (P-DRI) metric describes the probability that a trained human operator will successfully detect, recognize, or identify features from images provided by a UAV. One can maximize the overall P-DRI of specified objects by optimizing UAV flight paths. Several methods exist to estimate the probability of object detection, recognition, or identification. A geometric model describes a **seeability metric** which is defined by the UAV and camera position, terrain relief, viewing angle uniqueness, and the number of image frames for which a designated terrain point is visible from the camera. Computing the seeability metric requires little computational power. The seeability metric is a value bounded by 0 and 1 and is considered to be an estimate of the **probability of object identification** using the appearance data in an image frame. However, other parameters such as object size and lighting conditions are also important in estimating the probability of object identification from an image. An alternative formulation estimates the P-DRI, namely the **targeting task performance** (TTP) metric, which considers all spatial frequencies for which an object is detectable by an observer viewing a display whereas the **Johnson criteria** only considers the limiting spatial frequency for which the object is detectable. Once the targeting task performance metric is computed, the number of resolved spatial cycles on the object is estimated and then applied to a logistic function that describes a probability for detection, recognition, and identification of the imaged object. Targeting task performance enables to consider objects whose spatial frequency content is complex, but well represented using **Fourier se-**

ries coefficients. With accurate equations of motion, good state estimation, and 3-D terrain structure and lighting information, the parameters used to estimate the probability of identification can be predicted based on current state information to facilitate a path planner that maximizes the probability of identification through determination of an optimizing path [44].

In order to further reduce human involvement and speed up the process of finding causalities, the task of analyzing collected video can be delegated to an automated algorithm which analyzes the video footage in real time, online. An algorithm which can identify and geographically locate places where human bodies can be found is required to achieve such a task. In the task of observing and analyzing human appearance and movement, techniques can be categorized in many ways [49]:

1. One is the need for pre-processing, such as background subtraction, which can be achieved by frame differencing. Other factors include the types of features which are needed for describing human appearance e.g., shape, color, contour. A considerable amount of work is based on the idea of detecting humans by parts. For example, humans can be modeled as assemblies of parts which are detected separately and represented by co-occurrences of local features. A cascade of rejectors with variable size blocks of histograms of oriented gradients as features can also be used.

2. Another approach takes advantage of a classifier which is a cascade of boosted classifiers. Detecting humans in thermal imagery poses additional challenges such as lower resolution, halos around hot or cold objects, and smudging artifacts in case of camera movement.

Planning the trajectory of sensing agents to gather maximum possible information or reduce the uncertainty of the environment incorporate an information metric into the planning problem and achieve an information-rich trajectory. Uncertain environments can be represented as **Gaussian processes, Gaussian Markov random fields** or **Gaussian random fields**. The persistent monitoring problem can be formulated as an optimal control problem [25]. Density-based descriptors, such as color histograms, represent an attractive alternative for their low computational complexity and robustness to appearance changes. In the challenging UAV application context, strong simplifying assumptions are usually made in the vision algorithms. A color-based algorithm is used to track a fixed target and autonomously stabilize a UAV above it. Due to hardware limitations, the proposed tracking approach assumes that the target is clearly visible, without handling occlusions nor the presence of distractors of similar color. The visual tracking system aims to provide an estimate of the relative position and in-plane rotation between the UAV and the object. To achieve this in a robust way, a color-based representation is chosen and the tracking is performed in the particle filtering framework. Estimations of the relative position and translational velocity between the UAV and the object, obtained by the vision system, are used as an input to the proposed control law.

Definition 168 ***Geo-location*** *is the tracking of a stationary or moving point of interest using visual cameras for payloads. It is the process of using sensor data to develop statistical estimates of a point of interest on the ground. The geo-location system for a UAV requires the complex integration of several hardware components (camera, UAV, GPS, attitude sensors) and software components (camera image processing, inner-loop and path planning control, and estimation software) to develop accurate estimation of the object being tracked.*

The sensor biases and the unknown point of interest state can be estimated in a decentralized manner, while using the solution from the on-board navigation system to save significant computation. The joint estimation problem can be solved for multiple UAVs cooperating in a decentralized fashion such that the UAVs share information on the point of interest state and model only their local biases. This decentralized formulation saves computation as well as communication, while giving geo-location accuracy that is comparable with the centralized case. Further, this decentralized approach allows for effective cooperation not only among UAVs with potentially different biases, but also among different sensors altogether.

The environment affects sensor performance, and that variation in the environment throughout the search area causes search performance in some locations to be better than in other locations. Guidance algorithms for search missions can incorporate stochastic knowledge of the environment to improve search performance, and for the case the **environmental information** can be acquired in order to improve overall search performance:

1. Environmental characterization is performed prior to a search mission,
2. Environmental characterization is performed at the same time as the search mission by a separate asset than the search vehicle, and
3. Environmental characterization is performed at the same time as search by the same vehicle that performs the search task.

A decision-theoretic value function can be used that is associated with the accuracy of the estimate of the number of objects in the environment. Because search performance is dependent on the environment, knowledge of the environment can improve search performance due to better search plans. For example, one may choose to avoid searching areas that are known to contain excessive clutter and many false positives in favor of environments with few false positives. In situations where the environment is poorly known, efforts to acquire environmental information may lead to improved search effectiveness. Stochastic knowledge of the environment can be acquired, where the environment should be surveyed in order to improve overall search performance. One approach for selecting where to acquire environmental information is simply to characterize locations that yield the greatest reduction of uncertainty about the environment, maximizing change in **entropy** [64].

5.2.3 Human operator

UAVs provide essential support to human task forces in situation assessment and surveillance. A UAV's abilities, behavior, and possible achievements need to be transparent to a human operator: whether the UAV is doing something, what it is doing and why, and whether it has achieved a goal Humans perform under stress in complex environments. Situations, interactions, plans, and expectations change. As situations change, affecting the dynamics of the team, the UAV needs to adapt its behavior, and the way it presents that behavior to continue to provide adequate and effective transparency. In the system design, the human perspective is pervasive throughout the representations the UAV builds, and the way it determines its behavior. When it comes to human-UAV interaction and planning, humans must be explicitly modeled as actors, and action and interaction are planned in ways that conform to human operational practice. The UAV builds up a qualitative structure of dynamic space, and can make inferences about possible actions situated in that space. Mapping therefore builds up several layers of abstraction:

1. **Static target search and rescue**: a decision support method for search and rescue with a stationary target may be developed seeking to maximize the probability of mission success. Path planning is conducted via MILP for use in a rolling horizon optimization that allows for updates based on new information. The approach can be extended to multi-target and multi-agent search and rescue path planning. Both restrict attention to paths on a grid.

2. **Roaming target search and rescue**: their model and algorithms allow for Markovian movement.

3. **Constrained search and rescue**: the constraints serve to enforce visit order and camera detection range. A MILP planning model can be formulated.

4. **Fuel limited systems**: a single searcher with limited travel time is first assumed. The method uses a mixed linear/non linear integer programming model to optimize the tasks on a single trip or to generate task sequences for multiple fuel limited UAVs. Here, UAVs may return to the single start depot and cannot return to service. The first part can address resource selection, while the second part can determine task assignment for the vehicles allowing for multiple flights and vehicle loading capacity. There is no consideration of changing survival probabilities without considering multiple time periods.

An approach is developed in [62] for human-UAV team in disaster response where a planner agent gathers information with the assistance of UAVs and plans actions on behalf of fire responders. A scenario involves rescue tasks distributed in a physical space over which a radioactive cloud is spreading.

Tasks need to be completed by the fire responders before the area is completely covered by the cloud. As fire responders will die from radiation exposure, they need to know the radioactivity levels at the task sites and along the paths to the tasks. Crucially, such information should be collected by UAVs before fire responders decide their next actions. An algorithm is proposed that combines planning and sensing based on **Monte Carlo** simulation and value of information to compute policies both for fire responders and UAVs. By so doing, online planning is coupled with active sensing to achieve more effective rescue missions. In each time step, the human computes a policy and specifies their action, and the UAVs use this policy to update their sensing action. The UAVs' actions are dependent on what policy is chosen for the fire responder. In turn, they can update and enhance the policy based on information reported by UAVs as they fly over the areas chosen. The key procedures are

1. How to compute the fire responder's policy;
2. How to choose the best sensing action for the UAVs; and
3. How to update the belief state of the radioactive cloud given the observation from the UAVs.

A multi-modal interaction framework suitable for human-UAV interaction in search and rescue missions is presented in this paragraph. In the place of a fully dedicated operator, rescuers are deeply involved in the search and rescue mission, hence only able to provide fast, incomplete, sparse, although high-value, inputs to the UAVs. Since the human is not fully dedicated to the UAVs, the robotic system should support different control modes from an autonomous behavior to direct tele-operation, passing through the mixed-initiative mode, where the user can execute some operations while relying on the autonomous control system for the remaining ones. The human should focus his cognitive effort on relevant and critical activities, while relying on the autonomous system for specialized tasks. Moreover, the human should operate in proximity with the UAVs in hazardous scenarios. The multi-modal interaction should allow the operator to communicate with the robots in a natural, incomplete, but robust manner exploiting gestures, vocal, or tablet-based commands. In the framework proposed in [14], both command-based and joystick-based interaction metaphors can be exploited and smoothly combined to affect the UAVs' behavior. While a UAV is executing a task, the human can exploit gestures in joystick-based metaphor to adjust the generated trajectory or to directly tele-operate the selected UAVs. Indeed, the interpreted human interventions are continuously integrated within the robotic control loop by a mixed-initiative system that can adjust the UAV behaviors according to the operator intentions.

Performance metrics for human-robot interaction establish a way of describing and evaluating human-UAV task performance in this high-risk, time-critical domain. Human-robot interaction metrics focus on the work system as a whole, examining the UAV's effects on human task performance within the context of real-time human performance in field settings. Urban search and rescue task forces can be characterized as extreme teams who function in dynamic, high risk, and time critical environments. Team members must function in conditions which are often physically, mentally and emotionally taxing. The criteria for measures of robot-assisted human performance in urban search and rescue are search, rescue (extrication), structural evaluation, medical assessment and treatment, information transfer, command and control, and logistics [13]. As a continuous monitoring network, the best strategy for communication with a base station is to record data in batches and transmit relevant data in a hop by hop, store and forward nature. Limited energy reserves prohibit the transmission of every detected image through the network. It is, however, possible to perform automated image analysis to detect features of interest. Object detection is commonly accomplished using the Voila-Jones object detection framework applying filters to integral images that are calculated with one pass over the image. Each filter behaves as a weak classifying function that has an accuracy slightly better than random. A combination of several of these functions resulted in a detector that demonstrates high accuracy [40].

5.3 INFORMATION GATHERING

Problem 169 ***Informative motion planning*** *consists in generating trajectories for dynamically constrained sensing agents. The UAVs used as mobile agents to traversing the operating environment are typically subject to dynamic constraints. Obstacles in the operating environment can both constrain the UAV motion and occlude observations. Finally, the limitations inherent in the available sensing mechanism can further limit the informativeness of agent plans.*

Several solution strategies to effect information-rich path planning can be considered. Analytical solutions often use the Fisher information matrix to quantify trajectory information collection in an optimal control framework. Solutions seek to maximize, for example, a lower bound on the determinant of the **Fisher information matrix** (FIM) at the conclusion of the trajectory. While analytical solutions often have a simple form and perform optimally for low-dimension unconstrained problems, they typically are difficult to scale to complicated scenarios. The **information-rich rapidly exploring random tree** (IRRT) algorithm is proposed in [61] as an on-line solution method that by construction accommodates very general constraint characterizations for the informative motion planning problem. The IRRT method extends the RRT by embedding information collection, as predicted using the Fisher information matrix at the tree expansion and path selection levels, thereby modifying the structure of the growing feasible plan collection and biasing selection toward information rich paths. The Fisher information metric can be calculated once a probability distribution has been chosen. As the IRRT is a sampling-based motion planner, feasible solutions can be generated on-line.

Another approach is the **heuristic path shape** that performs well in steady state. For UAVs with side-mounted cameras, circular trajectories with optimal radii at a fixed altitude and varying altitude can be proposed. While these heuristically constrained trajectories may capture the physical and geometrical intuition of bearings-only target tracking, the gap between anticipated and realized informativeness of the motion plan can become arbitrarily large when operating under realistic dynamic and sensing constraints. The **partially observable Markov decision process** (POMDP) framework is a way of solving problems of planning under uncertainty with observations but there are also tractability issues. Belief space planning can be considered for both the target tracking problem and its inverse; that of localizing a vehicle through sensor measurements of perfectly known targets in a previously mapped environment. POMDP solutions are currently intractable for vehicle models with complex dynamics [18]. A **finite one-in-set traveling salesperson problem** (FOTSP) can be used to approximate a polygon-visiting TSP so that a roadmap is constructed. In the two-step approaches, the first step tries to obtain an optimal trajectory under the specific constraints. In the second step, the UAS has to be steered out of dangerous regions.

In [33], for the problem of path planning for multiple UAVs, the paths are planned to maximize collected amount of information from **desired regions** (DR) while avoiding **forbidden regions** (FR) violation and reaching the destination. The approach extends prior study for multiple UAVs by considering 3-D environment constraints. The path planning problem is studied as an optimization problem solved by a **genetic algorithm** (GA) with the proposal of evolutionary operators. The initial populations have been generated from a seed-path for each UAV. The seed-paths have been obtained both by utilizing the pattern search method and solving the **multiple traveling salesperson problem** (mTSP). This technique solves both the visiting sequences of desired regions and the assignment problem of which desired regions should be visited by which UAV. All of the paths in population in any generation of the GA have been constructed using the dynamical mathematical model of a UAV equipped with the autopilot and guidance algorithms. The algorithm consists of three main steps:

1. The problem is reduced to the multi-UAV path planning problems by solving the assignment problem of which desired region should be visited by which UAV. The desired regions visiting sequence for each UAV is also determined. **Pattern search algorithm** is utilized to find the distances between the centers of desired regions. The simplified form of the problem is modeled as a multiple traveling salesperson problem.

2. Instead of using randomly generated population for path search, seed paths are formed, for each UAV, to satisfy the physical constraints of the problem as much as possible. It provides a good starting point for path search.

3. Two new mutation operators are defined and implemented: **Ascend-To-**

EScape (ATES), and **Change ALTitude** (CALT). These operators also mimic the thinking process of a human path planner.

The main difficulty of the problem is due to the dynamic constraints of the UAVs. Otherwise, the first step (PatternSearch and mTSP) of the algorithm might have been sufficient to solve the problem. Even though using a full dynamic model introduces much more complex constraints to the problem, it makes the simulations more realistic, and it guarantees that the generated path does not violate dynamic limitations of UAV. In addition, the outputs of the controllers are saturated to handle the physical constraints of UAV actuators.

5.3.1 Detection

The challenges of **moving object detection** on-board UAVs include camera motion, small object appearances of only few pixels in the image, changing object background, object aggregation, and noise. In general, two modules are necessary for digital video stabilization: **global motion estimation** module and **motion compensation** module. A perfect motion correction needs an accurate global motion estimation. There are many methods that aim to estimate global motion accurately. A method for global motion estimation can be introduced by calculating the motions of four sub-images located at the corners of the image. A method based on circular blocks matching is also proposed in order to estimate local motion. The global motion parameters are generated by repeated least square. Global motion can be estimated by extracting and tracking corner features. When a UAV flies at hundreds of meters it provides a large region of surveillance, but with noisy and blurring images. Moving sensors in UAV surveillance system capture videos at low resolution and low frame-rate, which presents a challenge to motion detection, foreground segmentation, tracking, and other related algorithms. Scale and view variations and few pixels on region of interest are among these challenges. So the pre-processing is necessary to attain a better detection effect. Characteristics of the video itself, such as unmodeled gain control, rolling shutter, compression, pixel noise and contrast adjustment do not respect brightness constancy and geometric models. The background in the UAV surveillance platform changes frequently because of the high speed of the aircraft.

To identify traffic status and incidents, the complex background should be filtered and traffic features should be detected. **Scale-invariant feature transform** (SIFT) feature extraction can identify key points in order to be tracked over multiple frames of video and they are invariant to image translation, rotation and scale. The feature point tracking method is used to acquire a serial of features, which are then classified into three categories: undesired motion, moving object, and static object [57]:

1. Global motion estimation using **SIFT** feature point extraction and matching to eliminate camera vibration and noise.

2. Moving object detection using the **random sample consensus** (RANSAC) and **Kalman filtering**.
3. Motion compensation using **affine transformation**.

Two advanced path planning algorithms are described in [22]:

1. **Generalized Contour Search**: The ability to control the gimbaled camera to aim at a target point while orbiting enables a generalized contour search path planning algorithm. A queue of target points that follow the contours of the distribution of the missing person's likely locations can be created from which the algorithm interpolates and resamples at uniform distances. Lawnmower and spiral paths naturally emerge from the algorithm for uniform and Gaussian distributions, respectively, and they are the optimal paths. It is also possible to use the algorithm to follow the contours of steep terrain by aiming the camera out to the side of the UAV.
2. **Intelligent Path Planning**: The second path planning algorithm aims to maximize the accumulated probability for the path generated given a distribution, a starting point (optionally an ending point), and desired flight time. The camera footprint traverses a grid of probability nodes (enabled by the gimbaled camera), while the UAV approximates the path generated. Near-optimal flight paths are generated using an evolutionary approach, where seed paths are generated using various hill-climbing and **artificial potential fields** algorithms.

These advanced algorithms enrich the autonomy tool set for the UAV operator and can potentially be useful for the high priority search and exhaustive search techniques when systematic coverage is desired.

5.3.1.1 Agent model of a missing person

In **wilderness search and rescue** (WiSAR), a **missing person** (MP) must be found, before getting rescued or recovered. Search often involves observing large tracts of the environment to detect potential evidence of where the missing person has gone. Search algorithms can be posed as Bayesian decision problems. As the UAV explores its environment and monitors it with its sensors, the distribution of the missing person's location is updated through the use of **Bayes** rule. The decision of where to look next is determined through knowledge of the new posterior distribution. A WiSAR operation begins when an individual is reported missing. The first stage in the response is to construct a **missing person profile** (MPP), with information about the missing person's physical state, together with the time and point where the missing person was last seen, the direction the missing person was traveling, and information about where the missing person was intending to go. In addition, information about the environment (including elevation, vegetation and topography) is collected. From this, the search region A is constructed within which the search activities will take place. The distributions are non-Gaussian and a grid-based decomposition of A is used. Specifically, A is decomposed into a set of $M = |A|$ identically sized cells, where a is the ath cell. Suppose x is the cell which contains the missing person. $p(x_k = a)$ is the probability that the MP lies in cell a at time k. Bayes rule can be used to update this over time. To begin this process, the prior $p(x_{k_0} = a)$ must be chosen. However, this is strongly dependent upon the environment itself. The environment plays a fundamental role in the behavior of the missing person and the modeling of the search problem. The environment is characterized in three ways:

1. **Elevation model**, Γ. The elevation model describes the perceived slope of the environment. The slope is classified according to:

$$dom(\Gamma) = \left\{ \begin{array}{c} |S_l| < S_{lmin}, S_{lmin} < S_{lmed} \\ |S_l| < S_{lmed}, S_l > +S_{lmed}, \\ S_l < S_{lmed}, |S_l| < S_{lmax} \end{array} \right\} \tag{5.7}$$

   where S_l is the local slope, $S_{lmin}, S_{lmed}, S_{lmax}$ are minimum, medium, and maximum S_l thresholds.

2. **Topography classification model**, Ψ. This model classifies the search area according to its topography
   $dom(\Psi) = \{Obstacle, Water, Ground, Path\}$.

3. **Vegetation classification model**, Φ. This classifies the density of vegetation using the categories
   $dom(\Phi) = \{Sparse, Medium, Dense\}$.

With a **diffusion based model**, an occupancy grid is initialized with the distribution $p^*(x_0 = a)$ which is derived from the missing person profile. The diffusion model is run S times, where the kth iteration is

$$p^*(x_k = a|\Gamma, \Phi, \Psi) = \gamma_k \sum_{m \in A} p(x_k = a|x_{k-1} = m)p^*(x_{k-1} = m|\Gamma, \Phi, \Psi) \tag{5.8}$$

where $p(x_k = a|x_{k-1} = m)p^*(x_{k-1} = m|\Gamma, \Phi, \Psi)$ is the probability that a missing person transitions into cell a from cell m in a single time step. γ_k is the normalization constant. After S iterations have been completed, the prior for the search process is given by $p^*(x_{k_0} = a) = p^*(x_s = a)$ where k_0 is the time the search operation starts. The transition probability is affected by the physical constraints of the environment such as the gradient of the terrain, vegetation density, and the topography of the ground locally. The impacts of the different characteristics of the environment are assumed to be independent of one another. This approach fails to model the fact that a missing person is an intelligent entity whose internal state evolves over time. One example is view enhancing, in which a person will walk up to a high place, such as the top of a hill, to improve her knowledge of the surroundings [39]. Agent-model uses an internal state which can be used to describe complex behavior and human/environmental interactions within a spatial framework.

5.3.1.2 *Proximity relationship*

The approaches for the planning of search and rescue missions can be divided into two categories: probabilistic and non-probabilistic. When a probabilistic framework is considered, the mission assignment attempts to optimize the search by maximizing detection probability, minimizing the expected time for target detection, or number of detections. The related notion of exploration demands bounded range variables used to prioritize parts of the area according to the expected amount of exploration they require. The vehicle routing and multi-robot task allocation problems share the objective of efficiently allocating spatially distributed tasks to a team of agents. The task allocation problem is formulated as an optimization model where solution approaches are based on centralized control mechanism utilizing MILP, dynamic programming and genetic algorithms. For a complex mission where multiple UAVs are supposed to deal with moving targets, an efficient task allocation mechanism should combine search planning with a distributed setting [31]. In probabilistic physical search problems, there is a prior probabilistic knowledge regarding the price of the possible alternatives at each site, and traveling for the purpose of observing a price typically entails a cost. Furthermore, exploration of the item results in the expenditure of the same type of resource. The UAV's battery is used not only for performing the mission, but also for traveling from one potential location to another. Thus, the agent needs to carefully plan exploration and balance using the available budget between the exploration cost and the purchasing cost. Two variants of the problem can be considered:

1. **Max-probability** considers an agent that is given an initial budget for the task and needs to act in a way that maximizes the probability it will complete its task.

2. **Min-Budget** guarantees some pre-determined success probability, and the goal is to minimize the initial budget necessary in order to achieve the success probability.

As probabilistic physical search problems are hard on general graphs, approximations with guaranteed bounds or heuristics with practical running time can be considered [42].

The introduction of UAVs acting as data gatherers could potentially lead to an information overload toward the human search-and-rescue workers, if the data acquired by robotic tools are not managed in an intelligent way:

1. The different robotic systems require an interoperable framework in order to pass data from one to another and toward the unified command and control station.

2. A data fusion methodology combines the data acquired by the different heterogeneous robotic systems.

3. The computation needed for this process is published in a **software as**

a service (SaaS) model. The SaaS model helps in providing access to robotic data over ubiquitous Ethernet connections.

The system consists of interoperable and collaborative UGVs and UAVs, operated using field proven command and control tools. The data gathered by these unmanned systems are made publicly available in the form of a comprehensive dataset. These data can be processed in real time by a mobile data center to provide high-quality geo-referenced (3-D) data of the environment. **Open geospatial consortium** (OGC) standards have been used for storing and retrieving data from the **Geographic Information System** (GIS) Geoserver [6].

5.4 MOBILITY OF TARGETS

The basic discrete **search and research** (SAR) or optimal searcher path problem involving a **stationary target** is known to be NP-Hard (nondeterministic polynomial time). SAR may be generally characterized through multiple dimensions and attributes including: one-sided search in which targets are non-responsive toward a searcher's actions, two-sided describing target behavior diversity:

1. Cooperative, non-cooperative, or anti-cooperative,
2. Stationary versus moving target search,
3. Discrete versus continuous time and space search (efforts indivisibility/divisibility),
4. Static/dynamic, closed-loop decision models, and
5. Observation model, pursued objectives, target and searcher multiplicity and diversity.

Since the search problem is primarily concerned with the area to be searched, initial studies simplified the search problem to an area coverage problem. The introduction of the **probability of detection** along with advances in computational hardware led to more optimal allocation of search effort. The optimal search problem is a decision making process that involves two dynamic partakers: the searcher owns a sensor, and the target that is searched. This search problem can be modeled in a continuous way or in discrete form, depending on the searching region representation, the sensor type, and the searcher's possible actions. The case of a randomly located target can be discussed in the plane when the located target has symmetric and asymmetric distributions and with less information available to the searchers.

Another search plan in the plane divides the plane into identical cells. The searcher moves along spiral with line segment curve, finding the optimal value of the arcs and the target with minimum cost. When the searcher moves along parabolic **spiral curve**, the distance between the target position and the searcher starting point depends on the number of revolutions, where the complete revolution is done when $t = 2\pi$. A search model for a helix target motion in the space can be formulated by using a team of three searchers $S_i, i = 1, 2, 3$. Due to the importance of the target, the searching process is done by using more than one searcher at the same time to reduce the expected value of the first meeting time [20].

Recent taxonomies and comprehensive surveys on target search problems originate from search theory and artificial intelligence/distributed robotic control, and pursuit-evasion problem perspectives. Search-theoretic approaches mostly relate to the effort (time spent per visit) allocation decision problem rather than path construction. Based upon a mathematical framework, efforts have increasingly been devoted to algorithmic contributions to handle more complex dynamic problem settings and variants. In counterpart, many contributions on search path planning may be found in the area of robot motion planning and, namely, terrain acquisition and coverage path planning. Robot motion planning explored search path planning, primarily providing constrained shortest path type solutions for coverage problem instances. These studies typically examine uncertain search environment problems with limited prior domain knowledge, involving unknown sparsely distributed static targets and obstacles [12]. Because detection difficulties vary in different search sub-regions, flying patterns such as lawnmower and spirals do not guarantee optimal coverage. Integrating the task difficulty map into path planning adds another dimension of complexity and causes the performance of existing greedy-type algorithms to suffer. The path-planning problem can be modeled as a discrete combinatorial optimization problem and a heuristic, the **mode goodness** (MG) ratio can be proposed. This heuristic uses a **Gaussian mixture model** (GMM) to identify and prioritize search sub-regions. The hierarchical structure enables the algorithms to cluster probability volumes and prioritize search sub-regions at different levels of resolution [34].

5.4.1 Stationary target

The need for searching for hidden or lost objects arises in many applications. The searching device is subject to search-time constraints, fuel consumption and other factors. The typical objective of the search is to maximize the probability of detection or minimize the cost (or time) of the search. The choice of a search strategy strongly influences the search costs and losses incurred by the search as well as the probability of finding the target within a specific time limit. A search-and-detection process subject to the false-negative and false-positive inspection outcomes can be studied. The general resource-constrained problem being NP-hard, the study can be restricted to finding optimal strategies in a certain special scenario described below. The approach is to minimize the losses incurred during the search, using a greedy strategy in which, at each step, the on-board computer used by the UAV computes a current search effectiveness for each location and sequentially searches for a next location with the highest current search effectiveness. The local-optimal (greedy) strategy yields a global optimal. Being attractive due to its simplicity and computational efficiency, such local-search strategy guarantees finding an optimal search sequence with a pre-specified confidence level of target identification [52].

The case of N UAVs assumed to move either by first order or second order dynamics is presented. An appropriate metric is needed to quantify how well the trajectories are sampling a given probability distribution μ. It is assumed that μ is zero outside a rectangular domain $\mathbb{U} \in \mathbb{R}^n$ and that the UAV trajectories are confined to the domain $\mathbb{U}$. For a dynamical system to be **ergodic**, the fraction of the time spent by a trajectory must be equal to the measure of the set. Let $B(X,R) = \{R : \|Y - X\| \leq R\}$ be a spherical set and $\chi(X,R)$ be the indicator function corresponding to the set $B(X,R)$. Given trajectory $X_j : [0,t] \longrightarrow \mathbb{R}^n, j = 1..N$ the fraction of the time spent by the UAVs in the set $B(X,R)$ is given as:

$$d^t(X,R) = \frac{1}{Nt} \sum_{j=1}^{N} \int_0^t \chi(X,R)(X_j)(\tau) d\tau \tag{5.9}$$

The measure of the set $B(X,R)$ is given as

$$\bar{\mu}(X,R) = \int_{\mathbb{U}} \mu(Y)\chi(X,R)(Y)dY \tag{5.10}$$

For **ergodic dynamics**, the following relation must be verified:

$$\underbrace{lim}_{t \to \infty} d^t(X,R) = \bar{\mu}(X,R) \tag{5.11}$$

Since the equation above must be true for almost all points X and all radii R, this motivates defining the metric:

$$E^2(t) = \int_0^R \int_{\mathbb{U}} \left(d^t(X,R) - \bar{\mu}(X,R)\right)^2 dXdR \tag{5.12}$$

$E^2(t)$ is a metric that quantifies how far the fraction of the time spent by the UAVs in the spherical sets is from being equal to the measure of the spherical sets. Let the distribution C^t be defined as

$$C^t(X) = \frac{1}{Nt} \sum_{j=1}^{N} \int_0^t \delta\left(X - X_j(\tau)\right) d\tau \tag{5.13}$$

Let $\phi(t)$ be the distance between C^t and μ as given by the Sobolev space norm of negative index H^{-1} for $s = \frac{n+1}{2}$, i.e.,

$$\phi^2(t) = \left\|C^t - \mu\right\|^2_{H^{-s}} = \sum_K \Lambda_k |s_k(t)|^2 \tag{5.14}$$

where

$$s_k(t) = C_k(t) - \mu_k \qquad \Lambda_k = \frac{1}{\left(1 + \|k\|^2\right)^s} \tag{5.15}$$

$$C_k(t) = \left\langle C^t, f_k \right\rangle \qquad \mu_k = \langle \mu, f_k \rangle \tag{5.16}$$

Here, f_k are **Fourier basis functions** with wave number vector k. The metric $\phi^2(t)$ quantifies how much the time averages of the Fourier basis functions deviate from their spatial averages, but with more importance given to large-scale modes than the small-scale modes. The case is considered where the sensors are moving by first order dynamics; the objective is to design feedback laws so that the agents have **ergodic dynamics**. A **model predictive control** problem is formulated to maximize the rate of decay of the coverage metric $\phi^2(t)$ at the end of a short time horizon and the feedback law is derived in the limit as the size of the receding horizon goes to zero.

5.4.2 Moving target

5.4.2.1 Target capturability

The cyclic pursuit problem may be described as a problem where n bugs close in on each other, with each bug chasing its leader. In multi-agent systems simultaneous rendezvous, the agents must arrive at an agreement or consensus. At the same time, this task must be achieved in a decentralized fashion. If the communication graph containing the agents at the vertices and their connections as the edges is a connected graph, this consensus is achievable. This connectedness implies that there is a path that connects any agent i to any other agent j. The cyclic pursuit scheme is the simplest directed graph that remains connected at all times, while each agent shares information with exactly one other agent in the network. In general, the velocity of any agent i, in cyclic pursuit, is proportional to the distance separating the agent from its leader and is along the vector directed from agent i to agent $i+1$ [42]. The constant of proportionality is usually chosen to be the same for all the agents (homogeneous). For positional consensus, the agents must converge to a point, at steady state, and remain there. When the gains are homogeneous, the point of convergence (rendezvous point) is the centroid of the initial positions of the agents. By choosing heterogeneous gains for the agents, the point of convergence may be varied. The system will retain stability even when, at most, one of the gains is negative and subject to a lower bound. By choosing this negative gain, the set of reachable points could be significantly extended. However, some parts of the 2-D space are still not reachable for certain initial configurations of the agents, even with this negative gain. By choosing suitable angles of deviation for each agent, the reachability set could be expanded further to include points that were unreachable using heterogeneous gains. However, global reachability was not achievable even with these deviations.

The focus of cyclic pursuit is on performing formation maneuvers about certain designated goal points in space. Several variants and modified versions of the cyclic pursuit algorithm have been used to meet this objective. If this goal point is extended to a target trajectory, then target capture is possible using cyclic pursuit. The inherent advantages of cyclic pursuit, such as minimal communication requirements to retain connectivity, can then be gainfully used to track a moving target and possibly neutralize the threat.

In [17], the objective is to enclose the stationary target with n agents. It is assumed that each agent i has the information about the target's position and $(i+1)^{th}$ agent's position. The classical cyclic pursuit law for target enclosing problem is modified such that agent i, positioned at P_i, follows not only $(i+1)^{th}$ agent at P_{i+1} but also the target at P. This weighing scheme is mathematically equivalent to following a virtual leader located at the point $P'_{i+1} = \rho P_{i+1} + (1-\rho)P$ which is a convex combination of P and P_{i+1}, where the pursuit gain is given by $0 < \rho < 1$. Depending on their kinematics, the agents can form concentric circles or polygons.

5.4.2.2 Trajectory optimization

In the complex planning of moving target case, the timelines of the target and platform must be synchronized. The search-trajectory problem consists of the following aspects: the **probability map** for the target position, the target model, the sensor model, the platform model, and, finally, the search objective. The problem can be formulated as follows:

Problem 170 *The objective is to determine a search-trajectory* $o = (o_1, \dots, o_K)$ *maximizing the cumulative probability of detection over time period* K*,*

$$max_o PD(o) = \sum_{k=1}^{K} \sum_{c \in C} pd_k(o_k, c) \tag{5.17}$$

where $pd_k(o_k, c)$ *is the probability of detecting the target at time* k *in cell* c *and is calculated by:*

$$pd_k(o_k, c) = pc_k(c) pg_k(o_k, c) \tag{5.18}$$

The ***probability of containment*** $pc_k(c)$ *is calculated through the following equation:*

$$\begin{aligned} pg_k(o_k, c) &= Prob\left(Z_{c,k} = 1 | C_k = c, o_k\right) \\ pc_{k+1}(c) &= \textstyle\sum_{c' \in C} d(c', c) pc_k(c')(1 - pg_k(o_k, c')) \end{aligned} \tag{5.19}$$

In [48], a search-trajectory optimization is considering targets with a probabilistic motion model. It can be classified as K-step look ahead planning method, which allows for anticipation to the estimated future position and motion of the target, solved with a MILP that optimizes the cumulative probability of detection.

During search missions, efficient use for flight time requires flight paths that maximize the probability of finding the desired subject. The probability of detecting the desired subject based on UAV sensor information can vary in different search areas due to environmental elements like varying vegetation density or lighting conditions, making it likely that the UAV can only partially detect the subject. In [34], an algorithm that accounts for partial detection in the form of a difficulty map is presented. It produces paths that approximate the payoff of optimal solutions, the path planning being considered as a discrete optimization problem. It uses the **mode goodness ratio** heuristic that uses a Gaussian mixture model to prioritize search sub-regions. The algorithm searches for effective paths through the parameter space at different levels of resolution. The task difficulty map is a spatial representation of sensor detection probability and defines areas of different levels of difficulty.

5.4.2.3 Tracking a ground moving target

In airborne surveillance of moving ground targets, one important preliminary step is to continuously estimate UAV's state information. For instance, in order to analyze if the behavior of a vehicle in a surveillance region is normal, its position and velocity need to be extracted. For the continuous estimation of vehicle states, the **Kalman** or **particle filter** based on the Bayesian framework are widely applied. Considering a tracked object could maneuver with different movement types, the **interactive multiple model** (IMM) method was applied for state estimation with multiple state models. As the movement of a ground vehicle is not free but constrained by its operational terrain, some terrain information can be combined with certain filtering algorithms. The most widely used terrain information is the **road network** [65].

A UAV stabilization strategy based on computer vision and switching controllers is proposed in [23]. The main goal of this system is to perform tracking of a moving target on ground. The architecture implemented consists of a UAV equipped with an embedded camera which provides real-time video to a computer vision algorithm where images are processed. A vision-based estimator is proposed, which makes use of 2-D images to compute the relative 3-D position and translational velocity of the UAV with respect to the target. The proposed estimator provides the required states measurements to a micro-controller for stabilizing the vehicle during flight. The control strategy consists of switching controllers, which allows making decisions when the target is lost temporarily or when it is out of the camera's field of view.

The issue is achieving reliable target tracking for a UAV based on monocular camera sensing alone. Position, altitude and yaw of the UAV relative to the target are estimated based on images obtained through vision-based tracking alone. It is assumed that one dimension of the target is known. It could be a width of the road or length of a vehicle or another object of interest. Two modes of ground target are designed:

1. Static: This mode chases the target without changing the position established, just by rotating and tilting the camera.

2. Dynamic: This mode chases the target just by changing the pitch and roll maintaining the yaw and the camera's tilt constant.

The system is compounded by two different parts: a tracker and an **image-based visual servoing system** (IBVS). The tracker has the main function of detecting the selected target on the video signal and provide the position (X, Y) to the IBVS. The IBVS has the function of transforming the information provided by the tracker to a signal ready to be sent to the UAV in order to move it appropriately [40].

Tracking of a moving object in an unknown environment The task of chasing a **moving object** is considered in an unknown environment, and without any a priori model of the object. The reliability of the visual information is critical for the good realization of the vision-based control task. For autonomously performing such a task, one has to be able to robustly extract the object's location from images despite difficult constraints: large displacements, occlusions, image noise, illumination, and pose changes or image blurring [55]. A simple way to describe an object is the image template, which stores luminance or color values, and their locations. Describing how the object looks like pixel-wise, image templates can accurately recover a large range of motions. However, they are very sensitive to some modifications in the object's appearance due to pose changes, lighting variations, blurring, or occlusions.

Problem 171 *Consider a team of N UAVs, each one carrying a camera for detecting a moving target and having the sole information of the target position on the 2-D camera plane available. The aim is to track the moving target, while maintaining the UAVs in formation. The goal is to regulate the velocity of the flying cameras by exploiting the target position on the camera plane, keeping the target at the center of its field of view.*

The size of the camera plane is $\left[-\frac{W}{2}, \frac{W}{2}\right] \times \left[-\frac{H}{2}, \frac{H}{2}\right]$, Φ, f_L being respectively the angle of view and the focal length. The target trajectory point $x_t(k) \in \mathbb{R}^3$ is defined in the global coordinate system. The team is modeled by a graph $F(k) = (C(k), D(k))\,; C(k) = \{C_i(k)\}_{i=1}^N$ at time k where $C_i(k) = (x_i(k), R_i(k))$ is the state of the ith flying camera, $x_i(k) \in \mathbb{R}^3; R_i(k) \in SO(3)$ respectively the inertial coordinate system and rotation matrix from the body frame to the inertial frame. $D(k) \in \mathbb{R}^{N \times N}$ is the graph defining the team in terms of connectivity and distances among $C(k)$, D_i is the set of neighbors of $C_i(k)$ and $d_{i,j}(k)$ is the distance between $C_i(k)$ and $C_j(k)$. The thrust control, the velocity amplification to be applied in a certain direction, is divided into two steps:

1. **Gain**: the thrust gain is computed via a function $M : \mathbb{R}^2 \rightarrow \mathbb{R}$, an increasing function as the distance from the center of the camera plane increases:

$$M(\tilde{x}_{t,i}(k)) = m_i(k) = 1 - exp\left(-\frac{1}{2}\tilde{x}_{t,i}(k)\Sigma_m^{-1}\tilde{x}_{t,i}(k)\right) \tag{5.20}$$

Σ_m^{-1} being the covariance matrix.

2. **Direction**: the thrust direction is computed by projecting $\tilde{x}_{t,i}(k)$ on two virtual axes on the camera plane, used as a reference to infer the target dynamics.

$$\begin{aligned} a_{x,i}(k) &= sgn\left(\tilde{x}_{t,i}(k).e_{x,i}\right) m_i(k) \\ a_{y,i}(k) &= sgn\left(\tilde{x}_{t,i}(k).e_{y,i}\right) m_i(k) \end{aligned} \tag{5.21}$$

where $sgn()$ is the sign function, . is the scalar product, $e_{x,i} = \mathbf{R}_{c,i}e_1$, $e_{y,i} = \mathbf{R}_{c,i}e_2$ with $\mathbf{R}_{c,i}$ is the fixed orientation of the camera with respect to $\mathbf{R}_i(k)$, e_1, e_2 being the unitary vectors in the body-fixed plane.

In [46], the proposed vision-based servoing can deal with noisy and missing target observations, accounting for UAV oscillations. The flight direction of each camera is inferred via geometric derivation, and the formation is maintained by employing a distributed algorithm that uses the target position information on the camera plane and the position of neighboring flying cameras.

Moving uncooperative target tracking in wind The UAV's motion on the horizontal plane in the presence of wind can be described by the following equations:

$$\begin{aligned} \dot{x}_a &= v_a \cos\psi_a + v_w \cos\psi_w \\ \dot{y}_a &= v_a \sin\psi_a + v_w \sin\psi_w \\ \dot{v}_a &= u_a \\ \dot{\psi}_a &= \omega_a \end{aligned} \tag{5.22}$$

where $(x_a, y_a), v_a, \psi_a$ denote, respectively, the UAV's inertial position, airspeed and heading angle, u_a, ω_a are the acceleration and turning rate control inputs, v_w, ψ_w represent the speed and direction of the wind. The relative motion between a fixed-wing UAV and a ground moving target can be described as:

$$\dot{\ell} = v_a \sin\bar{\psi}_a - v_t \sin\bar{\psi}_t + v_w \sin\bar{\psi}_w \tag{5.23}$$

$$\dot{\lambda} = \frac{-v_a \cos\bar{\psi}_a + v_t \cos\bar{\psi}_t - v_w \cos\bar{\psi}_w}{\ell + R} \tag{5.24}$$

where $\bar{\psi}_* = \psi_* - \lambda$, R is a positive constant denoting the desired separation distance, v_t is the target velocity, ℓ denotes the distance error; $\lambda = \theta - \frac{\pi}{2}$ with

$\theta \in [0, 2\pi]$. The relative motion model allows a formulation to directly control the distance between the UAV and the ground moving target. The angle λ defines the orientation of the line connecting both vehicles. The target motion can be treated as the combined velocity of the target and the wind. The vehicle and wind speeds are assumed to satisfy $0 \leq v_t + v_w < v_a$ so that the tracking problem is feasible. The control goal of this tracking problem is to regulate the distance ℓ to zero.

For the single UAV-single target, the UAV speed is constant and the turning rate ω_a is the control input. A single UAV is sufficient for the tracking state if the target states are available. Multiple UAVs can perform **persistent tracking** cooperatively in a formation, although the limited sensor coverage may preclude them from performing this task individually. Further, multiple UAVs can carry different types of sensors that could complement each other. In addition, a tracking system using multiple UAVs has a higher level of redundancy and is more reliable against vehicle failures. In [67], the UAV team is required to maintain a circular formation with equal inter-vehicle angular separation. The resulting formation is a rigid equilateral N-polygon and its center represents the motion of the entire formation. The polygon center can be regarded as a virtual leader of the team and is denoted UAV_0. The $(N+1)$ UAV planar motion can be described by

$$\begin{aligned} \dot{x}_i &= v_i \cos\psi_i \\ \dot{y}_i &= v_i \sin\psi_i \\ \dot{v}_i &= u_i \\ \dot{\psi}_i &= \omega_i \end{aligned} \quad i = 0, \dots, n \tag{5.25}$$

Problem 172 *The problem now is to design control laws u_i and ω_i such that*

1. $\ell_i \to 0$ *as* $t \to \infty$
2. $(\lambda_{i+1} - \lambda_i) \to \frac{2\pi}{N}$ *as* $t \to \infty$

Assuming that all UAVs, including the virtual UAV, are subject to the same wind disturbance, the relative motion between UAV_0 and UAV_i can be expressed as:

$$\dot{\ell}_i = v_i \sin(\psi_i - \lambda_i) - v_0 \sin(\psi_0 - \lambda_i) \tag{5.26}$$

$$\dot{\lambda} = \frac{-v_i \cos(\psi_i - \lambda_i) + v_0 \cos(\psi_0 - \lambda_i)}{\ell_i + R} \tag{5.27}$$

where $\lambda_i = \theta_i - \frac{\pi}{2}$ and $\theta_i \in [0, 2\pi]$. It is assumed that each UAV can sense its states and has the information of the virtual UAV.

The variables $\lambda_i, i = 1, \dots, N$ need to be exchanged among UAVs in order to achieve equal inter-vehicle angular separation. The communication between two UAVs is bilateral and the communication graph topology is time-invariant. A Lyapunov approach method is used in [67] to solve this problem.

5.4.2.4 Moving source seeking

This paragraph focuses on moving source seeking using multiple UAVs with input constraints. There are various sources which can generate vector fields or scalar fields. Scalar fields may represent the temperature distribution throughout space, the strength of an acoustic signal, the concentration of a chemical agent, etc. In each scalar field, there may exist one or more sources which could be pointwise, area, or volume sources. Scalar field source seeking is to find the source using one or more agents based on the measured data. Many different approaches have been proposed to solve this problem in the literature. These works can mainly be classified into two categories:

1. For **single-agent source seeking**, if the gradient of the scalar field can be directly measured, then the source localization or seeking can be achieved by adopting a simple gradient climbing strategy. However, most agents can only measure the scalar field value instead of the gradient. In order to estimate the scalar field gradient, one commonly used method is to maneuver the UAV with sinusoidal inputs. An angular velocity controller for a non-holonomic agent can be proposed to achieve source seeking, while the forward speed is tuned to achieve source seeking with constant angular velocity. By using a sliding mode-based method, source seeking can also be achieved without gradient estimation. The proposed sliding mode-based turning rate controller can drive the UAV to the desired vicinity of the field maximizer in a finite time, and afterwards it remains in this region. However, since the change rate of the field value is required, the measurement noise may result in a significant error in the computation. A stochastic source seeking approach can be proposed for the UAV to converge to the unknown source. First, a stochastic trajectory is generated by extending simultaneous perturbation stochastic approximation technique, and, second, a simple source seeking controller is designed to follow this trajectory so that switching source seeking is achieved.

2. **Cooperation of multiple robots** for source seeking has several advantages over single robot, such as faster convergence, higher precision, and more robust performance. Mobile sensor networks are used to achieve gradient climbing in a sensed, distributed environment. The gradient climbing problem is decoupled into two tasks: **formation stabilization** and **gradient climbing** so that each task can be dealt with independently. A cooperative **Kalman filter** can be designed to estimate the gradient at the center of the formation. A geometric approach is adopted to design formation controllers, where the formation shape and orientation dynamics can be decoupled from the dynamics of the formation center based on the **Jacobi transformation**. A non-linear filter can also be proposed for target tracking with range measurement only. Based on the measured range, the proposed non-linear filter is used to estimate

the position, velocity, and acceleration of the target. However, this approach is not applicable to the problem of unknown scalar field source seeking since the structure of the field distribution is unknown. A gradient climbing problem can then be proposed to steer a group of UAVs to the extremum of an unknown scalar field distribution. The leader is first used to estimate the gradient by dithering its position, and after that the leader is controlled to achieve gradient climbing. The followers are controlled to follow the leader by using passivity-based coordination rules. The gradient is estimated by the leader. **Communication constraints** are taken into account in distributed source seeking. Based on the gradient estimated by the group of UAVs, a cooperative controller is proposed to achieve source seeking. A least-squares method is introduced to estimate the gradient of the scalar field at the leader UAV location based on the measurements of all UAVs. Since the moving source velocity is unknown, an adaptive estimator can be designed to obtain the velocity. Based on the estimated gradient and source velocity, a guidance law and a sliding mode-based heading rate controller are proposed for the leader UAV to achieve level tracking. Heading rate controller for each follower UAV is also developed to achieve circular formation around the leader UAV. Furthermore, the gradient estimation error is analyzed and its influence on moving source velocity estimation and level tracking accuracy is explored as well [69].

5.5 TARGET SEARCH AND TRACKING

Various setups can be considered when addressing the problem of **target search and tracking** [51].

5.5.1 Cooperative monitoring

In an automatic target search in a dynamic, noisy and uncertain environment, and in the presence of hostile zones, targets are scattered in a bounded domain and are capable of motion in $\mathbb{R}^2$. Information about target location or distribution is not available at the outset and targets have to be visited multiple times to ensure complete classification ($n_{seen} > 1$). Pop-up threats can become active, at random time instants during search. Cluster threats are groups of targets that pop-up simultaneously and are co-located. The UAV agents are capable of moving in $\mathbb{R}^3$ and equipped with localization equipment like GPS, wireless module for communication, cameras as sensors and sufficient on-board computational capacity [21].

5.5.1.1 *Optimal distributed searching in the plane*

A multi-target tracking and surveillance coordination approach that solves the task decomposition and the task allocation sub-problems concurrently is presented in [1]. It relies on the use of a region allocation tree for simultaneously dividing the environment into regions of interest based on the probability distributions of the target tracks and allocating the UAVs to these regions of interest based on their proximities. The region allocation tree is in the form of a **binary search tree** and therefore allows for a **divide-and-conquer approach** in achieving these functions in a single sweep. The computational complexity of the resulting coordination algorithm scales linearly with the number of agents and targets and logarithmically with the candidate regions encoded in the region allocation tree. In **Bayesian multi-target tracking**, the first design choice rests in deciding whether to estimate target states separately or in a joint state space. By assuming that target motions are mutually independent, target tracks are maintained separately. This assumption leads to a more conservative way of dealing with uncertainties, but simplifies the implementation and reduces the computational burden that arises in joint-state estimation. The multi-robot collaboration taxonomy uses a three-level classification: **knowledge level**, **coordination level**, and **communication level**.

Searching for an object on the plane with limited visibility is often modeled by a search on a lattice. The search agent identifies the target upon contact. An axis parallel lattice induces the Manhattan or L_1 metric on the plane. The distances traversed by the UAV can be calculated using this metric. Traditionally, search strategies are analyzed using the competitive ratio used in the analysis of on-line algorithms.

Definition 173 *For a single UAV, the* ***competitive ratio*** *is defined as the ratio between the distance traversed by the robot in its search for the target and the length of the shortest path between the starting position of the UAV and the target. The competitive ratio measures the detour of the search strategy as compared to the optimal shortest route.*

In the case of searches using 2 or 4 UAVs with symmetric paths, the following theorems have been proved in [35]:

Theorem 174 *Searching in parallel with $k = 2, 4$ UAVs for a point at an unknown distance n in the lattice is $\frac{2n+4+\frac{4}{3n}}{k} + o(1/n^2)$ competitive.*

This theorem can be generalized as:

Theorem 175 *Searching in parallel with k UAVs for a point at an unknown distance n in the lattice requires at least $\frac{2n^2+4n+\frac{4}{3}}{k} + o(1/n)$ steps, which implies a competitive ratio of at least $\frac{2n+4+\frac{4}{3n}}{k} + o(1/n^2)$.*

A generalization suggests a spiral strategy consisting of k nested spirals searching in an outward fashion. Another strategy is that each of the k UAVs covers an equal region of a quadrant.

5.5.1.2 Distributed estimation and control

This paragraph proposes a distributed estimation and control strategy for cooperative monitoring by swarms of UAVs. The focus is not on the boundary of the region spanned by the particles, but on controlling the (first and second order) geometric moments encoding an abstraction of the swarm to match the moments of the ensemble of particles observed by the UAVs. The word **particle** refers to any discrete entity belonging to a given ensemble, whose position in the plane has to be tracked over time. The agents have no access to a distribution density function, providing an a priori global measure of information or probability that some event takes place over the region of interest. Sensor measurements are used to learn the distribution of sensory information in the environment, each agent being equipped with a limited-footprint sensor, which allows it to detect only a fraction of the overall particles. Let the position of agent i denoted by $P_i(t) \in \mathbb{R}^3$, the configuration of the n agents is described by using a **swarm moment function** $f : (\mathbb{R}^3)^n \to \mathbb{R}^\ell$, assumed to be of the form:

$$f(P) = \frac{1}{n}\sum_{i=1}^{n} \Phi(P_i) \tag{5.28}$$

where the **moment generating function**: $\Phi : \mathbb{R}^3 \to \mathbb{R}^\ell$ is defined as:

$$\Phi(P_i) = \left[P_{ix}, P_{iy}, P_{iz}, P_{ix}^2, P_{iy}^2, P_{iz}^2, P_{ix}.P_{iy}, P_{ix}.P_{iz}, P_{iy}.P_{iz}, \dots\right] \tag{5.29}$$

with $\ell = \frac{1}{2}(r+1)(r+2) - 1$ where $r \in \mathbb{Z}^+$ is the maximum order of the moments appearing in relation (5.29) and that if ℓ moment constraints are specified on n agents, then there is in general a $(2n\ell)$-dimensional algebraic set of swarm configurations that satisfy them. The geometric moments of a team of UAVs are those of an ensemble of discrete particles describing the occurrence of some event of interest to be monitored. The primary objective of the agents is to move so that their final arrangement minimizes the error $f(p)f^*$, where the goal vector $f^* \in im(f)$ defines the desired shape of the formation. $f^*$is constant and a priori known to each agent. A proportional integrator average consensus estimator is run by each agent to locally estimate the desired moments of the swarm from the environmental data [41].

A network of mobile sensors has demonstrated the ability to enhance sensing flexibility and achieve the mission objective in a shorter time period. Decentralization among the sensors further provides **scalability**, **modularity**, and **redundancy** to the network. This reduces the vulnerability to central server failure, which further enhances the overall system robustness. Some solutions intrinsically exploit the **action-perception loop** to design autonomous cognitive agents based on recent computational and decision models of the human brain. The search problem can be formulated as a **probabilistic information gathering task**. Two kinds of probabilistic measures used as the team objective function can be distinguished:

1. **Entropy-related measures** such as the mutual information or the entropy itself;

2. **Detection-related measures** such as the probability of non-detection or the expected time of detection.

Once the target is found, the search mission is accomplished. Entropy-based measures usually have difficulties guiding the mobile sensors to find the target in scenarios with general (non-Gaussian) target probability distributions. For instance, the logarithmic probability of detecting the target can be implemented as the optimizing criterion. Therefore, instead of using entropy-related utility functions, a non-detection based objective function can be used to plan sensorial trajectories that minimize the probability of target non-detection, or, equivalently, that maximize the chance of detecting the target [33].

The motivating application of the scheme in [28] is that of hand-off of an object being tracked from one fixed-wing UAV to another in a team of UAVs, using on-board sensors in a **GPS-denied environment**. This estimation scheme uses optical measurements from cameras on-board a vehicle, to estimate both the relative pose and relative velocities of another vehicle or target object. It is obtained by applying the **Lagrange-d'Alembert principle** to a Lagrangian constructed from measurement residuals using only the optical measurements. This non-linear pose estimation scheme is discretized for computer implementation using the discrete Lagrange-d'Alembert principle, with a discrete-time linear filter for obtaining relative velocity estimates from optical measurements.

5.5.1.3 *Temporal planning approach*

The problem is to track a ground vehicle reactively while it is in view and plan a recovery strategy that relocates the target every time it is lost. The approach presented in [9] aims to handle big geographic areas, complex target motion models and long-term operations. The UAV is equipped with imaging systems allowing the observation of the target, prone to error and interference from terrain. The probability of observing the target on each observation cycle depends on how recently the target was last observed, the distance between the actual position of the target, the terrain and the mode of the imaging system. The camera has two modes:

1. **Wide-angle mode** used to increase the area being scanned when the target is not currently observed, at the cost of a lower probability of successfully observing the target in any specific part of the image.

2. **Narrow-angle mode** in which the viewing area is reduced, but the probability of detecting the target is higher.

The effect of terrain is to reduce the probability of spotting the target in urban, suburban, forested and mountainous areas, while in rough or open rural areas the probability is higher. A faster moving target in the viewing zone is considered easier to spot. The probabilistic approach relies on the use of **Recursive Bayesian Estimation** (RBE) techniques that recursively update and predict the probability density function of the targets state with respect to time, under the assumption that the prior distribution and the probabilistic motion model of the target are known. Another strategy can be used as the **orienteering problem with time windows** (OPTW). The set of search patterns corresponds to the set of vertices of the OPTW, whereas the time slots in which the search patterns are active correspond to the time windows. If the UAV observer loses the target beyond the short period for which it tracks the predicted location of the target, it must follow a search strategy such as spiral or Boustrophedon to attempt to rediscover the target.

In [10], the target is assumed to be located in 2-D space, characterized by a **road network** (RN), where each road is a sequence of connected line segments. Roads can be of different types, with different speed limits on the road. The target motion on each segment is assumed to follow a constant speed randomly and uniformly sampled in a interval $[\nu_{min}, \nu_{max}]$, where ν_{min}, ν_{max} are the minimum and maximum speed allowed in that segment depending on the road type. A circular sector centered on the target's **last known position** is taken as the optimal search area. This sector extends outwards with its symmetry axis aligned with the target's average bearing over the period it was observed. The radius of the sector is determined by considering both the target's travel speed and the time period over which the search is planned to be performed. The target motion is modeled as a continuous time stochastic process $X(t)$ that takes values on V and is described as follows:

1. The final destination cell $w \in W$ is sampled according to the probability distribution $\mu : W \rightarrow [0,1]$;

2. $X(t)$ moves from v_0 to w by following the shortest path $\gamma_w = (v_0, v_1, ..., v_l = w)$ and by jumping from v_k to v_{k+1} at the random time t_k;

3. The jumping time t_k is iteratively determined as: $t_{k+1}t_k = \delta/\nu_k$, where ν_k is a uniformly distributed random variable in the interval $[\nu_{min}(v_k, v_{k+1}), \nu_{max}(v_k, v_{k+1})]$.

Along with the flight action that allows a UAV to move from one way-point to another, actions correspond to the following search patterns: **spirals**, small and large **lawnmowers** and **contour searches**, used to get around obstacles. Small patterns cover limited portions of the map, but they do so with great accuracy. On the other hand, large patterns give a broader coverage, but are less accurate. The particular type of pattern to use to cover an area depends on the specific features of the area. Spirals, expanding square and sector search patterns, provide a more focused coverage than lawnmower searches; they are used in urban and suburban areas with a high road density. The UAV domain contains the following actions:

1. Taking-off;
2. Landing;
3. Hovering;
4. Flying between way-points;
5. Performing five types of patterns: parallel track, creeping line, expanding square, sector search and contour search;
6. Performing a re-localization procedure, planned for execution when the anticipated confusion of the UAV rises above a certain threshold.

In vision-based target tracking, image processing software determines the centroid pixel coordinates of a target moving in the image frame. Given these pixel coordinates, the intrinsic and extrinsic camera parameters, and the terrain data, one can estimate the 3-D location of the target in inertial coordinates and compute the associated error covariance. This is the process of geo-location for video cameras. The **geo-location error** is highly sensitive to the relative position of a UAV with respect to the target. When a UAV is far from the target, relative to its height above the target, the associated error covariance is elongated in the viewing direction. The smallest geo-location error comes when the UAV is directly above the target, in which case the associated covariance is circular. While a UAV would ideally hover directly above the target to minimize the error, the relative dynamics between a UAV and target typically preclude this viewing position from being held over a period of time. In the two-UAV scenario, the objective is to minimize the fused geo-location error covariance of the target position estimate obtained by fusing the individual geo-location measurements. The fused geo-location error is small when at least one UAV is close to the target and only slightly less when both UAVs are directly above the target. When both UAVs are far from the target relative to their altitudes, the fused geo-location error is greatly reduced when the UAVs have orthogonal viewing angles, though this error is still significantly greater than when at least one UAV is on top of the target. For two UAVs, a generally accepted practice is to have the UAVs orbit the target at a nominal standoff distance and maintain an angular separation of 90 degrees. The 90-degree separation angle minimizes the joint/fused geo-location (target localization) measurement error for the given standoff distance, as the individual measurement error ellipses are orthogonal. These principles give rise to cooperative (or coordinated) **standoff tracking**. When more than two UAVs are considered, the goal generally becomes having the group achieve a uniform angular separation on a circle centered at the target. Standoff tracking can be addressed using **Good-Helmsman steering**, **Lyapunov guidance vector fields**, **non-linear model predictive control**, **non-linear feedback**, and methods combining vector field guidance with adaptive control. Since multiple fixed-speed aircraft cannot maintain a uniform angular spread at a fixed distance from a constant-velocity target, works have explored the notion of spreading agents uniformly in time along a periodic trajectory at a fixed distance from the moving target. A number of approaches have employed **stochastic optimal control**, in the continuous time setting, modeling the target as a Brownian particle and the UAV as a deterministic Dubins vehicle. By minimizing the expected cost of the total squared distance error discounted over an infinite horizon, an optimal bang-bang turn-rate controller can be generated, that is highly robust to unpredictable target motion. Other authors studied the problem of having a single UAV optimally perform vision-based target tracking with a limited sensing region, wherein the cost objective was a function of the desired viewing geometry. A comparison was made between a **game theoretic approach** (addressing evasive target motion) and a **stochastic optimal control** approach (addressing random target motion). Others have employed optimal control to study optimal UAV coordination when the objective is to improve target state estimation.

The problem of multiple UAVs tracking multiple targets can also be formulated as a **partially observable Markov decision process** (POMDP) to present a new approximate solution. The problem of two variable-airspeed UAVs with bearing-only sensors tracking a stochastic ground target in the presence of packet losses in the communication with the base station, where target state estimation takes place, can be considered. The solution involved an **online receding horizon controller** that maximized the expected information (inverse covariance) of the target state estimate in an extended **information filter** over a short planning horizon, showing that one UAV will act as a relay to the base station when the target is far from the base. In the problem of optimally coordinating two camera-equipped Dubins vehicles with bang-off-bang turn rate control to maximize the geo-location information of a stochastic ground target over a short planning horizon, the results showed that a 90-degree separation in the viewing angle was essential in the case of terrestrial pursuit vehicles and less pronounced with airborne pursuit vehicles [47].

5.5.2 Communications

5.5.2.1 Asynchronous communication protocol

The environment in a disaster is inherently unpredictable, posing adverse conditions to the searchers. Moreover, the individuals that are to be rescued do not need to remain stationary. It is therefore necessary for the searchers to cooperate and hence they have to communicate. A controller selects actions according to its past knowledge of the state of both the system and environment in order for the system to satisfy a specification, and maintains a continuous interaction with the environment and can therefore be considered as a reactive system. A communication protocol is provided that is integrated in this setting. Specifications in **temporal logic** can easily be refined, so it can be used as an off-the-shelf building block for autonomous systems. A major restriction in most current applications of temporal logic synthesis is that cooperating agents are assumed to be perfectly synchronized and that communication is neglected. This asynchronous protocol overcomes this restriction by taking asynchrony into account in transmitting data across a reliable point-to-point channel. Moreover, reliable communication is subject to a set of complicating factors: the network itself is susceptible to data loss; communicating peers might fail, leading to protocol violations and unknown delays; and lack of synchronization might lead to observed stuttering and missed glitches. This has to be taken into account when proving correctness of the distribution of the specification. A controller is required to satisfy its guarantees only if the environment obeys certain assumptions. Such specifications are called **assumption/guarantee** (A/G) specifications, which are like contracts between the system and the environment, or between several sub-systems, which suggests that sub-systems satisfying A/G specifications can be composed to a larger system that satisfies a global specification. Since communication is between two agents that are physically separated, this notion of synthesis of a two-player game against the environment has to be extended to allow the synthesis of several controllers that together satisfy one global specification by coordinating in order to win against the environment. Such distributed systems are specified by a global specification ϕ together with an architecture defining how many **fair discrete systems** (FDSs) are to be synthesized and what their respective system and environment variables are. State-based transmission on reliable unidirectional links are assumed with unknown but finite delay, i.e., transmission is modeled via shared variables rather than via message-passing. These requirements and restrictions have to be taken into account when formulating the protocol specification. The control of the UAVs cannot be centrally coordinated, and instead local specifications are developed that together realize a reliable global search strategy. Interaction between UAVs is governed by the communication protocol, which is included in each UAV's specification as a standardized building block. The movement of the UAVs is modeled as transitions between discrete cells. Hence the UAVs can be viewed as moving on an underlying topology, represented by a strongly connected digraph G, i.e., it is possible to get from every cell to any other cell. All edges in G are initially contaminated. If a UAV moves along an edge $e = (v, v)$ from v to v while another UAV stays on (guards) vertex v, or v has no contaminated in-edges, e is cleared. An edge (v, v) becomes recontaminated if v is no longer guarded and has at least one contaminated in-edge. While a UAV moves between two cells, the targets may make an arbitrary number of moves. A UAV finds a target if both are on the same vertex or move along the same edge in opposite directions. Depending on the topology G, a minimum number of cooperating UAVs is required to reliably find the target. Each UAV is fitted with a transceiver that can both sense and set environment and system variables a_{in}, r_{in} and a_{out}, r_{out} respectively of the communication protocol. Also, each UAV can reliably detect whether it finds a target, and whether out-edges or in-edges of its current vertex are cleared or contaminated. Lastly, the UAV is equipped with actuators to move between vertices, one edge at a time. The local strategies consist of a combination of transmissions, receptions and moves between cells. The controller of a UAV R_i is then modeled as a fair discrete system R_i with sensor inputs being environment variables and actuators being system variables. The environment variables $a_{in,i}$ and $r_{in,i}$ as well as the system variables $aout, i$ and $r_{out,i}$ are added for communication. These are the only transmission variables shared between the fair discrete systems [61].

5.5.2.2 *Mobile ad-hoc network*

A **mobile ad-hoc network** (MANET) is defined as a distributed system that comprises of a group of similar mobile nodes, capable of dynamic and arbitrary movement. Such an autonomous collection of mobile nodes does not have a specific infrastructure, centralized administration or assigned base stations, thus the network topology is subjected to rapid changes where each node communicates over packet radios to other hosts that lie within the transmission range. Communication methodologies can enable preparedness, prior to a disaster by connecting emergency centers, broadcasting system and first responders. These techniques can help establish network on demand to alert and educate the victims and authorities about the likelihood of hazards involved. MANETs are generally applicable to ensure connectivity in disaster relief situation that are usually hampered by the absence of a network or communication infrastructure. The phases of a natural calamity can either be a post-disaster or pre-disaster. The pre-disaster phase comprises of mitigation and preparedness which allows to plan and react prior to a disaster. The post-disaster scenario consists of phases like response and recovery that are sequence of actions carried out after a disaster is struck [4].

The problem of localizing non-collaborative WiFi devices in a large region can be addressed. The main motive is to localize humans by localizing their WiFi devices. An active sensing approach is used that relies on UAVs to collect signal-strength measurements at informative locations. The approach is to localize WiFi devices measuring the received signal strength. As specified by the 802.11x standards, a **mobile-station** (MS) periodically broadcasts a management frame, called **probe request frame** (PRF), to actively scan the environment and discover access points. A WiFi device operating in monitor mode can receive and decode probe request frames, extracting the addresses of the source and measuring the corresponding **Received Signal Strength Indication** (RSSI). To localize the device, the UAV flies over the region X hoping to receive a measurement from the device. The measurements are received at random times. Assuming that by the time t, n_t measurements are received with time-stamps $t_1, t_2, \ldots; t_n$ such that $t_1 < t_2 < \cdots < t_{n_t} < t$. The location of the UAV is denoted by x_{t_i} when the ith measurement was made and y_{t_i} is the corresponding RSSI measurement. The RSSI measurement is a random process Y that depends on many factors such as the current position of UAV, the true location of the device and the time of measurement. The use of RSSI for localization purposes is problematic because there is no obvious connection between the measured signal strength and the distance between the transmitter and the receiver. The problem is challenging since the measurement is received at arbitrary times and they are received only when the UAV is in close proximity to the device. The Bayesian optimization approach based on **Gaussian process** (GP) regression can be used. The main characteristic of this data is that the measurement's strength is on average highest around the transmitting device [16].

If there is only one UAV performing the task of target tracking, the target will flee easily from the **field of view** (FOV) of camera, which is a small circular region and cannot cover the whole planning space. But team tracking will improve the sensor coverage by sharing information between UAVs. The actual flight environment of UAVs is usually very complicated, where in many cases UAVs should avoid various obstacles and track target at the same time. An effective target tracking method should be proposed to enhance the surveillance and reconnaissance ability of UAVs. Besides, there are dense static obstacles and moving threats in the planning space, so UAVs should avoid them safely during the process of target tracking. In addition, collision avoidance and cluster maintenance between UAVs should be considered for cooperative mechanisms. To ensure the feasibility of path, the planned path should meet UAV dynamic constraint. Overall, the cooperative path planning problem is actually an optimization problem under various constraints. The discrete way-points satisfying the certain optimization indexes under constraints should be generated quickly, and the planned path is finally obtained by connecting them [63].

The following function can be adopted to simplify the model of obstacle or threat:

$$\Gamma = \left(\frac{x - x_0}{a}\right)^{2p} + \left(\frac{y - y_0}{b}\right)^{2q} + \left(\frac{z - z_0}{c}\right)^{2r} \tag{5.30}$$

where a, b, c are the size parameters, p, q, r the shape parameters, x_0, y_0, z_0 the position of one obstacle center. Various kinds of obstacles, such as sphere and cylinder, can be obtained by choosing different parameters. $\Gamma = 1$ means the surface of obstacle; $\Gamma < 1$ is the danger area, i.e., the region inside of the obstacle; $\Gamma > 1$ refers to the safe area, i.e., the region outside of the obstacle. If the UAV is too close to the obstacle surface, UAV may not avoid it reactively in according to the maneuverability limitation.

Every search is initialized with the following inputs. The search space is limited to a bounded area of the plane Θ that contains a hidden, stationary subject with location ($\theta \in \Theta$). The search space is characterized by a pre-specified parameter $c(x, y)$, which corresponds to the **density of obstructions** in Θ at point (x, y). These could be buildings, fog, vegetation, etc. A distributed group of n human and k automated agents collaboratively control a set of m UAVs with mounted cameras for data collection. Each agent has an associated authority level $\alpha_i \in [0; 1]$ and maintains a PDF $P_{i,t}$ over Θ of the subject's location. In each iteration of a session, every agent specifies a rectangular frame to investigate further.

Definition 176 *A frame $f(x, y, z, t)$ corresponds to a rectangular sub-region of the search space, centered at point $(x; y)$, with zoom level z, and indexed by time t.*

Both human and automated agents submit requests in this unified format. In each iteration, the system must compute a set of m frames that maximize total satisfaction among the agents. Once data for a frame f is collected, the information is processed and a binary value $B(f)$ is returned indicating whether or not the subject is detected within f. $B(f)$ is a Bernoulli random variable that is more likely to return a correct answer when a frame is of high resolution and the density of obstructions is low. As the area spanned by an image and/or the density of obstructions increases, the quality of information decreases, a being the pre-specified termination threshold corresponding to the maximum acceptable probability of a false positive in a candidate frame. Then, a search session terminates when the sensor detects the subject in a frame with small enough area with probability $1-a$ that the sensor information is accurate. The maximum area of a terminating frame solely is a function of a and the average density of obstructions in the frame:

$$c_f = \frac{1}{Area(f)} \int_f c(x, y) dy.dx \tag{5.31}$$

Since time is a major factor determining the success of a search and rescue operation and hence the system, the goal of each agent is to minimize t, the number of iterations required to locate the subject. Models and algorithms are developed for the following four steps of the search process [12]:

1. **Agent Frame Request**: All agents generate frame requests based on their individual PDFs of the subject's location.
2. **UAV Frame Allocation**: Collects requests and computes an optimal frame assignment to the UAVs.
3. **Sensor Data Extraction**: Processes the resulting image data and specifies whether or not the subject was detected.
4. **Prior Distribution Update**: All agents update their probability distribution functions to incorporate the new data.

5.6 SURVEILLANCE

Surveillance is defined as a monitoring process (usually for people) of activities and various changing information with the aim to influence, manage, direct or protect. Most applications require the tasks to be performed in an optimal manner with specific timing constraints. The high level task specifications for these applications generally consist of temporal ordering of sub-tasks, motion sequencing and synchronization, etc. Some approaches mainly focus on **linear temporal logic** (LTL), which can specify tasks such as visiting goals, periodically surveying areas, and staying stable and safe. The main drawback of the LTL formulation is that it cannot specify time distance between tasks. In surveillance examples, a simple task may be to individually monitor multiple areas for at least x amount of time. Additionally, the LTL formulation commonly assumes the environment to be static. Traditional approaches commonly start with creating a finite abstraction of the environment including the dynamics, then combine it with the automata that is generated from the LTL specification. The cell decomposition performed in the abstraction process requires the environment to be static; but in most situations this is not the case. For example, the use of UAVs for surveillance in the commercial airspace needs to consider motion of other aircraft. The other weakness of the automata-based approach is that it is computationally expensive. Motion planning for surveillance in an airspace with finite time task constraints and safety guarantees should be considered. The other aircraft in the target area is assumed to be dynamic obstacles to the UAV. Further, the motions of these dynamic obstacles can be either predicted during the planning or are known a priori. Due to the limitations of the previous approaches, the method based on **metric temporal logic** (MTL) and an optimization problem formulation to solve the planning problem seem an interesting approach to solve this problem. MTL extends the LTL temporal operators so that it can express the requirements on time distance between events and event durations. This allows to describe the dynamic obstacle and survey durations. A path planning problem for surveillance under survey durations constraints can be considered for each region and overall temporal constraint to visit each region within given times. A path is generated that guarantees safety by avoiding static and moving obstacles in the workspace and the path is optimal in the sense that it minimizes a pre-defined cost function. There are specific time bounds associated with the regions, by which the surveillance has to be finished [68].

5.6.1 Stochastic strategies for surveillance

A team of UAVs are engaged in a surveillance mission moving in a unpredictable fashion. The analysis assumes that these UAVs can move forward and backward, as rotary wings vehicles do. Stochastic rules are used to guide the motions of the UAVs, minimizing centralized computation and communication requirements by focusing on local rules for each agent. The surveillance problem is formulated abstractly as a **random walk on a hypergraph**, where each node on a hypergraph corresponds to a section of the environment and where each edge of the graph is labeled with the probability of transition between the nodes. The problem of **parallel Markov chains** and **fastest mixing** when a team of agent moves among states on a graph is considered. A hypergraph is associated to a general class of search environments which are modeled as sets of line segments in a bounded domain in the plane. Since the states are directional segments between the way-points, the transition between states can be viewed as a set of turning probabilities. Given this representation, a Markov chain with the following transition probability matrix:

$$P = [P_{ij}] \tag{5.32}$$

$$\sum_i P_{ij} = 1 \tag{5.33}$$

$$0 \leq P_{ij} \leq 1 \tag{5.34}$$

where P_{ij} is a stochastic matrix, representing the probability of the agent going to state i from state j. Constraints (5.33) and (5.34) must hold since the sum of the probabilities must be one and all the probabilities must be non-negative.

The problem of defining surveillance strategies is parametrized. The following questions arise:

1. What **types of surveillance coverage** can the stochastic UAVs provide (i.e., what is the steady-state distribution for the system of UAVs with specified turning probabilities?)

2. What is the **rate of convergence** of the system of UAVs to this invariant distribution?

3. What are the appropriate **measures for comparing** different surveillance strategies?

4. How can **randomness** be captured in the motions of the UAVs?

5. How is the **trade-off between randomness and speed of convergence?**

The algorithms may use a triangular grid pattern, i.e., UAVs certainly go through the vertices of a triangular grid during the search procedure. A triangular grid pattern is asymptotically optimal in terms of the minimum number of UAVs required for the complete coverage of an arbitrary bounded area. Therefore, using the vertices of this triangular grid coverage guarantees complete search of all the region as well as better performance in terms of search time. A random triangular grid-based search algorithm may be proposed, where a UAV is located on a vertex of the common triangular grid. Consequently, it can explore the surrounding area using its sensors, and that depends on the sensing range of its sensors. After exploring that area, the UAV moves to another point which can be one of the six neighboring vertices in the triangular grid. To make sure that the whole area is explored by the team of the UAVs, each vertex in the triangular covering grid set of the area **W** must be visited at least one time by a member of the team. Consider **T** is a triangular covering grid of **W**, and also each vertex of **T** has been visited at least one time by a UAV of the team, ensuring that the area **W** has completely been explored by the multi-robot team. Since the UAVs do not have any map at the beginning, they need to complete their maps gradually. Whenever a UAV makes any changes to its map, it sends the new map to the other robots by transmitting packets. On the other side, whenever a UAV receives a packet of data, it extracts the new vertices from the received map and adds them to its map. Since the communication range of the UAVs is limited, if two UAVs are far from each other, they cannot directly communicate but can do it via other UAVs. Since there is a connected network of UAVs, each robot has the role of a hub in the network in order to share the maps among the robots. Therefore, all the UAVs that are connected and make a network have a common map. In the second phase of the algorithm when the robots have a common triangular grid map, a UAV can go far from the other robots and be disconnected from the team for a while. In this case, sharing maps between disconnected robot and the others is paused until the robot returns back to the communication range of the team again. In the next step, each robot randomly chooses one of the nearest neighboring vertices in the map and goes there. Since the map of a robot is a connected set, there exists always at least one neighboring vertex, and at most six vertices. If the robot reaches the target vertex, it marks that vertex as an explored vertex in its map and sends it to the other neighboring robots as well [8].

5.6.1.1 Analysis methods

The aim of this paragraph is to analyze the type of surveillance coverage and invariant distribution that the stochastic UAVs can provide. The probability distribution at each time $k+1$ is determined according:

$$\vec{p}_i^{k+1} = P_i \vec{p}_i^{k} \tag{5.35}$$

where $\vec{p}_i^{k}$ is the probability distribution for UAV i at time k and P_i is the transition probability matrix for UAV i. There is a unique invariant distribution for an **irreducible, aperiodic Markov chain** which is the eigenvector associated with the eigenvalue 1. This invariant distribution represents the steady-state probability of the UAV being at any state. The eigenvalues of the Markov chain can be sorted by magnitude:

$$1 = |\lambda_1|(P) \geq |\lambda_2|(P) \geq \ldots\ldots |\lambda_n|(P) \tag{5.36}$$

The **mixing rate** for the Markov chain is given by

$$\mu(P) = |\lambda_2|(P) \tag{5.37}$$

where $|\lambda_2|(P)$ is the eigenvalue which is second largest in magnitude. The smaller the mixing rate, the faster the Markov chain converges to its steady-state distribution. The expected composite distribution of a can be explicitly determined by

$$\vec{p} = \frac{\sum_{i=1}^{a} \vec{p_i^k}}{a} \tag{5.38}$$

The UAVs can move randomly and independently.

5.6.1.2 *Single UAV investigations*

The aim of this paragraph is to show how tuning problem parameters can affect the invariant distribution. The case of a probabilistic UAV walking on an $n-$node, one-dimensional lattice, taking steps to the right with probability ρ and steps to the right with probability $1-\rho$. For a given initial probability distribution $p_1^0, ..., p_n^0$, the probability distribution at each time evolves according to

$$\vec{p^{k+1}} = P\vec{p^k} \tag{5.39}$$

where

$$P = \begin{pmatrix} 0 & 1-\rho & 0 & \cdots & 0 \\ 1 & 0 & 1-\rho & \cdots & 0 \\ 0 & \rho & 0 & \cdots & 0 \\ \vdots & \vdots & \ddots & \ddots & \vdots \\ 0 & 0 & \cdots & \rho & 0 \end{pmatrix} \tag{5.40}$$

The steady-state, invariant distribution satisfies

$$\vec{p} = P\vec{p} \tag{5.41}$$

The components of this steady-state distribution can be found explicitly as solutions of recursive equations.

5.6.1.3 *Multi−UAV investigations*

A strategy for 1-D, $n-$node lattice is presented in this paragraph. By appropriate choice of parameters, it is possible to implement a probabilistic strategy in which UAVs disperse in the lattice domain as fast as possible. The turning parameters of this strategy are assigned by the following relation

$$\rho_i = \left\{ \begin{array}{cc} 0.9 & \text{for } k \leq \frac{a+1-i}{a+1} \times n \\ 0.5 & \text{Otherwise} \end{array} \right\} \tag{5.42}$$

where ρ_i is the probability that UAV i turns right, k is the number of steps, a is the number of UAVs, and n is the number of nodes in the lattice. This strategy disperses the UAVs along the graph before switching to equal turning probabilities. The UAVs do not have a uniform steady-state distribution prior to switching to the equal turning probabilities. After the UAVs switch to the equal turning probabilities, their steady-state distribution approaches the uniform distribution as the initial distributions are suppressed with time.

5.6.2 Urban surveillance

The task of both UAVs and UGVs is to make observations of terrain features and identify moving or stationary targets. A UAV is used to monitor a given area where a UGV is present to perform the inspection task. The images taken from the camera mounted on the UAV are sent continuously to the ground station. The images are processed in real time to provide localization information to the UGV to navigate in the area through the way-points selected by a human operator. To successfully inspect a given area within a reasonable frame of time, global and local coverages must be acquired simultaneously, which can be possible if UAVs and UGVs are deployed simultaneously. A human operator supervises the video flows, and selects progressively navigation way-points for the UGV. As the UGV navigates through the given way-points, the UAV follows continuously the UGV in order to keep it in its coverage view (the center of the image plan) using visual servoing. To successfully navigate in a given area, the relative pose of the UGV should be known [26]. **Scalability** provides group control over behavior-rich group with hundreds to thousands of group members. **Robustness** allows the controller agents (UAVs and UGVs) to generate control strategy even when the behaviors of the controlled group are not well modeled due to uncertainty. In order to achieve this objective, a planning and control framework based on **dynamic data driven, adaptive multi-scale simulation** (DDDAMS) simulation steers the measurement process for data update and system control [60]. UGVs can observe dynamic movements of individual agents in a sub-crowd within their observation range, with high resolution. UAVs can provide global but less accurate information on the overall dynamic movement of the crowd. By integrating the information from both UGVs and UAVs up to time t, the dynamics of the crowd region at future time horizon, e.g., time $(t+1)$ can be better estimated. During the planning stage, a decision planner, implemented in an agent-based simulation environment, will devise a set of best UAV/UGV control strategies for the evolution of real system. Various algorithms (e.g., UAV/UGV searching, detection, tracking, path planning) are implemented in the agent-based simulation, which helps evaluate alternative control strategies against different scenarios. For instance, UAV/UGV path planning algorithm involves selecting one from deterministic, probabilistic, and mixed strategies. Under a similar problem setting, the results may vary according to different choices of strategy. In addition, as different algorithms are integrated in the same simulation environment, strategy interactions and trade-offs should be considered for achieving the best system level performance in favor of individual algorithm performance. Next, each potential strategy combination is initialized as an instance for simulation-based evaluation. At the final step, statistical analyzes are performed for selecting the best control strategies due to the variations inherently in the simulation outputs.

In persistent surveillance, the environment can be abstracted into a graph, the movement of agents explicitly defined on the graph and the utility function to be optimized formulated as a cost of potential damage to the environment. There are two possible types of environment representation; both of them are discretization of continuous areas:

1. The environment may be represented by a tree graph $G = (N, E)$. The set of nodes N represents monitored parts of area. The length of the edge connecting these nodes equals l the diameter of the agent's sensor. Adjacent nodes parts in the original system are connected by a directed edge $e \in E$, where E is a set of all edges (i.e., for each pair of nodes n and m, a pair of edges $e(n, m)$ and $e(m, n)$) exist.

2. The environment may be tilled into a grid covering the to-be-surveyed system with hexagon tiles having diameter equal to the agent's sensor. The hexagon grid is represented by a directed graph $G_h = (N_h, E_h)$. Nodes $n \in N_h$ represent the centers of hexagon tiles. Directed edges $e \in E_h$ of length l connect nodes of neighboring tiles.

Complete graphs are considered and the associated turning parameters that yield the fastest mixing and approach the uniform steady-state distribution are investigated.

Theorem 177 *For a complete graph with n vertices, the probabilistic random walk having transition probability matrix*

$$p_{ij} = \left\{ \begin{array}{cc} \frac{1}{n-1} & \textit{if } i \neq j \\ 0 & \textit{Otherwise} \end{array} \right\} \tag{5.43}$$

has eigenvalues 1 of multiplicity one. The invariant distribution for this Markov chain is uniform, and the eigenvalue $\frac{-1}{n-1}$ *is smaller in magnitude than the eigenvalue of second largest magnitude corresponding to any other set of transition probabilities.*

In the general case, the structure of the problem can be considered as line segments on a bounded plane. A complete graph on its vertices is associated to each edge of rank ≥ 3 in $H(X, \epsilon)$ representing the set of segments in the search environment. The edges of these complete graphs represent the possible choices of transitions from one segment to another. Any general graph can be decomposed into a system of interconnected complete sub-graphs, and **cliques**. Cliques having two vertices are distinguished from those having more than two vertices. Those having two vertices correspond to transitions in which the only choices available to the UAV are to move ahead or to move backward. The intersections of a general graph can be thought of as a complete graph where the number of nodes in the complete graph is equivalent to the number of choices a UAV can make. With no restrictions on the movement of a UAV, the number of nodes in the complete graph is equal to the number of edges incident to the intersection in the graph representation. A hybrid strategy is obtained by combining the strategies for both linear graphs and complete graphs. It provides uniform coverage of a general graph while achieving this coverage quickly without a large sacrifice in the randomness in the behavior of the UAV. More details on this implementation can be found in [24].

5.6.3 Monitoring wildfire frontiers

The assumptions are as follows:

1. GPS and communication between the incident commander and the UAV are always available. The UAV can always localize itself and never needs to return to the starting location to transfer collected data.

2. The UAV always has the simulated fire frontier in order to find the hotspot locations.

3. The UAV is assumed to have unlimited endurance.

Each hotspot has a corresponding time since last tracked by the UAV and the maximum time it has been left untracked (ϕ) in the past. The sum of ϕ of all hotspots can be chosen as the metric to evaluate the effectiveness of an algorithm, thus fire-line intensity is used as the crucial information needed by the incident commander. The intensity is monitored through the clustering into hotspots, directly relating to the goal of providing the incident commander with up-to-date information about the fire progression:

$$J(t) = \sum_{i=0}^{hotspots} \phi_i \tag{5.44}$$

where ϕ is max time untracked. At mission start, the UAV must first find the fire and begin identifying the hotspot regions. Tracking a hotspot is done by calculating the distance between a previous set of hotspots relative to a new set. To determine when a hotspot moved as the fire progressed, a threshold is implemented. If a hotspot is not within the distance threshold of any previous hotspots, it is then classified as a new hotspot. Even after careful tuning, this approach can still lead to some untracked hotspots where the hotspot existence is too short for any response by the UAV. To identify hotspots, all points along the frontier with a fire-line intensity above a normalized threshold are parsed using a clustering technique called **mini-batch K-means. K-means** clustering was chosen because it directly relates the number of interest points (how active the fire is) to the number of cluster centers (hotspots). Algorithm 27 is an illustration of this approach.

The desirable amount of clusters or number of centers (K) changes as the fire evolves. the K value is actively determined for adaptive hotspot extraction

Algorithm 27 Baseline Algorithm

1. Inputs: UAV Location, frontier
2. for all points in frontier do
3. points.distance=$\sqrt{(points.x - UAVlocation.x)^2 + (points.y - UAVlocation.y)^2}$
4. end for
5. closest point = min(points.distance)
6. vector to nearest=([UAV location.x - closest.point.x], [UAV location.y - closest.point.y])
7. normalized vector = vector to nearest = distance to nearest
8. if distance to nearest > max distance to fire then
9. travel vector = vector to nearest
10. else if distance to nearest < min distance to fire then
11. travel vector = -vector to nearest
12. else
13. travel vector = (-vector to nearest.x, vector to nearest.y)
14. end if
15. path = travel vector

with:

$$K = \sqrt{N/2} \tag{5.45}$$

with N as the number of interest points. An algorithm A* path planning can be used for generating paths from the UAV to hotspot locations around the fire. A cost map is passed to the A* algorithm, and is generated by applying a blur to the map of the fire up to that point in time and assigning a high cost to areas within the fire as in Algorithm 28. This helps ensure the path generated for the UAV is not within dangerous proximity of the fire, but can still be navigated close enough to monitor the hotspots [52].

Algorithm 28 Weighted Algorithm

1. Inputs: hotspotslocation, time untracked, α, UAV location
2. for all h in hotspots do
3. h.path, h.path cost = ASTAR(h.location, UAV location)
4. h.score = h.time untracked $-\alpha*$ path cost(h)
5. if hotspot.score > target hotspot.score then
6. target hotspot = h
7. end if
8. end for
9. path = target hotspot.path

5.6.3.1 *Surveillance of risk sensitive areas*

A reactive motion-planning approach for **persistent surveillance** of risk-sensitive areas by a team of UAVs is sought. The planner seeks to maximize the area covered by sensors mounted on each UAV, maintain high sensor data quality, and reduce detection risk. To achieve the stated objectives, a cost function combines the detection risk with an uncertainty measure designed to keep track of the regions that have been surveyed and the times they were last surveyed. The uncertainty and detection risk is reduced by moving each UAV toward a low-cost region in its vicinity. By reducing the uncertainty, the coverage increased and persistent surveillance is provided. Moreover, a non-linear optimization formulation is used to determine the optimal altitude for flying each UAV in order to maximize the sensor data quality while minimizing risk [58].

In multi-UAVs surveillance missions, with local communication capabilities, one way to transmit the surveillance information to a base is streaming by the instantaneous data via multi-hop communications over a connected network. However, a connected communication network may become disconnected if some UAVs leave the surveillance area. A locally applicable, efficient, and scalable strategy that guarantees a connected communication network between the base and the agents in the face of any agent removal is presented in [3]. The proposed decentralized strategy is based on a sequence of local replacements, which are initiated by the agent leaving the network. The replacement sequence always ends with the relocation of an agent, for which the absence from its current position does not disconnect the network. Furthermore, the optimality of the proposed scheme is improved by incorporating a local **criticality** notion in the decision mechanism. In a multi-agent system, the connectivity of the communication network plays a significant role to achieve collaboration among the UAVs through some local interactions. UAVs can stream the surveillance data back to the base, if they have a connected communication network including the base. Even though UAVs begin a mission with a connected communication network, such systems have an inherent possibility of agent removal. In this case, the communication network may become disconnected. In networked systems, disconnection can be avoided or fixed by proactive or reactive approaches, respectively. In proactive approaches, a robust network topology is designed a priori such that the network can tolerate a certain number of agent removals.

Remark 178 *Relying only on proactive approaches can be impractical in applications where a large number of UAVs may eventually be removed from the network. In reactive approaches, a control strategy is developed for the network to reconfigure itself in the face of agent removals.*

A recovery process can be characterized as centralized or decentralized based on the information leveraged in the decision scheme. In a large-scale system, the availability of global information to individual UAVs is usually not feasible. Therefore, a decentralized strategy becomes more desirable than a centralized one due to **practicality** and **scalability** concerns.

A decentralized control strategy for a multi-UAV system that enables parallel multi-target exploration while ensuring a time-varying connected topology in cluttered 3-D environments is presented in [43]. Flexible continuous connectivity is guaranteed by building upon a recent connectivity maintenance method, in which limited range, line-of-sight visibility, and collision avoidance are taken into account at the same time. Completeness of the decentralized multi-target exploration algorithm is guaranteed by dynamically assigning the UAVs with different motion behaviors during the exploration task. One major group is subject to a suitable down-scaling of the main traveling force based on the traveling efficiency of the current leader and the direction alignment between traveling and connectivity force. This supports the leader in always reaching its current target and, on a larger time horizon, that the whole team realizes the overall task in finite time.

5.6.3.2 Cooperative surveillance

Identical UAVs are aimed at cooperative observation of a set of **areas of interest** (AoI) assigned by a human operator in an environment with obstacles. The obstacles may be physical obstacles such as building or no-fly areas. The surveillance of an AoI can be realized by a single UAV flying at a sufficiently low altitude or by several UAVs flying at a higher altitude with overlapping views of their surveillance cameras. In the swarm deployment method proposed in [50], a single optimization process is used to find a final swarm distribution in the environment together with feasible trajectories of all UAVs to reach these positions from the depot. This ensures that the surveillance locations are reachable by UAVs keeping the motion and relative localization constraints. In general, the surveillance region is divided into cells. The probability of targets or the uncertainty level of states is associated with the cells to represent the prior information. The probability of targets and uncertainty level of states may be time varying because of the movement of the targets. In the distributed receding horizon optimization, the UAV usually needs other neighboring agents' decisions in the motion planning in order to plan its optimal motion under the coupling constraints, for example, the collision avoidance. The UAVs may update their plans in sequence: only one UAV is allowed to plan its motion and share the plan results with adjacent UAVs at every step, while other UAVs hold their current plans. The iteration of the optimization can also be divided into two steps:

1. The UAV plans its presumed motion without taking the coupling constraints into account and shares the presumed motion with its neighbors.

2. The presumed motions are adjusted by taking the coupling constraints into account to obtain the feasible motion.

However, the method is based on the assumption that the difference between the presumed motion and the optimal feasible motion is small and the update goes fast enough. The potential field method can be utilized for coupling objectives in the distributed receding horizon control for the multiple agent stable flocking. The sufficient information sharing in the group is indispensable for the cooperation and information fusion, and thus, the network connectivity must be preserved. For discrete-time systems, the agent can choose control inputs from the allowable set (known as the connectivity constraint set) so that the connectivity can be maintained in the next step. The framework consists of:

1. A network topology control layer regulating the network topology and maintaining the network connectivity.

2. A motion planning layer planning the motion of UAVs using the distributed receding horizon optimization.

The model of the cooperative searching problem can be built based on the probability of targets and the detection history of UAVs over the region. The forgotten factor is introduced to drive the UAVs to revisit the areas that have been searched before. Furthermore, the tradeoff between the coverage enhancement and the network performance is achieved by taking into account the centrality of communication links in the deletion of communication links. The potential field design in the receding horizon optimization is presented to obtain the optimal motion of UAVs without violating the collision avoidance and network connectivity constraints [18].

If a team of UAVs is considered, the objective is usually to act cooperatively to gather as much information as possible from a set of **moving targets** in the surveillance area. This is a decision making problem with severe uncertainties involved. Relying on imperfect sensors and models, UAVs need to select targets to monitor and determine the best actions to track them. The positions of the targets are unknown and can only be observed with imperfect sensors; occlusions can occur due to elements in the scenario or due to other UAVs; the dynamic models for the UAVs and the trackers are not perfect. Considering these uncertainties when solving the tracking problem is key in order to ensure optimal solutions. The problem of target tracking has been extensively considered in [15] from the point of view of:

1. **Sensing** to maintain an estimation of the targets' positions and their associated uncertainties. For this purpose, many different stochastic filters integrating observations from sensors have been proposed. There also exist multi-robot filters that fuse information from all teammates in situations where several UAVs cooperate in the tracking mission. However, the problem is also a decision making problem as estimating the position of the targets with UAVs, but also controlling those sensors to optimize some objectives is also important. There are multiple targets and multiple UAVs, so it is needed to decide dynamically which UAVs track each target, and then, how they move in order to do so. The criteria to optimize may vary depending on the application, but they usually consider the percentage of time that each target is within field of view, the degree of uncertainty on the targets' positions, the fuel consumption, communication constraints, etc

2. **Decision making** is often formulated as a **stochastic optimal control problem** where a utility function is optimized. Many approaches have also been proposed to solve multi-target tracking with a team of several UAVs. Target estimations are uncertain and maximizing the information gathered from the targets is usually quite relevant. This information can be quantified by means of different metrics, such as **entropy** or **mutual information**. Many works assume Gaussian uncertainties and Kalman filters as the underlying estimation frameworks, defining utility functions based on these information metrics in order to determine the actuation. However, tracking applications can result in multimodal distributions when estimating the targets' positions. Hence, other works also consider alternative representations, such as discrete **Bayes**

filters or **particle filters**. Many works propose information-gathering approaches based on heuristics or rigid optimization problems. Those approaches lack usually adaptability to different scenarios and optimization criteria. POMDPs provide a mathematical framework for planning under uncertainties. Resulting policies reason about uncertainties and can combine multiple objectives, such as maximizing information and fuel consumption. Moreover, these policies do not constrain to specific scenarios, since they can be recomputed by adapting the models involved or the required objectives. However, often POMDPs are computationally expensive.

5.6.3.3 Cooperative relay tracking

The monitored 2-D space is divided into a number of regions: **Voronoi cells** with Voronoi diagrams. During the course of tracking, the tracking agents need to cooperate with each other and the relay scheme involves switching of topologies. Ignoring boundary effects and assuming each agent's sensing radius is R_s, then the corresponding sensing coverage area is πR_s^2. In a 2-D sensing field with an area S_c, the minimum number of agents N_m required to achieve 1-coverage is given by:

$$N_m = floor\left(\frac{S_c}{\pi\left(\frac{\sqrt{3}R_s}{2}\right)^2}\right) \tag{5.46}$$

Regions overlapped by k agents are called **k-coverage** regions. The **trilateration** algorithm can be used to determine the location of a target based on distance measurements by its neighbors. Trilateration is the process of calculating a 3-D target's position (x_T, y_T, z_T) based on measured distances $(d_{T_1}, d_{T_2}, d_{T_3})$ between three UAVs with known locations $(x_{a_1}, y_{a_1}, z_{a_1}), (x_{a_2}, y_{a_2}, z_{a_2}), (x_{a_3}, y_{a_3}, z_{a_3})$ and the target. Thus the following relationship of distances between the target and the agents is:

$$\begin{aligned} d_{T_1} &= \sqrt{(x_T - x_{a_1})^2 + (y_T - y_{a_1})^2 + (z_T - z_{a_1})^2} \\ d_{T_2} &= \sqrt{(x_T - x_{a_2})^2 + (y_T - y_{a_2})^2 + (z_T - z_{a_2})^2} \\ d_{T_3} &= \sqrt{(x_T - x_{a_3})^2 + (y_T - y_{a_3})^2 + (z_T - z_{a_3})^2} \end{aligned} \tag{5.47}$$

The position of the target is thus obtained:

$$\begin{bmatrix} x_T \\ y_T \end{bmatrix} = \begin{bmatrix} 2(x_{a_1} - x_{a_3}) & 2(y_{a_1} - y_{a_3}) \\ 2(x_{a_2} - x_{a_3}) & 2(y_{a_2} - y_{a_3}) \end{bmatrix}^{-1} \begin{bmatrix} x_{a_1}^2 - x_{a_3}^2 + y_{a_1}^2 - xa_3{}^2 + d_{T_3}^2 - d_{T_1}^2 \\ x_{a_2}^2 - x_{a_3}^2 + y_{a_2}^2 - ya_3{}^2 + d_{T_3}^2 - d_{T_2}^2 \end{bmatrix} \tag{5.48}$$

At least, three agents are required for localization, which means this monitoring area should be 3-coverage.

When the target moves into a new Voronoi cell, not only the topology switches but also one of the pursuers is replaced by the corresponding Voronoi site which induces jump of tracking errors. To maximize the coverage of a given region, distribution of agents should guarantee the coverage of the whole region [19]. The relay tracking algorithm is given in Algorithm 29.

Algorithm 29 Cooperative Relay Tracking

1. Calculate the required number of agents and initialize the deployment guaranteeing the monitoring area is 3-overlapped.

2. When a target goes into the monitoring area, the 3-neighboring agents $a_i, i = 1, \ldots, 3$, send their detected distances to the Voronoi site agent where the target resides.

3. Using trilateration relation, the **Voronoi** site agent estimates the target's location through three measurement distances.

4. Then the Voronoi site agent and $N_f - 1$ other nearest agents start tracking the target, becoming tracking agents.

5. If the tracking agent is a monitoring agent, in order to guarantee at least 3-coverage of this area, the nearest redundant UAVs move to the monitoring agent's previous location immediately. If the tracking agent is one of the redundant agents, the other agents stay at their original positions, then recreate a new Voronoi diagram, in which the target and the tracking agents are exclusive.

6. If the target is not captured in the first Voronoi cell, it will go to the next Voronoi cell. The new corresponding Voronoi site agent will become a tracking agent, meanwhile, one of the original tracking agents will quit tracking based on the distance discipline.

7. When a target is caught, it stops moving and releases the corresponding tracking agents to become redundant agents, this is called a release policy, with which there are more redundant agents in the domain. Repeat all the procedures until there is no more new and uncaptured targets.

5.6.3.4 *Path planning with temporal logic constraints*

The UAV motion in the environment is assumed to be modeled as a transition system obtained by partitioning the environment into regions, using triangulations and rectangular partitions [53]. The UAV task is to collect rewards that dynamically change their values in the regions and that can be sensed only within a certain vicinity of the UAV's current position. An approach to this kind of problem, i.e., an optimization problem on a dynamically changing plant is **model predictive control** or **receding horizon control** based on iterative re-planning and optimization of a cost function over a finite horizon.

This paragraph focuses on the integration of a receding horizon control with a synthesis of a path that is correct with respect to a given temporal logic formula. While the satisfaction of a linear temporal logic mission is required, the target is also to collect maximal rewards within a given horizon. The trade-offs between both goals are allowed to be partially driven by user-defined preferences that may change dynamically during the execution of the UAV motion. In particular, an LTL mission includes surveillance of a set of regions and a user definite preference function expressing the desired trade-off between the surveillance and the reward collection given the history of the UAV motion. Considering a UAV moving in a 2-D partitioned environment modeled as a weighted deterministic transition system, a dynamically changing non-negative real-valued reward is associated with each state of the transition system. The UAV senses the rewards in its close proximity and collects the rewards as it visits the regions of the environment, i.e., the states of the transition system change. Moreover, the UAV is given a high level linear temporal logic mission. A user-defined preference function is considered that, given a history of the UAV's movement, expresses whether moving closer to a region under surveillance or collecting rewards is prioritized. Then an arbitrary reward dynamic is considered and the concrete reward dynamics assumption is captured through a state potential function. The problem is to design a control strategy that guarantees the satisfaction of the mission locally optimizes the collection of rewards, and takes into account the preference function and the reward dynamics assumptions.

Problem 179 *The UAV motion in the environment is given as a transition system :* $\mathbb{T} = (Q, q_0, T, \prod, L, W)$*. The reward can be sensed at time* t_k *within the visibility range* $v \in \mathbb{R}^{*+}$ *from the UAV's current position* q_k*.*

$$V(q_k) = \{q | W^*(q_k, q) \leq v\} \tag{5.49}$$

The set of states that are within the ***visibility range*** v *from* q_k*, assuming that* $q \in V(q_k), \forall (q_k, q) \in \mathbb{T}$ *and by* $R : Q \times Q^+ \rightarrow \mathbb{R}^+$ *the reward function, where* $R(q, q_0, ..., q_k)$ *is the reward sensed in the state* q *at time* t_k *after executing the run prefix* $q_0, ..., q_k$*, defined if and only if* $q \in V(q_k)$ *and it is known only at time* t_k *and later, not earlier.*

A user-defined planning horizon and a state potential function can be employed to capture the user's assumptions about the reward dynamics and its interests. For instance, the rewards may appear according to a probabilistic distribution or their changes might be random. The rewards might or might not disappear once they are collected by the UAV. The user might be interested in the maximal, expected or minimum sum of rewards that can be collected from a given state during a finite run whose weight is no more than the planning horizon. The horizon is $h \in \mathbb{R}^{*+}, h \geq min_{(q,q') \in T} W(q, q')$ and the state potential function is $pot : Q \times Q^+ \times \mathbb{R}^{*+} \rightarrow \mathbb{R}^{*+}$ where $pot(q, q_0, \ldots, q_k, h)$ is the potential of the state q at time t_k. The value of $pot(q, q_0, \ldots, q_k, h)$

is defined for all q, where $(q_k, q) \in \mathbb{T}$ and captures the rewards that can be collected after execution of the run prefix $q_0, \dots, q_k$ during a finite run $\rho_{fin} \in P_{fin}(q, q_k, h)$ where $P_{fin}(q, q_k, h) = \{\rho_{fin} | \rho_{fin} \text{is a finite run of } \mathbb{T}\}$ such that

1. ρ_{fin} starts at q,
2. $W(\rho_{fin}) + W(q_k, q) \leq h$,
3. The states that appear in ρ_{fin} belong to $V(q_k)$.

There is a set of regions labeled with a surveillance proposition $\pi_{sur} \in \prod$ and a part of the mission is to periodically visit one of those regions. The missions are then LTL formulas of the form:

$$\phi = \varphi \vee GF\pi_{sur} \quad \text{where } \varphi \text{ is an LTL formula over } \prod \tag{5.50}$$

The formula $GF\pi_{sur}$ states that π_{sur} has to be visited always eventually, i.e., infinitely many times. The operator can partially guide whether the UAV should collect high rewards or whether it should make a step towards the satisfaction of π_{sur} through a preference function. Formally, the function $pref : Q^+ \rightarrow \mathbb{R}^+$ assigns a non-negative real value to each executed run prefix $q_0, \dots, q_k$ of $\mathbb{T}$, possibly taking into account the current values of the state potential function. A shortening indicator function I indicates whether a transition leads the UAV closer to a state subject to surveillance: $I : \mathbb{T} \rightarrow \{0, 1\}$ is defined as follows:

$$I(q, q') = \left\{ \begin{array}{ll} 1 & min_{q_\pi \in Q_\pi} W(q', q_\pi) < min_{q_\pi \in Q_\pi} W(q, q_\pi) \\ 0 & \text{Otherwise} \end{array} \right\} \tag{5.51}$$

where $(q, q') \in \mathbb{T}$ and $Q_\pi = \{q_\pi | \pi_{sur} \in L(Q_\pi)\}$.

Problem 180 *Given the UAV motion model* $\mathbb{T} = (Q, q_0, T, \prod, L, W)$, *the surveillance proposition* $\pi_{sur} \in \prod$, *the visibility range* v, *the reward* $R(q, q_0, \dots, q_k)$ *at time* $t_k, \forall q \in V(q_k)$, *the planning horizon* h, *the state potential function pot, the linear temporal logic formula* ϕ *over* $\prod$, *in* (5.50) *and the preference function pref, find a control strategy such that:*

1. *The run generated by* C *satisfies the mission* ϕ
2. *Assuming that* $q = C(q_0, \dots q_k)$, *the cost function*

$$pot(q, q_0, \dot{;} q_k, h) + I(q_k, q).pref(q_0, q_k) \tag{5.52}$$

is maximized at each time t_k

In general, the satisfaction of the second condition may cause violation of the first objective. The goal is thus to guarantee accomplishment of the mission and to maximize relation (5.50) *if possible.*

The solution consists of two separate steps:

1. An **offline preparation** before the deployment of the system, involves the construction of a **Buchi automaton** for the given linear temporal logic mission and its product $\mathbb{P}$ with the transition system. The offline algorithm assigns two Boolean indicators to each transition of the product which indicate whether the transition induces a progress to a sub-goal, i.e., a surveyed state of the transition system and both a surveyed state and an accepting state of the product, respectively.

2. An **online feedback algorithm** determines the next state to be visited by the UAV is iteratively run. In each iteration, attractions of the states of $\mathbb{P}$ are computed.

5.6.4 Probabilistic weather forecasting

Weather conditions such as thunderstorms, icing, turbulence and wind have great influence on UAV safety and mission success. It is therefore important to incorporate weather forecasting into path planning. The recent development in numerical weather prediction makes possible the high-resolution ensemble forecasting. In ensemble forecasting, different weather models with model inputs, initial conditions and boundary conditions are being slightly changed for each run. Each single run contains a different number of ensemble members and generates a prediction spectrum. This allows to build an objective and stochastic weather forecast that supports statistical post-processing [54]. Based on this information, a probabilistic weather map can be constructed. With continuous ensemble forecasting updating at a rate of once per time unit, this analysis provides an online 4-D weather map. In the probabilistic 3-D weather map, the path can be optimally planned. The problem is defined as follows.

Problem 181 *Given a UAV in an area of operation described by non-uniform grids, each way-point is assigned with a probability of adverse weather that is updated periodically, find a path from start to destination with minimum cost on defined terms and meeting the constraints on mission failure risk. The cost function is defined as:*

$$Min \ (J = w_{time}T_{time} + w_{wea}W_{wea}) \tag{5.53}$$

subject to:

$$R_{mission} < R_{critical} \qquad T_{time} < T_{max} \tag{5.54}$$

where w_{time}, w_{wea} are the weighting factors on mission duration and weather condition respectively with: $w_{time}+w_{wea} = 1$, $R_{mission}$ is the risk of the mission and $R_{critical}$ is the critical risk level defined by users. T_{time} is the mission duration, T_{max} is the maximum mission duration allowed, and W_{wea} is the weather condition along the flight route.

Incremental search algorithm makes an assumption about the unknown space and finds a path with the least cost from its current location to the goal. When a new area is explored, the map information is updated and a new route is replanned if necessary. This process is repeated until the goal is reached or it turns out that goal cannot be reached (due to obstacles for instance). When the weather map is updated, the weather in cells close to the UAV is more certain than those in the grids far away from the UAV. In this sense, the weather uncertainty in grids is proportional to its distance to the UAV. When the uncertainties become larger, the weather condition can be regarded as unknown. Therefore, the weather map is not completely known.

The **mission risk** evaluation and management can be improved by integrating an uncertainty factor in path planning. The uncertainty factor for a grid denoted as $U_{un}(x)$ can be defined as a Gaussian function:

$$U_{un}(X) = 1 - \exp^{\left(-\frac{(X-X_0)^2}{2\sigma^2}\right)} \tag{5.55}$$

where σ is an adjustable parameter, X_0 is the UAV current location and X are the centers of the grids. Everytime the weather forecasting is updated, this uncertainty factor is recalculated to obtain a new set of uncertainty so that the impact of adverse weather to mission success is also updated. The probability of adverse weather in each grid is then weighted by the uncertainty factor.

$$P_{ad-un} = P_{ad}(i)\left(1 - U_{un}(i)\right) \tag{5.56}$$

where P_{ad-un} is the probability of adverse weather adjusted by the uncertainty factor, $P_{ad}(i)$ is the probability of adverse weather in i^{th} grid, and $U_{un}(i)$ is the probability of adverse weather adjusted by the uncertainty factor.

To evaluate the mission risk of a planned path, the probability of adverse weather in each grid cell needs to be converted to the probability of UAV failure as it traverses the cell. The **Weibull distribution** can be used to calculate the probability of failure. The inputs consist of the probability of adverse weather occurring in each cell along the path and the time that the UAV spends flying through each of the cells. In the study presented by [66], the Weibull scale factor is calculated as:

$$\alpha = \frac{\mu_{fail}}{\Gamma\left(1 + \frac{1}{\beta}\right)} \tag{5.57}$$

where $\Gamma(.)$ is the Gamma function, μ_{fail} is the average time to failure for the aircraft in each cell. Then a Weibull distribution can be established to calculate the probability of UAV failure.

5.7 CONCLUSION

Search, tracking and surveillance are very important in everyday life and vital in disasters. Search and tracking is the problem of searching for a mobile target and tracking it once it is found. Multiple UAVs, moving targets, constrained resources, and heterogeneous UAV capabilities can be considered. In surveillance tasks, complex sequences of behaviors plan generally consist of gathering as much information as possible given the constraints and communicating findings to human operators. Observers usually operate in unpredictable environments with rapidly changing information. They must decide what action to perform and how to coordinate with other observers almost instantaneously and be highly trained to react quickly. At the same time observers have limited resources and need to be strategic in deciding what course to follow, looking ahead at their remaining lifespan and fitting their objectives within this time-frame.

Bibliography

[1] Adamey, E.; Oguz, A. E.; Ozguner, U.: *Collaborative Multi-MSA Multi-Target Tracking and Surveillance: a Divide and Conquer Method Using Region Allocation Trees*, Journal of Intelligent and Robotic Systems, pp. 1-15, DOI 10.1007/s10846-017-0499-4, 2017.

[2] Adams, S. M.; Friedland, C. J.: *A Survey of Unmanned Aerial Vehicle (UAV) Usage for Imagery Collection in Disaster Research and Management*, publisher not identified, 2011.

[3] Aksaray, D.; Yazicioglu, A. Y.; Feron, E.; Mavris, D. N.: *Message-Passing Strategy for Decentralized Connectivity Maintenance in Multi-agent Surveillance*, AIAA Journal of Guidance, Control and Dynamics, vol. 38, pp. 542-555, 2015.

[4] Anjum, S. S.; Noor, R. M.; Anisi, M. H.: *Review on MANET Based Communication for Search and Rescue Operations*, Wireless Personal Communications, 1-22, DOI 10.1007/s11277-015-3155-y, 2015.

[5] Babel, L.: *Curvature-constrained traveling salesman tours for aerial surveillance in scenarios with obstacles*, European Journal of Operational Research, doi:10.1016/j.ejor.2017.03.067, 2017.

[6] Balta, H.; Bedkowski, J.; Govindaraj, S.; Majek, K.; Musialik, P.; Serrano, D.; Alexis, K.; Siegwart, R.; De Cubber. G.: *Integrated data management for a fleet of search and rescue robots*, Journal of Field Robotics, vol. **34**, pp. 539-582, 2017.

[7] Banaszuk, A.; Fonoberov, V. A.; Frewen, T. A.; Kobilarov, M.; Mathew, G.; Mezic, I; Surana, A: *Scalable approach to uncertainty quantification and robust design of interconnected dynamical systems*, IFAC Annual Reviews in Control, vol. **35**, pp. 77-98, 2011.

[8] Baranzadeh, A.; Savkin, A. V.: *A distributed control algorithm for area search by a multi-robot team*, Robotica, pp. 1-21, DOI: http://dx.doi.org/10.1017/S0263574716000229, 2016.

[9] Bernardini, S.; Fox, M.; Long, D.; Bookless, J.: *Autonomous Search and Tracking via Temporal Planning*, In Proceedings of the Twenty-Third International Conference on Automated Planning and Scheduling, pp. 353-361, 2013.

[10] Bernardini, S.; Fox, M.; Long, D.: *Combining temporal planning with probabilistic reasoning for autonomous surveillance missions*, Autonomous Robots, vol. **41**, pp. 181-203, 2017.

[11] Bernaschi, M.; Lulli, M.; Sbragaglia, M.: *GPU based detection of topological changes in Voronoi diagrams*, Computer Physics Communications, vol. **213**, pp. 19-28, 2017.

[12] Bitton, E.; Goldberg, K.: *Hydra: A framework and algorithms for mixed-initiative UAV-assisted search and rescue*, In IEEE International Conference on Automation Science and Engineering, pp. 61-66, 2008.

[13] Burke, J. L.; Murphy, R. R.; Riddle, D. R.; Fincannon, T.: *Task performance metrics in human-robot interaction: Taking a systems approach*, Proceedings of the Performance Metrics for Intelligent Systems Workshop, Gaithersburg, MD, pp. 1-8, 2004.

[14] Cacace, J.; Finzi, A.; Lippiello, V.: *Multimodal Interaction with Multiple Co-located Drones in Search and Rescue Missions*, arXiv preprint arXiv:1605.07316, 2016.

[15] Capitan, J.; Merino, L.; Ollero, A.: *Cooperative Decision-Making Under Uncertainties for Multi-Target Surveillance with Multiples UAVs*, Journal of Intelligent and Robotic Systems, pp. 1-16, DOI 10.1007/s10846-015-0269-0, 2015.

[16] Carpin, M.; Rosati, S.; Khan, M. E.; Rimoldi, B.: *UAVs using Bayesian Optimization to Locate WiFi Devices*, arXiv preprint arXiv:1510.03592, 2015.

[17] Daingade, S.; Sinha, A.; Borkar, A. V.; Arya, H.: *A variant of cyclic pursuit for target tracking applications: theory and implementation*, Autonomous Robots, vol. **40,** pp. 669-686, 2016.

[18] Di, B.; Zhou, R.; Duan, H.: *Potential field based receding horizon motion planning for centrality-aware multiple UAV cooperative surveillance*, Aerospace Science and Technology, vol. **46**, pp. 386-397, 2015.

[19] Dong, L.; Chai, S.; Zhang, B.; Nguang, S. K.; Li, X.: *Cooperative relay tracking strategy for multi-agent systems with assistance of Voronoi diagrams*, Journal of the Franklin Institute, vol. **353**, pp. 4422-4441, 2016.

[20] El-Hadidy, M. A.: *Optimal searching for a helix target motion*, Science China Mathematics, vol. **58**, pp. 749-762, 2015.

[21] Gade, S.; Joshi, A.: *Heterogeneous UAV swarm system for target search in adversarial environment* In IEEE International Conference on Control Communication and Computing (ICCC), pp. 358-363, 2013.

[22] Goodrich, M. A. et al.: *Supporting Wilderness Search and Rescue with Integrated Intelligence: Autonomy and Information at the Right Time and the Right Place*, 24th AAAI Conference on Artificial Intelligence, pp. 1542-1547, 2010.

[23] Gomez-Balderas, J. E.; Flores, G.; Garcia Carrillo, L. R.; Lozano, R.: *Tracking a ground moving target with a quadrotor using switching control*, Journal of Intelligent and Robotic Systems, vol. **70**, pp. 65-78, 2013.

[24] Grace, J.; Baillieul, J.: *Stochastic strategies for autonomous robotic surveillance*, In Proceedings of IEEE Conference on Decision and Control, pp. 2200-2205, 2005.

[25] Ha, J.S.; Choi, H.L.: *Periodic sensing trajectory generation for persistent monitoring*, In Proceedings of 53rd IEEE conference on Decision and Control, Los Angeles, pp. 1880-118, 2014.

[26] Harik, E. H. C.; Gurin, F.; Guinand, F.; Breth, J. F.; Pelvillain, H.: *A decentralized interactive architecture for aerial and ground mobile robots cooperation*, In IEEE International Conference on Control, Automation and Robotics (ICCAR), pp. 37-43, 2015.

[27] Hosseini, S.; Mesbahi, M.: *Energy aware aerial surveillance for a long endurance solar-powered unmanned aerial vehicle*, AIAA Journal of Guidance, Control and Dynamics, vol. **39**, pp. 1-13 , 2016.

[28] Izadi, M.; Sanyal, A. K.; Barany, E.; Viswanathan, S. P.: *Rigid Body Motion Estimation based on the Lagrange-d'Alembert Principle*, arXiv preprint arXiv:1509.04744, 2015.

[29] Kassem, M. A. E. H.; El-Hadidy, M. A. A.: *Optimal multiplicative Bayesian search for a lost target*, Applied Mathematics and Computation, vol. **247**, pp. 795-802, 2014.

[30] Khaleghi, A. M.; Xu, D.; Wang, Z.; Li, M.; Lobos, A.; Liu, J.; Son, Y. J.: *A DDDAMS-based planning and control framework for surveillance and crowd control via UAVs and UGVs*, Expert Systems with Applications, vol. **40**, pp. 7168-7183, 2013.

[31] Kim, M. H.; Baik, H.; Lee, S.: *Response threshold model based UAV search planning and task allocation*, Journal of Intelligent and Robotic Systems, vol. **75**, pp. 625-640, 2014.

[32] Klodt, L.; Khodaverdian, S.; Willert, V.: *Motion control for UAV-UGV cooperation with visibility constraint*, In 2015 IEEE Conference on Control Applications (CCA) pp. 1379-1385, DOI: 10.1109/CCA.2015.7320804, 2015.

[33] Lanillos, P.; Gan, S. K.; Besada-Portas, E.; Pajares, G.; Sukkarieh, S.: *Multi-UAV target search using decentralized gradient-based negotiation with expected observation*, Information Sciences, vol. **282**, pp. 92-110, 2014.

[34] Lin, L.; Goodrich, M. A.: *Hierarchical heuristic search using a Gaussian mixture model for UAV coverage planning*, IEEE Transactions on Cybernetics, vol. **44**, pp. 2532-2544, 2014.

[35] Lopez-Ortiz, A.; Maftuleac, D.: *Optimal Distributed Searching in the Plane with and without Uncertainty*, International Workshop on Algorithms and Computation. Springer International Publishing, pp. 68-79, 2016.

[36] McCune, R.; Purta, R.; Dobski, M.; Jaworski, A.; Madey, G.; Wei, Y.; Blake, M. B.: *Investigations of dddas for command and control of uav swarms with agent-based modeling*, In Proceedings of the IEEE Winter Simulation Conference: Simulation: Making Decisions in a Complex World, pp. 1467-1478, 2013.

[37] McCune, R. R.; Madey, G. R.: *Control of artificial swarms with DDDAS*, Procedia Computer Science, vol. **29**, pp. 1171-1181, 2014.

[38] Meng, W.; He, Z.; Teo, R.; Su, R.; Xie, L.: *Integrated multi-agent system framework: decentralised search, tasking and tracking*, IET Control Theory and Applications, vol. **9**, pp. 493-502, 2015.

[39] Mohibullah, W.; Julier, S. J.: *Developing an Agent model of a missing person in the wilderness*, In IEEE International Conference on Systems, Man, and Cybernetics, pp. 4462-4469, DOI: 10.1109/SMC.2013.759, 2013.

[40] Montserrat, D. M.: *Ground target chasing with a UAV*, Master thesis, Universtat politecnica de Catalunya, Barcelona, Spain, 2015.

[41] Morbidi, F., Freeman, R. A.; Lynch, K. M.: *Estimation and control of UAV swarms for distributed monitoring tasks*, In Proceedings of the IEEE American Control Conference, pp. 1069-1075, 2011.

[42] Mukherjee, D.; Ghose, D.: *Target capturability using agents in cyclic pursuit*, AIAA Journal of Guidance, Control and Dynamics, vol. **39**, pp. 1034-1045, 2016.

[43] Nestmeyer, T.; Giordano, P. R.; Bulthoff, H. H.; Franchi, A.: *Decentralized simultaneous multi-target exploration using a connected network of multiple robots*, Autonomous Robots, pp. 1-23, DOI: 10.1007/s10514-016-9578-9, 2016.

[44] Niedfeldt, P.; Beard, R.; Morse, B.; Pledgie, S.: *Integrated sensor guidance using probability of object identification*, In Proceedings of the 2010 American Control Conference, pp. 788-793, 2010.

[45] Pace, P.; Aloi, G.; Caliciuri, G.; Fortino, G.: *A Mission-Oriented Coordination Framework for Teams of Mobile Aerial and Terrestrial Smart Objects*, Mobile Networks and Applications, pp. 1-18, DOI 10.1007/s11036-016-0726-4, 2016.

[46] Poiesi, F., Cavallaro, A.: *Distributed vision-based flying cameras to film a moving target*, In IEEE International Conference on Robotics and Automation, vol. **3**, pp. 2453-2459, 2015.

[47] Quintero, S. A.; Ludkovski, M.; Hespanha, J. P.: *Stochastic optimal coordination of small UAVs for target tracking using regression-based dynamic programming*, Journal of Intelligent and Robotic Systems, vol. **82**, pp. 135-162, 2016.

[48] Raap, M.; Zsifkovits, M.; Pickl, S.: *Trajectory optimization under kinematical constraints for moving target search*, Computers and Operations Research, 2017.

[49] Rudol, P.; Doherty, P.: *Human body detection and geolocalization for UAV search and rescue missions using color and thermal imagery*, In IEEE Aerospace Conference, pp. 1-8, 2008.

[50] Saska, M.; Vonsek, V.; Chudoba, J.; Thomas, J.; Loianno, G.; Kumar, V.: *Swarm distribution and deployment for cooperative surveillance by micro-aerial vehicles*, Journal of Intelligent and Robotic Systems, vol. **84**, pp. 469-492, 2016.

[51] Senanayake, M.; Senthooran, I.; Barca, J. C.; Chung, H.; Kamruzzaman, J.; Murshed, M.: *Search and tracking algorithms for swarms of robots: A survey*, Robotics and Autonomous Systems, vol. **75**, pp. 422-434, 2016.

[52] Skeele, R. C.; Hollinger, G. A.: *Aerial vehicle path planning for monitoring wildfire frontiers*, Field and Service Robotics, pp. 455-467, 2016.

[53] Svorenova, M.; Tumova, J.; Barnat, J.; Cerna, I.: *Attraction based receding horizon path planning with temporal logic constraints*, IEEE 51th Control and Decision Conf., pp. 6749-6754, 2012.

[54] Sydney, N.; Paley, D. A.: *Multiple coverage control for a non stationary spatio-temporal field*, Automatica, vol. **50**, pp. 1381-1390, 2014.

[55] Teuliere, C.; Eck, L.; Marchand, E.: *Chasing a moving target from a flying uav*, In IEEE/RSJ International Conference on Intelligent Robots and Systems (IROS), pp. 4929-4934, 2011.

[56] Vitte, T.: *Optimization of the range for continuous target coverage*, AIAA Journal of Aircraft, vol. **52**, pp. 896-902, 2015.

[57] Walha, A.; Wali, A.; Alimi, A. M.: *Video stabilization with moving object detecting and tracking for aerial video surveillance*, Multimedia Tools and Applications, vol. **74**, pp. 6745-6767, 2016.

[58] Wallar, A.; Plaku, E.; Sofge, D. A.: *Reactive motion planning for unmanned aerial surveillance of risk-sensitive areas*, IEEE Transactions on Automation Science and Engineering, vol. **12**, pp. 969-980, 2015.

[59] Wanasinghe, T. R.; Mann, G. K.; Gosine, R. G.: *Relative localization approach for combined aerial and ground robotic system*, Journal of Intelligent and Robotic Systems, vol. **77**, pp. 113-133, 2015.

[60] Wang, Z.; Li, M.; Khaleghi, A. M.; Xu, D.; Lobos, A.,; Vo, C.; Lien, Y. M.; Son, Y. J.: *DDDAMS-based crowd control via UAVs and UGVs*, Procedia Computer Science, vol. **18**, pp. 2028-2035, 2013.

[61] Wiltsche, C.; Ramponi, F. A.; Lygeros, J.: *Synthesis of an asynchronous communication protocol for search and rescue robots*, In European Control Conference, pp. 1256-1261, 2013.

[62] Wu, F.; Ramchurn, G.; Chen, X.: *Coordinating human-UAV teams in disaster response*, soton.ac.uk, 2016.

[63] Yao, P.; Wang, H.; Su, Z.: *Cooperative path planning with applications to target tracking and obstacle avoidance for multi-UAVs*, Aerospace Science and Technology, vol. **54**, pp. 10-22, 2016.

[64] Yetkin, H.; Lutz, C.; Stilwell, D.: *Environmental Information Improves Robotic Search Performance*, arXiv preprint arXiv:1607.05302, 2016.

[65] Yu, M.; Oh, H.; Chen, W. H.: *An improved multiple model particle filtering approach for maneuvering target tracking using airborne GMTI with geographic information*, Aerospace Science and Technology, vol. **52**, pp. 62-69, 2016.

[66] Zhang, B.; Tang, L.; Roemer M.: *Probabilistic weather forecasting analysis of unmanned aerial vehicle path planning*, AIAA Journal of guidance, control and dynamics, vol. **37**, pp. 309–312, 2014.

[67] Zhang, M.; Liu, H. H.: *Cooperative tracking a moving target using multiple fixed-wing UAVs*, Journal of Intelligent and Robotic Systems, vol. **81**, pp. 505-529, 2016.

[68] Zhou, Y.; Maity, D.; Baras, J. S.: *Optimal mission planner with timed temporal logic constraints*, In European Control Conference (ECC), pp. 759-764, DOI: 10.1109/ECC.2015.7330634, 2015.

[69] Zhu, S.; Wang, D.; Low, C. B.: *Cooperative control of multiple UAVs for moving source seeking*, Journal of Intelligent and Robotic Systems, vol. **74**, pp. 333-346, 2014.

CHAPTER 6

General Conclusions

Between the progressing regulatory boundaries and the introduction of innovative technologies, the world of commercial UAVs evolves quickly. It is a new industry with quickly changing rules. Civilian applications of UAS have increased considerably in recent years due to their greater availability and the miniaturization of sensors, GPS, inertial measurement units, and other hardware. This technology enables UAVs ranging in scale from sub-meter to full scale to autonomously inspect, map, survey, and transport with applications in infrastructure maintenance, agriculture, mining, emergency response, cargo delivery, etc.

This book allows the reader to examine different mission types and provide new insight into the rationale dealing with work-processes and optimization methods overlapping technical and human-centered disciplines. The reader discovers the real problems that need to be resolved, independently of the application. A common perspective of autonomy is to segregate functions according to the nature of the tasks involved, such as those that are unique to the UAV as compared to those that are applicable to the mission-level activities. Aircraft level autonomy includes vehicle stabilization and flight control, maneuvering flight and basic auto-land. Mission level autonomy encompasses functions such as auto-navigation, route planning, mission objective determination, flight plan contingencies, dynamic trajectory management, task execution and collision/obstacle avoidance. Generic robotic approaches such as orienteering, coverage, deployment, patrolling and foraging are used in the mission level. In the formal definition of UAS, control station and data link, there is no explicit mention of mission preparation and execution. However, an essential component of UAS is the mission planning and management subsystem. Although the mission planning and real-time management component is usually an integrated part of commercial UAS, there are open issues that still lead users to repeat acquisition campaigns simply because of flawed mission design or execution.

In the near future, rapid innovation is expected in advanced sensors, real-time situational navigation, artificial intelligence, and continued integration in air-space currently occupied by manned aircraft. Expanding UAV battery life is extremely important for UAS. Flight time continues to be a challenge for UAVs, especially those being used by companies for beyond visual line-of-sight operation. For example, delivery companies want UAVs with enhanced battery life, and are testing delivery of parcels with UAVs to see how far they can go. UAV technology also supports the development of new solutions such as the provision of communication access using UAVs in category of high-altitude long endurance or/and high-altitude pseudo satellite. Technical requirements refer, for example, to the remote identification of UAVs. Operational requirements refer among others to geo-fencing, to ensure the UAVs does not enter a prohibited zone. Big data analytics related to services such as industrial preventative maintenance, precision agriculture and research purposes are still in their early phases of development and must continue to evolve for UAV applications to successfully transform businesses and processes. Advances are still needed in detect and avoid, datacom technology, air traffic management, security and cyber reliance along with the availability of authorized and safe testing environments.

The future business opportunities are in the commercial, beyond visual line-of-sight operations that operate autonomously or semi-autonomously. To be highly reliable, organizations must conduct consistent, sustainable and low error operations based on informed high-quality decision making and practices. Based on the rapid evolution over recent years, unforeseen missions may in fact act as a significant multiplier to the value and demand of UAVs. Value added services are expected to represent the largest market opportunity in the value chain. **UAVs-as-a-service** and the insight procured by leveraging UAV technologies is the real intention of commercial users and stakeholders.

CHAPTER 7

Acronyms

2-D	Two-dimensional
3-D	Three-dimensional
4-D	Four-dimensional
AANET	Aeronautical ad-hoc network
ABS	Agent-based navigation
ACO	Ant colony optimization
ACT	Adaptive control of thought rationale
ADS-B	Automatic dependent surveillance broadcast
AFCS	Automatic flight control system
AFF	Artificial force field
AGL	Above ground level
AGP	Art gallery problem
AGS	Adaptive group size
AI	Artificial intelligence
AIA	Adaptive intelligent agents
ANO	Air navigation order
APM	Autopilot module
ASRM	Aviation system risk model
ATOL	Automatic take-off and landing

ATC	Air traffic control
ATSP	Asymmetric traveling salesman problem
BCD	Boustrophedon cellular decomposition
BDI	Belief desire intention
BIM	Building information model
BN	Bayesian network
BVLOS	Beyond visual line-of-sight
BVP	Boundary value problem
C3	Command, control, communication
CA	Control allocation
CAA	Civil aviation authority
CAGR	Compound annual growth rate
CAS	Collision avoidance system
CBAA	Consensus-based auction algorithms
CCF	Common cause failures
CCPP	Capacitated Chinese postman problem
CD	Complex dependencies
CDT	Center of democracy and technology
CDV	Coverage driven verification
CLARATy	Coupled layered architecture for robotic autonomy
CM	Cognitive model
COA	Certificate of authorization
CONOPS	Concept of operations
CPP	Chinese postman problem
CSM	Crop surface model
CSP	Constraint satisfaction problem
CT	Coordinated turn
DAC	Data analysis center

DAG	Directed acyclic graph
DBA	Distributed bidding algorithm
DDDAMS	Dynamic data driven application multi-scale system
DDDAS	Dynamic data driven application system
DDL	Domain description language
DEM	Digital elevation model
DISO	Direct sensor orientation
DME	Distance measurement equipment
DMVP	Dynamic map visitation problem
DOF	Degrees of freedom
DRPP	Dynamic rural postman problem
DSL	Domain specific language
DSM	Digital surface model
DSPP	Dubins shortest path problem
DSS	Decision support system
DTM	Digital terrain model
DTMC	discrete-time Markov chain
DTRP	Dynamic traveling repairman problem
DTSP	Dynamic traveling salesman problem
DWI	Driving while intoxicated
EA	Evolutionary algorithm
EA-DDAS	Energy aware-dynamic data driven application system
EASA	European aviation safety agency
EEDI	Ergodic exploration of distributed information
EFC	Equitable facility configuration
EFCS	Electronic flight control system
EID	Expected information density
EIF	Extended information filter

EKF	Extended Kalman filter
ELP	Equitable location problem
ELOS	Equivalent level of safety
EP	External pilot
ESD	Event sequence diagram
ESS	Expanding square search
ESSW	Early-site specific weed management
EST	Expansive search trees
ETSP	Euclidean traveling salesman problem
ETTD	Expected time to detection
EVLOS	Extended visual line-of-sight
EVS	Enhanced vision system
EXIF	Exchanged image format
EWS	Early warning system
FAA	Federal aviation administration
FAR	Federal aviation regulations
FCC	Flight control computer
FCFS	First-come first-serve
FDD	Fault detection and diagnosis
FDI	Fault detection isolation
FDS	Fair discrete system
FDSS	Forestry decision support system
FFA	Functional fault analysis
FIFO	First-in first-out
FL	Flight level
FLOA	Functional level of autonomy
FMEA	Failure modes and effects analysis
FOM	Figures of merit

FOTSP	Finite-one in set traveling salesman problem
FOV	Field-of-view
FPV	First person view
FRBS	Flight rule-based system
FU	Footprint of uncertainty
GA	Genetic algorithm
GCP	Ground control points
GCS	Ground control station
GIS	Geographic information system
GMM	Gaussian mixture model
GNC	Guidance, navigation and control
GNSS	Global navigation satellite system
GPS	Global positioning system
GUI	Graphical user interface
GVD	Generalized Voronoi diagram
GVRP	Generalized vehicle routing problem
HAS	Human-agent collaboration
HCAS	Hazard classification and analysis system
HDRA	Hybrid deliberative reactive architecture
HMI	Human-machine interaction
HRI	Human-robot interface
HSC	Human supervisory control
IA	Instantaneous assignment
IAI	Intelligent adaptive interface
IBVS	Image-based visual servoing
IC	Incident commander
ICAO	International civil aviation organization
ID	In-schedule dependencies

IF	Information fusion
IFR	Instrument flight rule
ILS	Instrument landing system
IMU	Inertial measurement unit
INS	Inertial navigation system
INSO	Indirect sensor orientation
IRRT	Information-rich rapidly exploring random tree
IRS	Indoor residual spraying
IRU	Inertial reference unit
ISA	International standard atmosphere
ISHM	Integrated system health management
ISO	Integrated sensor orientation
ISR	Intelligence, surveillance, reconnaissance
JAR	Joint aviation regulations
KF	Kalman filter
KPA	Key performance attributes
KSA	Knowledges, skills and attitudes
LAN	Local area network
LKP	Last known position
LOA	Levels of autonomy
LOS	Line-of-sight
LRU	Line replaceable unit
LRV	Least recently visited
LSM-LC	Larval source management via larviciding
LSS	Logistic service station
LTL	Linear temporal logic
LTID	Linear temporal incremental deployment
MA	Maneuver automaton

MANET	Mobile ad-hoc network
MAS	Multi-agent system
MATP	Maximum area triangulation problem
MBFS	Model-based fuzzy system
MC	Monte Carlo
MCDA	Multi-criteria decision analysis
MCMC	Markov chain Monte Carlo
MCS	Motion capture system
MCS-OPTW	Maximum coverage stochastic orienteering problem with time windows
MDL	Mission description language
MDP	Markov decision process
MFP	Mission flight planning
MILP	Mixed integer linear programming
MIP	Mixed integer programming
MIMO	Multi-input multi-output system
ML	Machine learning
MLD	Master logic diagram
MP	Mission priority
MR	Mission risk
MRTA	Multi-robot task allocation
MRTP	Minimum relay triangulation problem
MSN	Mobile sensor network
MST	Minimum spanning trees
MTSP	Multiple traveling salesmen problem
MTOM	Maximum take-off mass
MTOW	Maximum take-off weight
NAAA	National agricultural aviation association

NAS	National air space
NBC	Nuclear, biological, chemical
NCRSTF	National consortium for remote sensing in transportation flows
NDVI	Normalized difference vegetation index
NDM	Naturalistic decision making
NETD	Noise equivalent temperature difference
NHTSA	National highway traffic safety administration
NIR	Near infrared
NNI	Nitrogen nutrition index
NP	Non-polynomial
OLS	Orthogonal least squares
OOBN	Object-oriented Bayesian network
OODA	Observe, orient, decide, act
ORS	Outdoor residual spraying
OS	Orientation system
OTP	Optimal travel path
PbD	Privacy by design
PCTL	Probabilistic computation tree logic
PDC	Parallel distributed compensator
PDF	Probability distribution function
PDP	Pick-up and delivery problem
PID	Proportional integral derivative
PLS	Point last seen
PN	Proportional navigation
POC	Probability of containment
POI	Point of interest
POMDP	Partially observable Markov decision process

PRF	Probe request frame
PRA	Probabilistic risk assessment
PRM	Probabilistic road map
PTS	Parallel track search
PV	Photovoltaic
PVDTSP	Polygon visiting Dubins traveling salesman problem
PWM	Primary way-points mission
QV	Quantization vector
RANSC	Random sample consensus
RBE	Recursive Bayesian estimation
RBF	Radial basis function
RC	Radio control
RCA	Root cause analysis
RF	Radio frequency
RGB	Red, green, blue
RHTA	Receding horizon task assignment
RL	Relative localization
ROM	Rough order of magnitude
ROW	Right of way
RPA	Remotely piloted aircraft
RPAS	Remotely piloted aircraft system
RPDM	Recognition primed decision making
RRT	Rapidly exploring random tree
RS	Robust stabilizability
RSA	Random sequential adoption
RSSI	Received signal strength indication
RT	Risk tolerance
SAA	Sense and avoid

SAR	Search and rescue
SAROPS	Search and rescue optimal planning system
SCA	Swarm control agent
SCADA	Supervisory control and data acquisition
SCP	Set covering problem
SCS	Stochastic coverage scheme
SDL	Scenario description language
SEC	Software enabled control
SF	Scaling factor
SFC	Specific fuel consumption
SFDIA	Sensor fault detection, isolation, accommodation
SFM	Structure from motion
SHM	System health management
SIFT	Scale invariant feature transform
SISO	Single-input single-output
SLAM	Simultaneous localization and mapping
SoS	System of systems
SP	Synchronization and precedence
SPOI	Sensed point of interest
SPP	Shortest path problem
SSE	Sum square errors
SSWM	Site specific weed management
ST	Single task
STSP	Symmetric traveling salesman problem
TA	Terrain avoidance
TAA	Technical airworthiness authority
TAP	Task assignment problem
TCAS	Traffic collision avoidance system

TCG	Temporal causal graph
TDL	Team description language
TF	Terrain following
TGP	Task graph precedence
TPR	Third party risk
TSP	Traveling salesman problem
TSPN	Traveling salesman problem with neighborhoods
TTF	Temporal transition to failure
TW	Time window
UAS	Unmanned aerial system
UAV	Unmanned aerial vehicle
UCA	Uniform cost assignment
UGS	Unattented ground sensor
UKF	Unscented Kalman filter
USAR	Urban search and rescue
VFR	Visual flight rules
VI	Vegetation index
VLL	Very low level
VLOS	Visual line-of-sight
VNS	Variable neighborhood search
VOR	VHF omni-directional range
VSN	Visual sensor network
VR	Virtual reality
VRT	Variable rate technology
VRTW	Vehicle routing problem with time windows
VTP	Virtual target point
WAMI	Wide area motion imagery
WG	Way-point generation

WiSAR	Wilderness search and rescue
WLC	Weighted location coverage
WLS	Weighted least squares
WSN	Wireless sensor network
WUI	Wildland urban interface
ZVD	Zermelo-Voronoi diagram

Index